George M. Hopkins'
Experimental
SCIENCE

Volume One

1906

Elementary Practical and Experimental Physics

reprinted by Lindsay Publications Inc

PLATE I.

An Electrical Gyroscope.

Experimental Science

ELEMENTARY

PRACTICAL AND EXPERIMENTAL

PHYSICS

BY
GEORGE M. HOPKINS

TWENTY-FIFTH EDITION
(Revised and Enlarged)

NEW YORK:

1906

Experimental Science
Volume One

by George M Hopkins

Originally published by
Munn & Co

Original copyright 1902
by Munn & Co

Reprinted by
Lindsay Publications Inc
Bradley IL 60915

All rights reserved.

ISBN 0-917914-49-X paper
0-917914-51-1 cloth

1 2 3 4 5 6 7 8 9 0

1997

WARNING

Remember that the materials and methods described here are from another era. Workers were less safety conscious then, and some methods may be downright dangerous. Be careful! Use good solid judgement in your work, and think ahead. Lindsay Publications Inc. has not tested these methods and materials and does not endorse them. Our job is merely to pass along to you information from another era. Safety is your responsibility.

Write for a complete catalog of unusual books available from:

    Lindsay Publications Inc
    PO Box 12
    Bradley IL 60915-0012

PREFACE.

THE design of this work is to afford to the student, the artisan, the mechanic, and in fact all who are interested in science, whether young or advanced in years, a ready means of acquiring a general knowledge of physics by the experimental method. One of its principal purposes is, also, to furnish to the teacher suggestions in experimentation, which will be helpful in making class-room work interesting and attractive, rather than dry and monotonous.

Most of the apparatus here illustrated and described may be constructed and used by any one having ordinary mechanical skill. Simple and easily made devices have been chosen for physical demonstration.

With scarcely an exception the experiments described were performed at the time of writing, to insure fullness of detail, and to avoid inaccuracies. The reader can therefore be assured that by following the instructions, success will be certain.

Mathematics has been almost entirely excluded. The few problems presented are capable of arithmetical solution. The importance of mathematical knowledge in all branches of science is fully recognized, but the majority of students have little taste for the intricacies of numbers. Faraday was an illustrious example of a scientific man without great mathematical proclivities.

The late Clerk Maxwell, one of the most eminent mathematicians and electricians of the present century, said: "A few experiments performed by himself will give the student a more intelligent interest in the subject, and will give him a more lively faith in the exactness and uniformity of nature, and in the inaccuracy and uncertainty of our observations, than any reading of books, or even witnessing elaborate experiments performed by professed men of science."

A large proportion of the material of this work consists of original articles published from time to time in the *Scientific American*. These have been revised or rewritten, with copious additions of text and engravings. Very few of the conventional illustrations of the text books have been used. Most of the engravings are now for the first time given in book illustration.

The leading principles of physics are here illustrated by simple and inexpensive experiments. The endeavor has been to make the explanations of both apparatus and experiment plain and easily understood.

If what is here written shall induce any who are now indifferent to the subject to begin the study of physics experimentally, so as to gain even a faint conception of the marvelous perfection of the physical world, or if anything in these pages proves helpful to those who instruct, or who seek scientific information, the end sought by the writer will have been gained.

GEORGE M. HOPKINS.

NEW YORK, January, 1890.

PREFACE TO EDITION OF 1898.

THE seventeenth edition of Experimental Science contained an appendix including much new matter, but, in the four years which have elapsed since the publication of this edition, several startling physical discoveries have been made, among which are the X-Ray and its phenomena, Wireless Telegraphy, the Liquefaction of Air, and Acetylene Gas. These have been included in the present edition. Besides these, a number of additional experiments are given, some of which are new and original. The book has been considerably enlarged by the additions, and it has been revised so that it is in accord with recent ideas of the subjects treated.

The new matter added will prove acceptable to such as seek information on the more recent scientific discoveries.

GEORGE M. HOPKINS.

September 7, 1898.

PREFACE TO THE TWENTY-FIFTH EDITION

IN order to broaden the scope of this work, the author has relaxed the rather rigid rule heretofore adhered to, which called for the trial by himself of every piece of apparatus described in its pages, and has now availed himself of the experience of others. He is therefore able to present to the readers of the twenty-fifth edition, a full explanation of the Polyphase Generator, Induction Motors, and Rotary Transformers, also to give accurate information regarding the construction of modern direct current motors for 110 volts pressure.

A full description of Edison's New Storage Battery is introduced, also some interesting experiments by Prof. John Trowbridge, and some Electrical Measuring Apparatus by N. Monroe Hopkins. Wireless telegraphy is brought up to date, and other recent discoveries are noticed.

The new edition, owing to the great amount of new matter, is published in two volumes. It presents the more recent developments in modern science, and gives information which assists the reader in comprehending the great scientific questions of the day.

GEORGE M. HOPKINS.

New York, June, 1902.

CONTENTS.

CHAPTER I.—PROPERTIES OF BODIES.

PAGES

Extension and Impenetrability—Cotton and Alcohol Experiments—Solution of Sugar in Water—Reduction of Volume of Alcohol and Water Mixture—Mixture of Sulphuric Acid and Water—Divisibility—Example of Extreme Divisibility—Porosity—Physical and Sensible Pores—Porosity of Wood—Mercurial Shower—Porosity in Nature — Porosity in the Arts — Compressibility — Pneumatic Syringe—Elasticity—Gases and Liquids Perfectly Elastic—Elasticity of Flexure—Elasticity of Torsion—Experiment showing the Elasticity of Glass.. 1 to 7

CHAPTER II.—REST, MOTION, AND FORCE.

When a Body is at Rest—All Bodies continually changing Position—Absolute Rest Possible—Inertia—Force—Matter Incapable of changing from Rest to Motion, or the Reverse—Equalizing Effect of Fly Wheels—Persistent Rotation due to Inertia—Action of Projectiles, Hammers, Drop Presses and the Hydraulic Ram, due to Inertia—Inertia Locomotive—Friction due to Roughnesses—The Effect of a Lubricant—Silding Friction—Rolling Friction—Roller and Ball Bearings—Centrifugal Force—Centrifugal Railway—Normal Path of a Moving Body a Straight Line—Spiral Railway—Effect of Centrifugal Force on Air—Choral Top—Effect of Centrifugal Force on Liquids—The Glass Top—Effect of Centrifugal Force on Liquids of Different Densities contained in the Same Vessel—A Scientific Top—Persistence in maintaining Plane of Rotation—Gyroscopic Action—Examples of Centrifugal Action—Oblate Spheroid—Centrifugal Hero's Fountain... .... 8 to 18

CHAPTER III.—THE GYROSCOPE.

Toy Gyroscope—A Large Gyroscope—Gyroscope with Friction Driving Gear—Pneumatic Gyroscopes—Electrical Gyroscope—Steam Gyroscope—Gyroscopes for showing the Rotation of the Earth—Equatorially Mounted Electrical Indicator—Bursting of Fly Wheels by Gyroscopic Action—Flexible Fly Wheel....................... 19 to 37

CHAPTER IV.—FALLING BODIES—INCLINED PLANE—THE PENDULUM.

In a Vacuum All Bodies fall with Equal Rapidity—Effect of Resistance on Falling Bodies—Water Hammer—Swiftest Descent Apparatus

vi CONTENTS.

PAGES

—Inclined Plane—Concave Circular Curve—Cycloidal Curve—Isochronal Curve—Method of describing the Cycloid—Dropped and Projected Balls—Gun for dropping and projecting Balls—Oscillating and Conical Pendulums—Variations in Length of Seconds Pendulum at Different Places—Galileo's Discovery—Isochronism of the Pendulum—Length of Pendulum—Foucault's Experiment showing the Rotation of the Earth—Pendulum with Audible Beats—Kater's Reversible Pendulum—Measurement of Time by the Pendulum—Huyghens' Invention—Torsion Pendulum—Hooke's Invention—The Balance—Flying Pendulum......................38 to 55

CHAPTER V.—MOLECULAR ACTIONS.

Cohesion—A Demonstration of Cohesion—Strain—Prince Rupert's Drops—Bologna Flask—The Breaking of Lamp Chimneys and Water Gauge Tubes—Adhesion—Surface Tension—Surface Tension exhibited in Water Drops—Oil Globule suspended in Equilibrium—Capillarity—Capillary Elevation and Depression—Designs on Wire Cloth—Absorption of Gases—Absorption of Carbonic Acid by Charcoal—Preparation of Carbonic Acid Gas—Diffusion of Gases—Endosmose—Exosmose—Simple Way of showing the Diffusion of Gases—Pressure by Endosmose—Vacuum by Exosmose—The Law governing the Diffusion of Gases—Endosmometer..56 to 71

CHAPTER VI.—LIQUIDS—PRESSURE EXERTED BY LIQUIDS.

Pascal's Law of the Pressures of Liquids—Demonstration of Pascal's Law—Pascal's Experiment—Equilibrium in Communicating Vessels—Principle of the Hydraulic Press—Hypothetical Hydraulic Press—Simple Hydraulic Press—Lateral Pressure—Rectilinear and Rotary Motion produced by Reaction—Hydraulic Ram—Superposition of Liquids of Different Densities—Vial of Four Liquids—Effect of Liquids of Different Densities—Cartesian Diver........72 to 84

CHAPTER VII.—GASES.

Gases are Elastic Fluids—Expansion of Gases—The Air in a State of Equilibrium—Expansibility of Air—Dilatation of Balloon in a Vacuum—Weighing of Gases—Wheel operated by Gas—Determination of the Weight of Air—Hand Glass—Effects of Air Pressure—Crushing Force of the Atmosphere—The Weight lifted by the Air Pressure—The Barometer—Mercurial Column supported by Atmospheric Pressure—Torricelli's Experiment—Pascal's Experiment—Simple Air Pump—Testing the Air Pump—Water boiling in Vacuo—Rarefied Air a Poor Conductor of Sound—Bell in Vacuo—Destruction of Life by Removal of Air—Desiccation by Removal of Air—The Ball Experiment—Card Experiment—Atom-

CONTENTS. vii

PAGES

izing Petroleum Burner—Aspirators for Laboratory Use—Bunsen Filter Pump—Elongation of Discharge Pipe of Bunsen Filter Pump necessary to Best Effects—Chapman's Metallic Aspirator—Principle of the Giffard Injector—Experiment with the Aspirator—Exhausting a Geissler Tube—Blast produced by the Aspirator—Plate and Receiver for Aspirator—Mouth Vacuum Apparatus—Hero's Fountain—Wirtz's Pump—Inertia of Air—The Flight of Birds—The Operation of Windmills and Propulsion of Sailing Vessels due to Inertia of Air—Aerial Top—The Fly Wheel—Mechanical Bird—The Boomerang—Vortex Rings............................85 to 115

CHAPTER VIII.—SOUND.

Toys as Experimental Apparatus—Sound a Sensation of the Ear—Sound due to Irregular Vibrations—Musical Sounds due to Rapid and Uniform Vibrations—The Cricket or Rattle—The Buzz as Savart's Wheel—Vibrating Rods—Tranverse Vibration of Rods—The Zylophone—Tuning the Zylophone—The Metallophone—The Musical Box a Reed Instrument—Mouth Organ or Harmonica an Example of a Reed Instrument—Tuning Reed Instruments—The Bugle—Longitudinal Vibrations of Rods—Of a Steel Rod—Longitudinal Vibrations of Wooden Rods—Marloye's Harp—Stopped Pipe—Pandean Pipes—Open Pipes—Flageolet—Ocorina—Stringed Instruments—Lateral Vibrations of Strings—Zither—Division of Strings into Vibrating Segments—Vibrations of Strings by Sympathy—Conduction of Sound—The String Telephone—Harmonic Vibrations—Cumulative Effects of Harmonic Vibrations—Vibration of Railroad Bridges—Slow Vibratory Period of the East River Bridge—The Breaking of an Iron Girder by Bombardment of Pith Balls—Steel Bar vibrated by Drops of Water—By Magnetic Impulses—Sound Recorder—Tracings of the Motion of a Telephone Diaphragm—Vibrating Flames—Simple Device for showing Vibrating Flames—The Speaking Flame—Annular Burner for producing Vibrating Flames—Manometric Flames—Composition of Vibrations—Optical Method of studying Sonorous Vibrations—Apparatus for producing Lissajous' Figures—Re-enforcement of Sound—Resonance studied by Simple Apparatus—Selective Power of a Resonant Vessel—Bell and Resonator—Mouth used as a Resonator—Experiment with the Jew's Harp—Tuning Forks and Resonant Tubes—Musical Flames—Apparatus for the Production of Sounding Flames—Analyzing Vibrating Flames by a Revolving Mirror—A Simple Phonograph—The Perfected Phonograph—Edison's New Phonograph—Edison listening to the First Phonogram from England—The Phonographic Record—Reflection and Concentration of Sound—Adjustable Sound Reflector—Reflection of Light and Sound—Trevelyan Rocker—Refraction of Sound—Sound Lens—The Sensitive Flame—Apparatus for producing Gas Pressure for Sensitive Flame—Sensitive Flame

PAGES

with Gas at the Ordinary Pressure—Determining Speed by Resonance—Siren for measuring Velocities........................116 to 172

CHAPTER IX.—EXPERIMENTS WITH THE SCIENTIFIC TOP.

Siren applied to the Top—Savart's Wheel—Gyrating Perforated Disk —Gyrating Disk with Polished Beads—Chameleon Top—Changes of Hue by the Shifting of the Cover Disk—Phantom Forms—Revolving Mirror—Koenig's Manometric Flames................... 173 to 180

CHAPTER X.—HEAT.

Heat the Manifestation of Rapid Vibratory Motion of Molecules—A Heated Mass can impart Vibratory Motion to Ether—Heat partially or wholly balances Molecular Attraction—Expansion—A Metallic Thermometer—Simple Thermostat—Air Thermometer—Pulse Glass —Thermoscopic Balance—Electric Meter on the Principle of the Thermoscopic Balance—Wollaston's Cryophorus—Freezing by Rapid Evaporation—The Radiometer—Tyndall's Experiment on Radiant Heat—Action of Radiant Heat on Different Gases—Reflection and Concentration of Heat—Conduction of Heat—Conductivity of Different Metals—Heat due to Friction—Heat due to Pressure and Compression—Pneumatic Syringe—Force of Steam—Candle Bomb—Steam Engine—Fifty Cent Steam Engine—Ascensional Power of Heated Air—Hot Air Motor—Hygrometry—Toy Hygroscope—Sensitive Leaf—Chemical Thermoscope—Hydroscopic and Luminous Roses .. ..181 to 199

CHAPTER XI.—LIGHT.

Theories of Light—The Emission Theory—Undulatory Theory—Comparison of Sound and Light Waves—Sound propagated by Compression and Rarefaction—Vibrations of Light at Right Angles with its Line of Progression—Ether—Reflection—Refraction—Huyghens' Explanation of Refraction—Prisms—Course of Light through a Prism—Polyprism—Lenses—Hypothetical Lens—Forms of Lenses —Converging or Magnifying Lenses—Principal Focus of a Convex Lens—Concave Lens—Converging Rays with a Convex Lens—Diverging Rays with a Concave Lens—Real and Diminished Image— Real and Magnified Image—Virtual Image with Convex Lens— Water Bulb Magnifier—Mirrors—A Convex Cylindrical Mirror— Concave Cylindrical Mirror—Caustics—Convex Spherical Mirror— Concave Spherical Mirror—Phantom Bouquet—Multiple Reflection —The Kaleidoscope—Analysis and Synthesis of Light—Rocking Prism—The Spectrum—Simple Method of producing the Spectrum —Apparatus for producing the Spectrum—Chromatrope—The Blending of Surface Colors—Persistence of Vision—Zoetrope—Irradiation—Examples of Irradiation—Intensity of Light—The Light

PAGES

of the Sun—The Light of the Moon—Measurement of Light—Photometer—Optical Illusions—Illusion from Engineering Drawings—Apparent Deviation by Oblique Lines—Apparent Displacement of a Single Oblique Line—Curious Optical Illusions—Prof. Thompson's Optical Illusions—Webster's Optical Illusions—Rapieff's Optical Illusions.. ....200 to 232

CHAPTER XII.—POLARIZED LIGHT.

Glass, Single Refracting—Double Refracting Bodies—Iceland Spar—The Investigation of Newton on the Properties of Light--Course of Light through Iceland Spar—Nicol's Prism the Most Perfect Instrument for Polarizing—Tourmaline Crystals—Polarization by Reflection and Refraction—Angle of Polarizing for Glass—Stewart's Explanation of Polarized Light by Reflection—Arrangement of Polarizer and Analyzer—Simple Experiment in Polarized Light—Polarizing by Reflection from Blackened Glass—Analyzing by Bundle of Glass Plates—Strained Glass—Glass strained by Pressure—Glass strained by Heat—Polarizing and analyzing with a Single Bundle of Plates—Norremberg Doubler—Double Polarization with a Single Glass Plate—Mica Objects for the Polariscope—Mica Semi-cylinder—Mica Semi-cylinders crossed—Mica Cone—Maltese Cross—Mica Wheel—Star, Fan, and Crossed Bars of Mica—Polariscopes—Simple Norremberg Doubler—Half and Quarter Wave Films—Wide-angled Crystals—Hoffman's Improvement—Polariscope for exhibiting Wide-angled Crystals—Examination of Various Crystals with the Polariscope—Tourmaline Tongs—The Polariscope a Test for Quartz Lenses—Polariscope for Large Objects—Examination of Glassware, etc., by Polariscope—Simple Polariscope for Microscopic Objects—Construction of Simple Polarizer and Analyzer—Method of holding Cover Glasses for cleaning—Practical Applications of the Polariscope—Wheatstone's Polar Clock—Suggestions in Decorative Art—Various Crystals and Combinations of Crystals................233 to 277

CHAPTER XIII.—MICROSCOPY.

Microscopic Objects—Microscopy in Chemistry and Mineralogy—The Microscope a Necessity to the Physicist—Inexpensive Microscope—Water Lens Microscope—Water Lens Microscope with Stand—Compound Microscope—Accessories for the Compound Microscope—Diaphragm and Fine Adjustment—Substitute for the Revolving Table—Illumination of Microscopic Objects—A Modern Microscope—Light Modifier—Iris Diaphragm—Sub-stage Condenser—Gathering Microscopic Objects—Various Books on Microscopy—Implements for gathering Microscopic Objects—Various Microscopic Objects—Transferring Objects to the Slide—Compressor—Microscopic Examination of Ciliated Objects by Intermittent Light—Light Interrupter for the Microscope—Circulation in Animal and Vegetable Tissues—Simple Frog Plate—Circulation of Blood in a

Fish's Tail—Quick Method of Mounting Dry Objects—Dr. Stiles' Wax Cell—Microscopic Examination of the Phenomenon of Colors in Thin Plates—Newton's Rings for Microscopic Examination—Microscopic Examination of Soap Films—Of Mica Plates—Of Vibrating Rods—Simple Polariscope for the Microscope—Objects for the Polariscope..278 to 308

CHAPTER XIV.—THE TELESCOPE.

Inexpensive Telescope—Terrestrial and Celestial Eyepiece for the Telescope—Collimation—Objects to be examined by the Telescope—Simple Telescope Stand—Compact Telescope...................309 to 317

CHAPTER XV.—PHOTOGRAPHY.

Manipulative Skill in Photography—Dry Plates—The Lens—The Camera Box—The Plate Holder—Focusing Cloth—Exposures—Management of a Camera—Timing the Exposure—Copying—Development of the Plate—Treatment for Overexposure—For Underexposure—Beach's Pyro-Potash Developer—Washing and Clearing or Fixing—The Fixing Solution—Hydrochinon Developer—Lantern Slides—Photographic Printing—Toning—Solution for Black Tones—Solution for Brown Tones—Fixing Bath—Mounting Prints—A Pocket Camera—Simple Photographic and Photo-Micrographic Apparatus—Arrangement of Microscope and Camera for Photo-Micrography—Daguerreotypy—The Invention of Daguerre—Scouring the Plate—Buffing the Plate—Sensitizing—The Dark Room—The Operating Room—Developing the Plate—Fixing—Gilding or Toning—Mounting..318 to 346

CHAPTER XVI.—MAGNETISM.

Magnetism by Induction from the Earth—By Torsion—Magnetization of Straight and U-shaped Bars—Motion produced by a Permanent Magnet—Effect of the Armature—Effect of Permanent Magnet on a Bar magnetized by Induction—Neutralizing Effect of an Opposing Pole—Neutral Point between Unlike Poles—Consequent Pole—Formation of Magnetic Curves—Magnetic Curves in Relief—Arborescent Magnetic Figures—Floating Magnets—Mayer's Floating Needles—Rolling Armature—Magnetic Top...................347 to 358

CHAPTER XVII.—FRICTIONAL ELECTRICITY.

Action of Frictional Electricity on Pith Balls—Electric Pendulum—Electroscope—Masked Electricity—Ano-Kato—Mutual Repulsion of Electrified Threads—Self-luminous Buoy—Electrical Machines—Electrophorus—Winter's Electrical Machine—Modified Wimshurst Electrical Machine—Attachment of Leyden Jar—Distribution of Electricity on the Wimshurst Plates—Experiments with the Induction Machine—Various Phases of the Electrical Discharge—Length-

PAGES

ening the Spark—Diversion of the Discharge by Moisture—Glow at the Positive Collector—Glow at the Negative Collector—Discharge through a Geissler Tube—Franklin's Plate—Leyden Jar—Measuring Jar—Disruptive Effect of the Discharge—Electrical Chime—Electric Fly—Fly on Inclined Plane—Jointed and Universal Dischargers—Insulating Stool—Insulated Sphere—Cylindrical Conductor—Gas Pistol—Electric Mortar—Dancing Pith Balls—Gassiot's Cascade—Pith Ball Electroscope...................359 to 391

Chapter XVIII.—Dynamic Electricity.

Generator of the Electric Current—Experimental Battery and Galvanometer—Polarization—Single-Fluid Batteries—Smee's Battery—Grenet Battery—Simple Plunge Battery—Large Plunge Battery—Forming the Lining for Battery Cells—Chloride of Silver Cell—Leclanche Battery—Dr. Gassner's Dry Battery—Caustic Potash Battery—Two-Fluid Batteries—Daniell Battery—Gravity—Grove—Chromic Acid—The Fuller Cell—Mechanical Depolarization of Electrodes—Application of Air Jets to Depolarization—Mechanical Agitator for Depolarizing—Secondary Battery—Roughening the Plate—Method of Connecting the Plates—Forming the Cell—Thermo-Electric Battery—Electrical Units—Arrangement of Battery Cells—Galvanometers—Deprez-D'Arsonval Galvanometer—Arrangement of Galvanometer, Lamp, and Scale—Tangent Galvanometer—Circuit of the Tangent Galvanometer—Electrical Measurements—Wheatstone's Bridge—Resistance Box Connections—Bridge Key—Branch Circuits—Joint Resistance of Branch Circuits—Expansion Voltmeter—Ammeter—Recording Voltmeter—Electro-Magnets—Magnet for Experimentation—Magnet and Switch—Inexpensive Magnet—Form for the Coils—Foucault's Experiment—Experiments with the Electro-Magnet—Diamagnetism—Experiments illustrating the Principle of the Dynamo—Magnetization of a Steel Bar—Magneto-electric Induction—Magnetic Induction—Induced Currents from Induced Magnetism—Simple Current Generator—Simple Motor—Fifty Cent Electric Motor—Gramme Machine for Illustration—Armature and Magnetic Fluid—Drum Armature—Magneto Electric Machines—Principle of the Bell Telephone—Magnetic Key—Polarized Bells—Annunciator—Hand Power Dynamo—Details of Construction of Hand Power Dynamo—Electro-plating Dynamo—Connections of Plating Dynamo—Simple Electric Motor—Details of Construction of Simple Electric Motor—Circuit of the Simple Electric Motor—Cast Iron Magnet..392 to 509

Chapter XIX.—Quarter-Horse Power Electric Motor.

Quarter-Horse Power Electric Motor—Field Magnet—Armature Core—The Winding—The Commutator—The Journal Box—Starting Box—Motor as a Dynamo—Table of Tangents................510 to 523

EXPERIMENTAL SCIENCE.

CHAPTER I.

PROPERTIES OF BODIES.

Extension, impenetrability, divisibility, porosity, compressibility, elasticity, inertia, and gravity are general properties common to all bodies, whether solid, liquid, or gaseous, while some bodies possess specific properties, such as solidity, fluidity, tenacity, malleability, color, hardness.

EXTENSION AND IMPENETRABILITY.

To all matter must be attributed two essential qualities: first, that in virtue of which it occupies space, and which is

FIG. 1.

A Hatful of Cotton in a Tumblerful of Alcohol.

known as extension, and, second, that which allows only one particle or atom of matter to occupy a given space—the

2 EXPERIMENTAL SCIENCE.

property known as impenetrability. That matter occupies space is appreciated by our senses, and needs no particular proof, but that two portions of matter cannot occupy the same space at the same time sometimes seems anomalous, as is shown by some of the following experiments.

Into a tumbler filled with alcohol may be crowded a hatful of loose cotton without causing the alcohol to overflow.* The success of the experiment depends upon the slow intro-

FIG. 2.

Solution of Sugar in Water.

duction of the cotton, allowing the alcohol to invest the fibers, before they are fairly plunged beneath the surface of the alcohol.

In this experiment the penetration of the alcohol is only apparent; the fibers displace some of the alcohol, but the quantity is so small as not to be observable. If the cotton were compressed to the smallest possible volume, it would be found to occupy but very little space. So small a body

* See also chapter on projection.

would be incapable of raising the level of the alcohol enough to be appreciable by an ordinary observer.

A more puzzling experiment consists in slowly introducing some fine sugar into a tumblerful of warm water. A considerable quantity of sugar may be dissolved in the water without increasing its bulk appreciably.

Here the physicist is forced to acknowledge that either the water is penetrated or its atoms are so disposed as to receive the sugar between them, possibly in the same way as a scuttle filled with coal might contain also a bucketful of sand. This latter view is adhered to. The atom or ultimate particle is held to be impenetrable.

In the case of the mixture of water and alcohol, or water

Fig. 3.

Representing Volume of Unmixed Alcohol and Water.

Fig. 4.

Reduction of Volume of Alcohol and Water Mixture.

and sulphuric acid, a curious phenomenon is presented. Take alcohol and water for example. Equal volumes of alcohol and water, when mixed, occupy less space than when separate. If the sum of the volumes of the two separate liquids is 100, the volume of the mixture will be only 94. In the case of the mixture of sulphuric acid and water, the difference is greater.

An easy way to perform this experiment is to fill a narrow-necked flask up to a line which may conveniently be marked by a rubber band around the neck, then removing one-half of the water, measuring it exactly, and replacing it with a volume of alcohol exactly equal to that of the water removed. It will be found that when the liquids are mixed, the mixture will not fill the flask up to the original mark.*

The only reasonable explanation of this phenomenon is that the molecules of the two liquids accommodate themselves to each other in such a manner as to reduce the pores, and thus diminish the volume of the mixture.

DIVISIBILITY.

The property of a body which admits of separating it into distinct parts, and which is known as divisibility, is possessed by all matter. An example of extreme divisibility is found in the coloring of a pail of water with a minute particle of aniline.

POROSITY.

There are two kinds of pores, viz., physical or intermolecular pores and sensible pores. In the case of the former, the interspaces are so small that the molecules are within each other's influence and may attract or repel each other. Expansion by heat, contraction by reduction of temperature, and reduction of volume by compression are among examples of phenomena rendered possible by the existence of physical pores.

Sensible pores are small cavities or spaces, across which molecular forces are unable to act.

The experiment illustrated by Fig. 5 shows the existence of sensible pores. In the neck of an Argand chimney is inserted a plug of Malacca wood, which is sealed around the periphery with wax or paraffine. In the top of the chimney is inserted a stopper, through which projects a short glass tube, having its upper end bent over or capped with a small test tube. To the outer end of the glass tube is applied a rubber tube. When the chimney is in an inverted position, as shown in the engraving, a quantity of mercury is placed in the larger part of the chimney, and the air is partly exhausted from the chimney, by applying the mouth to the

---

* See also chapter on projection.

rubber tube and sucking. The mercury readily passes through the porous wood and falls in a shower. By employing an air pump for producing the partial vacuum, the mercury may be drawn through a plug of pine. These experiments show in a striking manner the porosity in a longitudinal direction of these pieces of wood.

Wood, vegetable, and animal tissues, sponge, pumice stone, and many other substances have sensible pores that

FIG. 5.

Mercurial Shower.

may readily be seen. Physical pores cannot be seen even by the aid of the most powerful microscope; but their existence is proved by the fact that all bodies may be compressed or diminished in volume.

Sensible pores play an important part in the operations of nature, especially in the vegetable and animal kingdoms.

The property of porosity is utilized in the arts, in the

filtration of liquids, in the absorption of liquids and gases, in electrolytic processes, in assaying, etc.

COMPRESSIBILITY.

The property by virtue of which a body may be diminished in volume, by pressure, without losing weight, is known as compressibility. This property is possessed in the greatest degree by gases, which may be reduced by compression to from one-tenth to one-hundredth their original volume.

The simplest piece of apparatus for showing the compression of a gas is a well-made toy popgun, such for example as that shown in Fig. 6. By closing the mouth of this gun by means of a piece of sheet metal or mica, and oiling

FIG. 6

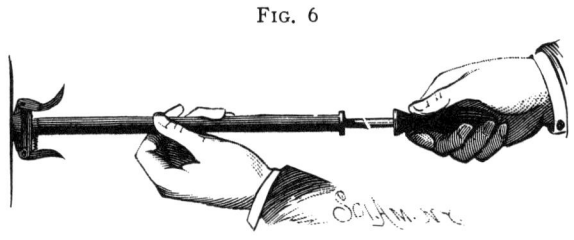

The Popgun used as a Pneumatic Syringe.

the piston well with a heavy oil, to prevent the escape of air from the barrel, it may readily be shown that the air contained by the barrel may be greatly reduced in volume by simply pushing in the piston.

ELASTICITY.*

When a body resumes its original form or volume after distortion or compression, it possesses the property of elasticity, and is therefore known as an elastic body. Elasticity may be shown by pressure, by bending, by torsion or twisting, or by tension or stretching. Gases and liquids are perfectly elastic. When compressed and afterward allowed to

* See also chapter on projection.

return to their original pressure, they are found to possess exactly their original volume.

Among solids, glass is apparently perfectly elastic. A plate of glass bent under pressure and allowed to remain under stress for twenty-five years, when released and carefully tested for any permanent set, was found to have returned to exactly its original shape. Elasticity by flexure or bending is seen in various springs, such as carriage springs, gun-lock springs, etc.

The elasticity of torsion is exhibited by door springs of certain forms, spiral springs, and by twisted threads of cotton, linen, and other material. The elasticity of tension is shown in the strings of all stringed musical instruments, and notably in soft rubber in its various forms.

CHAPTER II.

REST, MOTION, AND FORCE.

A body is said to be at rest when its position is not being changed, but this statement needs some qualification, since any rest known to us is only relative. All bodies with which we are acquainted are continually changing their position either in relation to adjacent objects or along with adjacent objects relatively to distant objects. For example: a bowlder is said to be at rest when it maintains its position relative to the earth's surface, but since the earth itself is not at rest, it is evident that whatever is fixed on the face of the earth cannot be at rest.

On the other hand, if the bowlder were rolling down a declivity, it would be changing its position relative to the earth's surface as well as to all other objects, and would therefore be said to be in motion; but a body may be apparently in motion while in reality absolutely at rest. If we were to suppose a body projected from the earth into space with a velocity equal to that of the earth, but in a direction opposite that of tne earth's motion and uninfluenced by heavenly bodies, the body, although having apparently a high velocity relative to the earth, would be absolutely at rest.

INERTIA.

No body is of itself able to change from a state of rest to a state of motion, neither can a body in motion change its direction or pass unaided to a state of rest. That which causes or tends to cause a body to pass from a state of rest to one of motion, or accelerates or retards the motion of a body, or changes its direction, is known as Force. The incapability of matter to change from rest to motion, or the reverse, is a negative property known as Inertia.

To inertia is due the equalizing effect of flywheels; when

set in motion, they tend to maintain their revolution in opposition to considerable resistance. If sufficient force is applied to the flywheel to counteract the resistance, a practically equable motion is secured, even though the force applied be an intermittent one.

The top is an example of persistent rotation due to inertia. To inertia is due the action of projectiles, hammers, drop-presses, also the hydraulic ram.

The property of inertia, the storage of power, the transfer of power by friction, and the conversion of rotary into rectilinear motion are illustrated by the toy locomotive shown in the annexed engraving. The flywheel, A, is mounted on the shaft, B, which rests on the supporting and driving wheels, C. The wheel, A, is spun by means of a string in the same manner as a top. By virtue of its inertia, the wheel, A, tends to continue its rotary motion. If unaffected by outside influences, it would run on forever; but the friction of its bearings and of the air and other causes combine to bring it to rest.

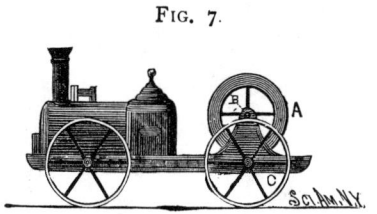

FIG. 7.

Inertia Locomotive.

The power imparted to and stored in the wheel, A, is given out in turning the wheels, C, overcoming friction, and propelling the machine forward.

FRICTION.

The resistance caused by the moving of one body in contact with another is known as friction. No perfectly smooth surface can be produced, all surfaces having minute projections or roughnesses, so that when the surfaces of any two bodies are moved in contact with each other, the projections of one body engage the projections of the other body, thus offering resistance to the free motion of the bodies. When the surfaces are covered with a lubricant, their inequalities are filled and smoothed over and the friction is lessened.

10 EXPERIMENTAL SCIENCE.

The friction developed by the sliding of one body upon another is known as "sliding friction," and the kind developed by the rolling of a body upon another is "rolling friction." Rolling friction absorbs much less power than sliding friction. Owing to this fact, the journals and steps

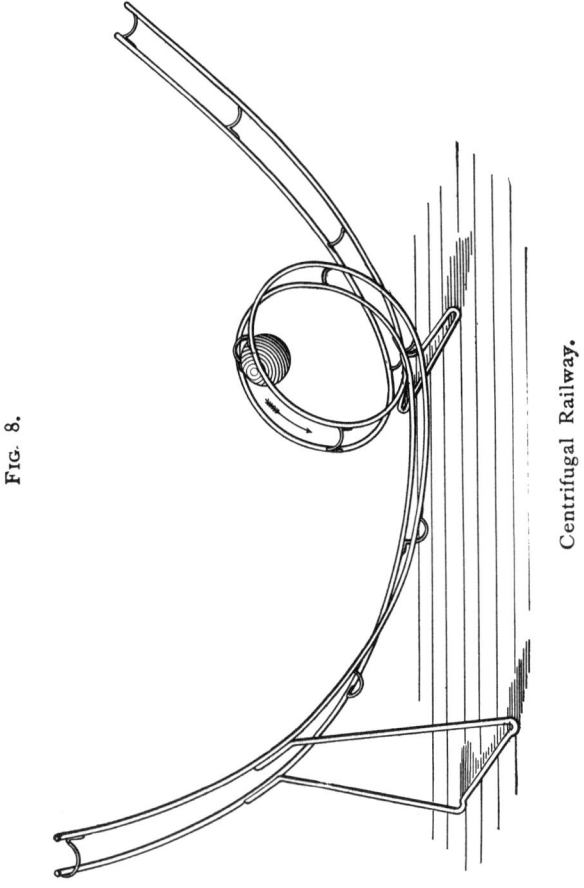

FIG. 8.

Centrifugal Railway.

of many kinds of machinery are provided with roller or ball bearings, thus substituting rolling for rubbing surfaces. An example of bearings of this kind is found in the pedals and shafts of bicycles and tricycles, which are provided with ball bearings.

CENTRIFUGAL FORCE.

The normal path of any moving body is a straight line; the body can be made to move in a curved path only by restraining it sufficiently to counteract its tendency to leave a circular path and move in a straight line. This tendency is called centrifugal force. When a body moving in a circular path is released, it does not fly off radially, but on a line tangent to the circular path. The fact that a body traveling in a circular path, when released from all restraint, will move in a straight line, proves that the normal path of a moving body is a straight line. The centrifugal railway represented in Fig. 8 shows with what force a restrained body tends to fly from a circular path.

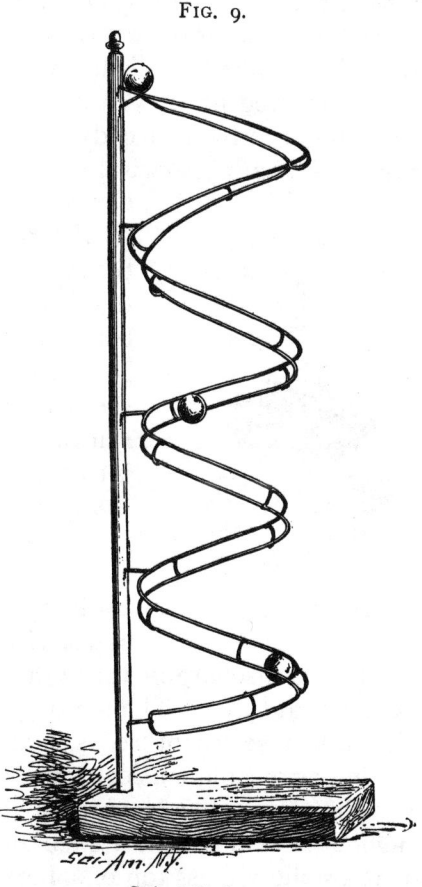

FIG. 9.

Spiral Railway.

This railway is made in the same manner as the swiftest descent apparatus described on another page. Two wires are bent into spiral loops around a cylinder, and the extremities are curved upwardly as shown. The two curved wires are connected together by curved wire cross pieces fastened by soldering, and two wire feet are attached to complete the apparatus. No particular rule is required for the construction of the centrifugal railway. The only precaution necessary is to see that the height of the higher end of the railway is to the height of the circular part in a greater ratio than 5 to 4.

A ball started at the higher end of the railway follows the track to the opposite end, and at one point in its travel it is held by centrifugal force against the under side of the track in opposition to the force of gravity.

In Fig. 9 another example of centrifugal action is exhibited by a spiral railway upon which a ball rolls down upon a track consisting of two rails arranged vertically one over the other. The track is formed of two wires bent spirally and connected by curved cross pieces, as in the case of the centrifugal railway already described. The upper convolution of the spiral is twisted so that the ball may start on a horizontal track. During its descent on the twisted portion of the track, the ball acquires sufficient momentum to cause it to follow the vertical track, being held outwardly against the rails by centrifugal force. The descent of the ball is accelerated. The spiral railway represented in the engraving is two feet high, six inches in diameter, the rails being ¾ inch apart.

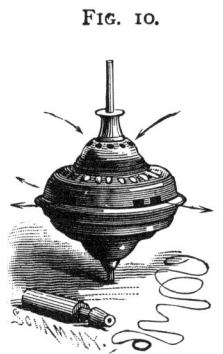

FIG. 10.

The Choral Top.

The effect of centrifugal force on air is beautifully exhibited by the ordinary choral top. As the top spins, air, which enters the holes at the top, is discharged through the holes at the equator by centrifugal force. The air, in going through the top, passes through a series of reeds, setting them in vibration, producing agreeable musical sounds.

The annexed engraving shows a very simple but effective device for exhibiting the effect of centrifugal force on liquids. It is a hollow glass top of spherical form, having a tubular stem, and a point on which to spin.

These tops are filled with various liquids, some of them containing two or more. The one shown at Fig. 11 is filled partly with water and partly with air. When the top is spun, the water flies as far from the center as possible, leav-

REST, MOTION, AND FORCE.   13

ing in the center of the sphere an air space, which at first is almost perfectly cylindrical, but which gradually assumes the form of a parabola as the velocity of the top diminishes.

At 2 is shown a top having a filling consisting of air,

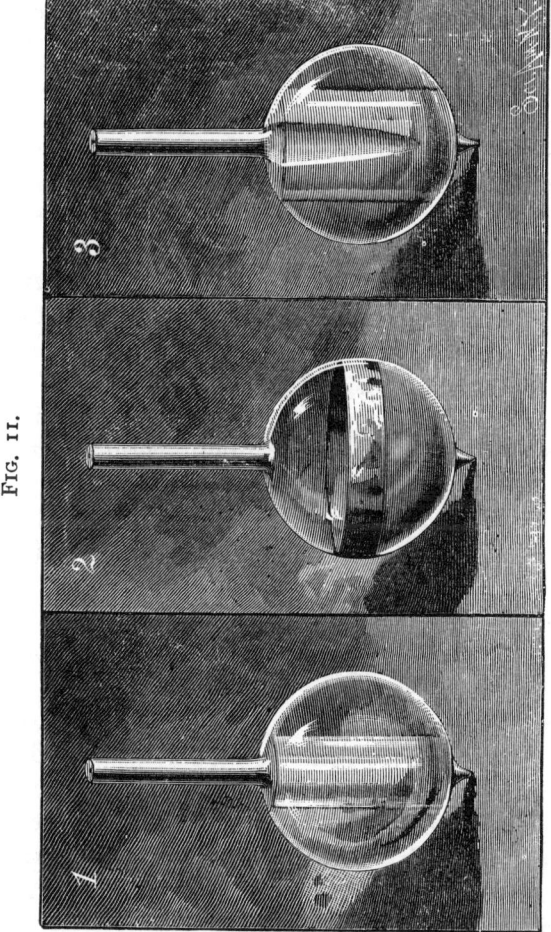

FIG. 11.

Top for Showing the Action of Centrifugal Force on Liquids

water, and a small quantity of mercury. The water acts as above described, and the mercury forms a bright band at the equator of the sphere.

At 3 is shown a top containing water and oil (kerosene).

The water, being the heavier liquid, takes the outside position, the oil forming a hollow cylinder with a core of air.

The top, after being filled, is corked and sealed. It is spun by the hands alone or with a string and the ordinary handle. The diameter of the top is $1\frac{1}{2}$ inches. It is made of considerable thickness, to give it the required weight and strength.

A SCIENTIFIC TOP.

Every street urchin can spin a top, and get an unending amount of amusement out of it; but it would seriously puzzle the majority of "boys of larger growth" to satisfactorily explain all the phenomena of this simplest of toys.

Why does it continue to revolve after being set in motion? Why does its motion ever cease? Why does it so persistently maintain its plane of rotation? When its axis is inclined to the vertical, why does it revolve slowly around a new axis while turning rapidly upon its own axis? And when so inclined, why does it gradually right itself until it rotates in a horizontal plane? Why does it not revolve proportionately longer when its speed is increased? These and many other questions arise when we begin the examination of the action of the top. They have all been answered so far as it is possible to answer them, still it is difficult to reach far beyond the mere knowledge of the actions themselves.

The top has already risen to some importance as a scientific toy, but it is worthy of being elevated to the dignity of a truly scientific instrument. To give it that eminence, three things are necessary: first, a considerable weight, and in consequence of this an easy and effective method of spinning, and finally it requires a good bearing, having a minimum of friction.

The top illustrated has these three requisites. It weighs $3\frac{1}{2}$ pounds, and its weight might be increased somewhat with advantage. It has a frictional spinning device by which a velocity of 3,000 revolutions per minute may readily be attained. It is provided with a hardened steel pivot which

REST, MOTION, AND FORCE. 15

PLATE II.—A SCIENTIFIC TOP.—1. The Top. 2. Persistence in Maintaining Plane of Rotation. 3. Gyroscopic Action. 4, 5, 6. Examples of Centrifugal Action. 7. Formation of Oblate Spheroid. 8, 9, 10, 11. Examples of Centrifugal Action on Liquids. 12. Centrifugal Hero's Fountain.

turns on an agate or steel step.* It is almost perfectly balanced, and the friction of its bearing is very slight. When unencumbered, it will run for over 42 minutes in the open air with once spinning, and its motion may at any time be accelerated without stopping, by a new application of the friction wheel.

The brass body of the top is 6 inches in diameter, and $\frac{5}{8}$ inch thick in the rim. Its steel spindle is $\frac{3}{8}$ inch in diameter and has a tapering longitudinal hole which is $\frac{1}{4}$ inch in diameter at its larger end. To this tapering hole is fitted the tapered end of a rod supporting the stud on which the friction driving wheel turns. The upper end of the rod is provided with a handle, and to the boss of the friction wheel is secured a crank.

A sleeve fixed to the spindle of the top is furnished with an elastic rubber covering which is engaged by the beveled surface of the driving wheel. After imparting the desired speed to the top, by turning the driving wheel, the wheel and the rod by which it is supported may be withdrawn from the top, without interfering in any way with its action.

A large number of interesting experiments may be performed by means of a top of this character. Most demonstrations possible with the whirling table may be adapted to this top, and, besides, many phenomena peculiar to the top itself may be exhibited. A few of the more striking experiments are illustrated.

By suddenly pressing upon one side of the top with a small rubber-covered wheel, as shown in Fig. 2 (Plate II.), it will be found impossible to change its plane of rotation by the application of any ordinary amount of force. In fact, the side of the top to which the pressure is applied will rise rather than yield to the pressure.

By placing the step of the top on an elevated support, such as a tumbler, as shown in Fig. 3 (Plate II.), and gently pressing against one side of the spindle, the axis of the top will be gradually inclined, and a gyroscopic action will be

---

* An agate mortar of the smallest size, about $1\frac{1}{4}$ inches in diameter, mounted in a wooden base, forms a very good step, but a steel disk, having a concave upper surface, and made as hard as possible, is preferable.

set up. The top will swing around with a very slow, majestic movement, traveling six or eight turns per minute around a vertical axis while revolving rapidly on its own axis, and it will slowly regain its original position.

As the peripheral speed of the top is almost a mile a minute, a little caution is necessary in handling it while in rapid motion, as any treatment that will cause it to leave its bearings will be sure to result in havoc among the surroundings, besides being liable to injure the operator.

Several methods of showing centrifugal action are illustrated, the simplest being that shown in Fig. 4 (Plate II.) A small Japanese umbrella, about 20 inches in diameter, is arranged to be rotated by the top, by applying to its staff a tube which fits over the spindle of the top. In this, as well as the other experiments, the top is set in motion before the object to be revolved is applied. The tube attached to the umbrella having been placed on the revolving spindle, the arms are thrown up by centrifugal action, thus spreading the umbrella.

Fig. 5 (Plate II.) shows a ring formed of two pieces of heavy rubber tubing secured to two metallic sleeves fitted to a rod adapted to the tapering hole of the top spindle. The lower sleeve is fixed, and the upper one is free to slide up or down on the rod. Normally, the rubber forms a ring, as shown in dotted lines, but, when rotated, the centrifugal force reduces it to a flat ellipse. A similar experiment, in which two elastic rings are secured on opposite sides of the rod, is shown in Fig. 6 (Plate II.); the rings being circular when stationary and elliptical when revolved.

In Fig. 7 is shown a device for illustrating the formation of an oblate spheroid. A tube, closed at the lower end and fitted to the hole in the top spindle, is provided near its lower end with a fixed collar and a screw collar, between which the lower wall of a hollow flexible rubber sphere is clamped. The upper wall of the sphere is clamped in a similar way between collars on a sleeve arranged to slide on the tube. The tube is perforated above the lower pair of collars to admit of filling the hollow ball with water. When the ball is filled or partly filled with water, and rotated, it becomes flattened at the poles and increases in diameter at the equator, perfectly illustrating the manner in which the earth received its present form.

The glass water globe represented in motion in Fig. 8 exhibits a cylindrical air space extending through it parallel with the axis of rotation, the water having been carried as far as possible from the center of rotation by centrifugal action.

When the speed of the globe is reduced, gravity asserts itself and the air space assumes a parabolic form, as shown in Fig. 9 (Plate II.)

In the globe represented in Fig. 10 the filling consists of water and mercury. The rotation of the globe causes the mercury to arrange itself in the form of a narrow band at the equator of the globe.

Fig. 11 shows a globe filled with air, oil, and water, which, when the globe is revolved, arrange themselves in the order named, beginning at the center of the globe.*

A Hero's fountain, operated by centrifugal force instead of gravity, is shown in Fig. 12 (Plate II.) The metallic vessel contains three concentric compartments. The jet tube extends downward into the central compartment and is bent laterally, so that it nearly touches the wall of the compartment. The intermediate compartment communicates with the outer compartment, and the outer and central compartments are connected by an air duct. The central and intermediate compartments are filled with water, and as the vessel is revolved the water in the intermediate compartment is carried by centrifugal action into the outer compartment, and, compressing the air contained in that compartment, drives it through the air duct, with a force due to the centrifugal action, into the central compartment, where it exerts a pressure on the water sufficient to cause it to be discharged through the jet.

* See also chapter on projection.

CHAPTER III.

THE GYROSCOPE.

This instrument has always been a puzzle to physicists. Its phenomena seems to be incapable of explanation in a popular way. In view of the complicated nature of the calculations involved, no attempt will here be made to explain the action of the gyroscope mathematically,* the object of the present article being merely to describe a few modifications of the instrument and to mention peculiarities noticed in the performance of some of these modified forms.

Fig. 12.

Toy Gyroscope.

The difficulty of securing a high speed in a large gyroscope led to the application of a friction driving device, as shown in Figs. 13 and 13a, by means of which an initial velocity of from 4,500 to 5,000 revolutions per minute may readily be attained.

The instrument, after being set in motion, behaves like other gyroscopes not provided with means for maintaining the rotary motion of the wheel, but its size and the facility with which it may be operated render it very satisfactory.

The gyroscope wheel is 6 inches in diameter, $\frac{5}{8}$ inch thick, and, together with its shaft, weighs $3\frac{1}{2}$ pounds. The annular frame weighs $1\frac{3}{4}$ pounds. So that $5\frac{1}{4}$ pounds must be sustained by gyroscopic action when the counterbalance is not applied.

The driving wheel is $7\frac{3}{4}$ inches in diameter. Its face is

---

* For a mathematical explanation see "Rotary Motion as applied to the Gyroscope." by Gen. J. G. Barnard.

¾ inch wide. Its shaft is journaled in an arm pivoted to the base, with its free end adapted to enter a recess in the edge of the annular frame, for supporting the gyroscopic wheel while motion is being imparted to it. Upon the shaft of the

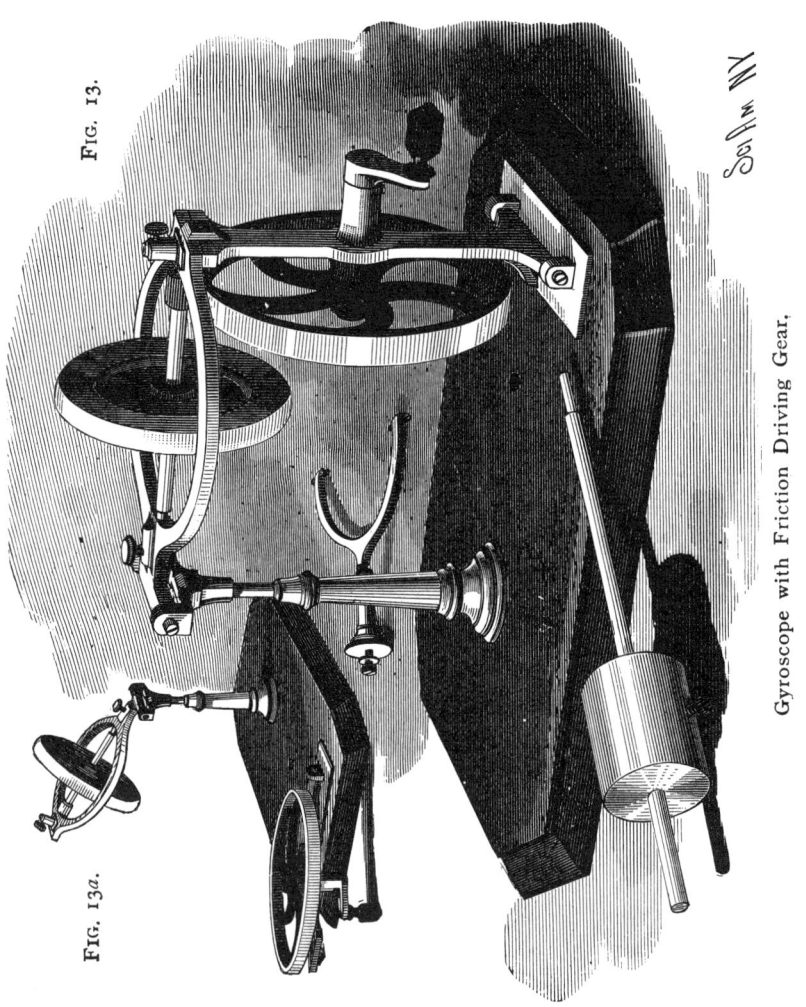

FIG. 13.
FIG. 13a.
Gyroscope with Friction Driving Gear.

gyroscope wheel is secured a soft rubber tube having an external diameter of nine-sixteenths inch. This shaft makes 13.84 revolutions to one turn of the drive wheel, so that when the drive wheel is turned six times per second, the

THE GYROSCOPE.

gyroscope wheel will make very nearly 5,000 turns per minute (4,982).

This gyroscope may be arranged as a Bohnenberger apparatus by removing the tall standard and attaching the shorter one to the center of the base by means of a bolt. The annular frame of the instrument is suspended on pivotal screws in the extremities of the semicircular support, which is capable of turning on the upper end of the short standard. In the engraving the short standard, together with the semicircular support, is shown lying on the table. The usual counterbalance is also shown lying on the table. Fig. 13 shows the drive wheel in position for imparting motion to the gyroscopic wheel, and Fig. 13a shows the driving wheel withdrawn and the gyroscope in action.

As this instrument does not differ from the ordinary one, except in the application of the driving mechanism, it will be unnecessary to go into particulars regarding its performance.

In Figs. 14, 15, and 16 are shown pneumatic gyroscopes, and Fig. 17 represents a steam gyroscope.

The pneumatic gyroscope shown in Fig. 14 consists of a heavy wheel provided with flat arms arranged diagonally, like the vanes of a windmill. The wheel is pivoted on delicate points in an annular frame having an arm pivoted in a fork at the top of the vertical support. The arm of the annular frame carries a tube, which terminates near the vanes of the wheel in an air nozzle which is directed toward the vanes at the proper angle for securing the highest velocity. The opposite end of the tube is prolonged beyond the pivot of the frame.

The support of the annular frame, shown in vertical section in Fig. 15, consists of an inner and outer tube, the inner tube having a closed upper end terminating in a pivotal point. The lower end of this tube communicates with the horizontal tube, through which air is supplied to the machine.

A sleeve, closed at its upper end and carrying the fork in which the arm of the annular frame is pivoted, is inserted in the space between the inner and outer tubes, and turns

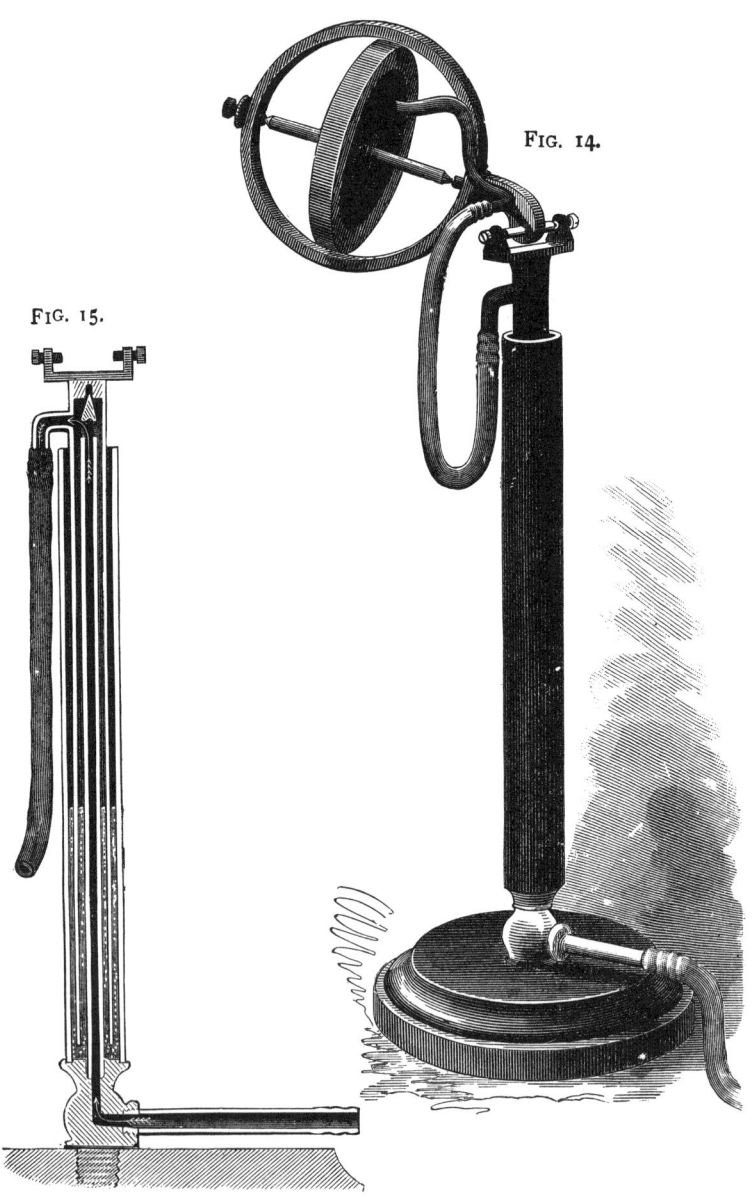

Pneumatic Gyroscope.

on the pointed end of the inner tube. The inner tube is perforated near its pointed end, to permit of the escape of air to the interior of the sleeve, and the lower end of the sleeve is sealed by a quantity of mercury contained by the space between the inner and outer tubes. The air pipe carried by the annular frame communicates with the upper end of the sleeve by a flexible tube. When air under pressure passes through the inner pointed tube, through the sleeve, and through the air nozzle, and is projected against the vanes of the wheel, the wheel rotates with great rapidity, and the gyroscope behaves in all respects like the electrical gyroscope referred to.

The gyroscope shown in Fig. 16 is adapted to the standard just described, but the heavy wheel is replaced by a very light paper ball, whose rotation is maintained by two tangential air jets, which play upon it on diametrically opposite sides, and nearly oppose each other, so far as their action on the surrounding air is concerned. The rotary motion is produced solely by the friction of the air on the surface of the ball. The upwardly turned nozzle is arranged to deliver an air blast which is a little stronger than that of the lower nozzle, so that a slight reactionary force is secured, which assists the gyroscope in its movement around the vertical pivot sufficiently to cause the ball to maintain its horizontal plane of rotation continuously. In fact, this gyroscope will start from the position of rest, raise itself in a spiral course into a horizontal plane, and afterward continue to rotate in the same plane so long as air under pressure is supplied.

It may be questioned whether this machine is a true gyroscope. However this may be, it is certain that the reactionary power of the stronger air jet is of itself insufficient to produce the motion about the vertical pivot; neither is there a sufficient vacuum at the top of the ball to produce any appreciable lifting effect.

The steam gyroscope shown in Fig. 17 hardly needs explanation. It differs from all the others in generating its own power within its moving parts. The boiler is supported by trunnions resting in a fork arranged to turn on a fine vertical pivot. The engine is attached to the boiler, so that both engine and boiler swing on the trunnions in a vertical plane. The wheel of the engine is made disproportionately large and heavy, to secure the best gyroscopic action.

The performance of the steam gyroscope is like that of

FIG. 16.

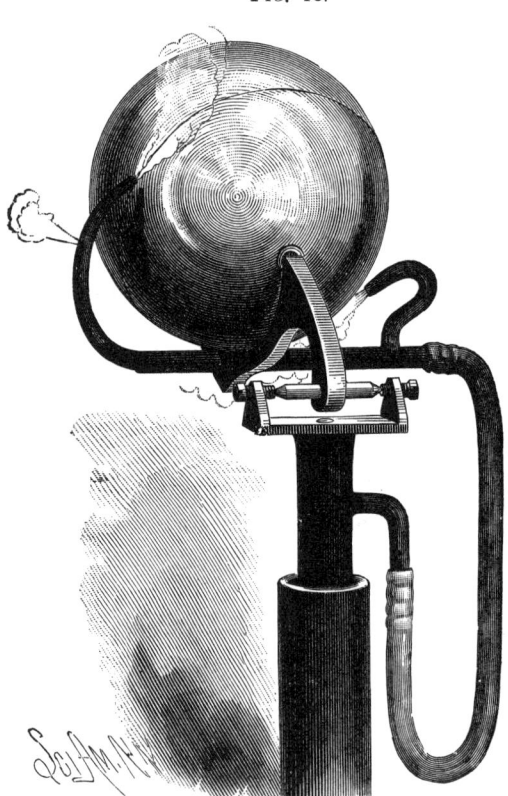

Pneumatic Gyroscope having Continuous Action.

the other power-propelled gyroscopes, and needs only a reactionary jet of steam or some other slight force to keep up the rotation around the vertical pivot, and thus render the action of the instrument continuous.

AN ELECTRICAL GYROSCOPE.

To render the operation of the gyroscope as nearly con-

FIG. 17.

Steam Gyroscope.

tinuous as possible, so that its movements may be more thoroughly studied, electricity has been applied as a motive agent.

The gyroscope illustrated in Plate I. (frontispiece) and in Fig. 18 has a weighted base piece, from which projects a pointed standard that supports the moving parts of the instrument. The frame, of which the electro-magnets form a part, has an arm in which is fastened an insulated cup, that rests upon the point of the standard. One terminal of the magnet coil is connected with this cup, and the other terminal is connected with the yoke connecting the cores of the two magnets.

To the top of the yoke is secured a hard rubber insulator, which supports a current-breaking spring arranged to touch a small cylinder on the wheel spindle twice during each revolution of the wheel.

The wheel, whose plane of rotation is at right angles with the magnet cores, carries a soft iron armature, which turns very near the face of the magnet, but does not touch it. The armature is arranged in such relation to the contact surface of the current-breaking cylinder that twice during each revolution, as the armature nears the magnet cores, it is attracted, but immediately the armature comes directly opposite the face of the magnet cores, the current is broken, and the acquired momentum is sufficient to carry the wheel forward until the armature is again within the influence of the magnet.

The current-breaking spring is connected with a fine copper wire, that extends backward as far as the pointed standard, and is coiled several times to render it very flexible, and is finally bent downward so as to dip in mercury contained in an annular vulcanite cup placed on the pointed standard near the base piece.

The base piece is provided with two binding posts for receiving the battery wires. One of the binding posts is connected with the pointed standard, and the other communicates by a small wire with the mercury in the vulcanite cup.

The wheel, magnet, and parts connected therewith are

free to move in any direction on the point of the standard. When two large or four small Bunsen cells are connected with the gyroscope, the wheel revolves with enormous velocity, and upon letting the magnet go (an operation requiring some dexterity), the wheel sustains not only itself, but also the magnet and other parts between it and the point of the standard, in opposition to gravity.

Fig. 18.

Electrical Gyroscope.

The wheel, besides rotating rapidly on its axis, sets up a slow rotation about the pointed standard in the direction in which the under side of the wheel is moving.

By attaching the arm and counterbalance shown in the engraving, so as to exactly balance the wheel and magnets on the pointed standard, the whole remains stationary. By overbalancing the wheel and magnets, the rotation of the apparatus around the standard is in an opposite direction, or in the direction in which the top of the wheel is turning.

This gyroscope illustrates the persistency of a rotating body in maintaining its plane of rotation. It also exhibits the result of the combined action of two forces tending to produce rotations about two separate axes lying in the same plane, one force being gravity.

The rotation of the wheel upon its axis, produced in this instance by the electro-magnet, and the tendency of the wheel to fall, or rotate in a vertical plane about a second horizontal axis at right angles to the first, results in a tendency to continually rotate about a new horizontal axis intermediate between the two. The continual adaptation to this new axis implies rotation of the whole mass additionally around a vertical axis which is coincident with that of the pointed standard.

ELECTRICAL GYROSCOPE FOR SHOWING THE ROTATION OF THE EARTH.

Although the apparent displacement of the plane of vibration of the pendulum had long been noticed, it was not until the year 1852 that the fact was coupled with the diurnal rotation of the earth. In September of that year M. Foucault, the distinguished French physicist, suspended a ball, by means of a fine wire, from the dome of the Pantheon at Paris, and for the first time in the history of the world made visible the rotation of the earth. The pendulum thus formed, after receiving an impulse, vibrated for many hours, and preserved its plane of vibration while the earth slowly turned under it. This splendid experiment was subsequently repeated at the Capitol at Washington, and at other places.

Soon after the pendulum experiment, Foucault, to illustrate the same thing, constructed a gyroscope which was a modification of Bohnenberger's machine. This gyroscope received a rotating impulse from the hand of the operator, and the momentum of the disk was depended on to continue the rotation for a sufficient length of time to exhibit the movement of the earth.

FIG. 19.

Gyroscope for showing the Earth's Rotation.

To furnish a more practicable means of making visible the diurnal movement of the earth, the action of the gyroscope is made continuous by applying electricity as a propelling power.

In Fig. 19 (which represents the machine arranged for the purpose named) the rectangular frame which contains the wheel is supported by a fine and very hard steel point, which rests upon an agate step in the bottom of a small iron cup at the end of the arm supported by the standard.

The wheel spindle turns on carefully made steel points. Upon the spindle are placed two cams—one at each end—which operate the current-breaking springs.

The horizontal sides of the frame are of brass, and the vertical sides are iron. To the vertical sides are attached the cores of the electro-magnets, and the wheel is provided with two armatures—one on each side—which are arranged at right angles to each other. The two magnets are oppositely arranged in respect to polarity, to render the instrument astatic.

An insulated stud projects from the middle of the lower end of the frame to receive an index that extends nearly to the periphery of the circular base piece and moves over a graduated semicircular scale. An iron point projects from the insulated stud into a mercury cup in the center of the base piece, and is in electrical communication with the platinum-pointed screws of the current breakers. The current-breaking springs are connected with the terminals of the magnet wires, and the magnets are in electrical communication with the wheel-supporting frame.

One of the binding posts is connected by a wire with the mercury in the cup, and the other is connected with the standard. A drop of mercury is placed in the cup that contains the agate step, to form an electrical connection between the iron cup and the pointed screw. The instrument is covered with a glass shade to exclude air currents, and the base piece is provided with leveling screws.

The current breaker is contrived to make and break the current at the proper instant, so that the full effect of the magnets is realized, and when the binding posts are con-

THE GYROSCOPE.

nected with four or six Bunsen cells, the wheel rotates at a high velocity.

The wheel will maintain its plane of rotation, and when it is brought into the plane of the meridian, the index will appear to move toward the right of a person facing northward with the index pointing northward in front of him. To a person in New York, therefore, the index seems to turn *toward the east*. To a person at the north pole, where

FIG. 20.

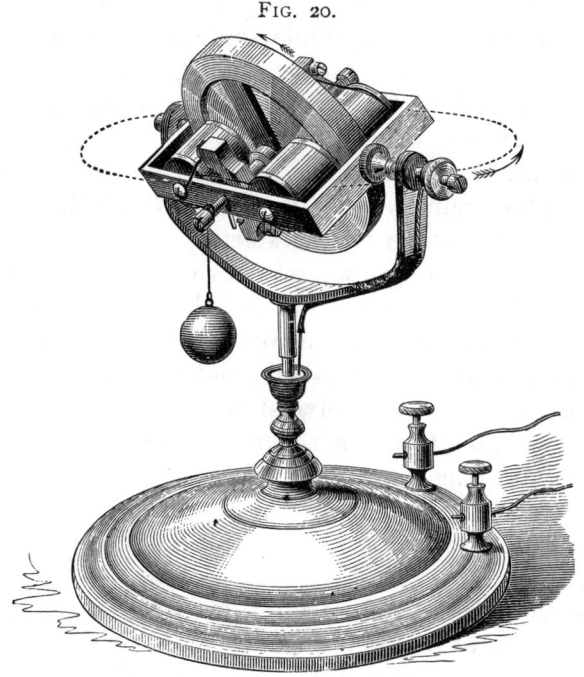

Electrical Gyroscope.

north is up and east is left, the hourly deviation is 15° *rightward*, or *westward*. At the equator there is, of course, no deviation.

It makes no difference whether the index points northward or southward, its apparent motion is always toward the right, thus affording visible evidence that the earth rotates.

The instrument thus described may be easily modified,

so as to illustrate other interesting phenomena of rotary motion.

By removing the index and point from the insulated stud at the lower part of the frame and unscrewing the supporting piece from the top of the frame, the frame may be suspended in a horizontal position upon pointed screws in a fork which is supported upon a vertical pivot, as shown in Fig. 20.

The pointed screw entering the insulated stud is itself insulated, and communicates, by an insulated wire, with mercury contained in an annular vulcanite cup on the fork-supporting pivot. One of the binding posts is connected with the pivot of the fork and the other communicates with the mercury in the vulcanite cup.

When the instrument is connected with a battery, the wheel revolves rapidly, and if undisturbed will remain in the position in which it was started. If a small weight, such as a key, be hung upon one of the pivot screws of the wheel spindle, the frame containing the wheel does not turn quickly on its pivots, as might be expected, or as it would if the wheel were not revolving, but the entire apparatus immediately begins to revolve slowly on the vertical pivot, while the weighted side of the frame descends almost imperceptibly. Transfer the weight to the opposite pivot, and while the wheel still revolves in the same direction, the apparatus will turn on the vertical pivot in the opposite direction.

By removing the weight from the pivot screw and turning the apparatus on the vertical pivot, the converse of what has just been described will result; that is, the wheel besides revolving on its own axis will turn in a plane at right angles to its plane of rotation.

If the apparatus be turned on the vertical pivot in the opposite direction, the rotation of the wheel on its new axis will be reversed, and by oscillating the apparatus on the vertical pivot the wheel and frame will revolve rapidly on the pointed screws that support the frame.

The law controlling these movements is as follows: "Where a body is acted upon by two systems of forces,

THE GYROSCOPE.   33

tending to produce rotations about two separate axes lying in the same plane, the resultant motion will be rotation about a new axis situated in the same plane between the directions of the other two."

By means of this continuously operating gyroscope Dr. Magnus' experiments showing some of the causes of deviation of projectiles may be exhibited.

EQUATORIALLY MOUNTED ELECTRICAL INDICATOR.

In Fig. 21 a gyroscope is shown which is suspended with the axis of the wheel-supporting frame, C, at right angles

FIG. 21.

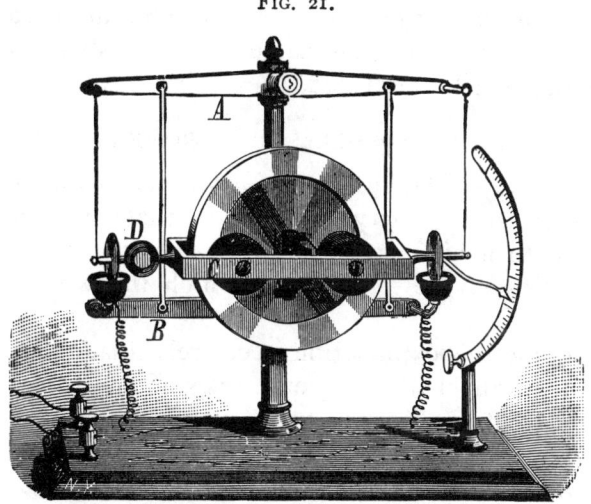

Electrical Indicator.

to the plane of the equator and parallel with the polar axis of the earth. The frame, C, is suspended by silk threads from studs that project from the beam, A. Two vulcanite mercury cups are supported by the beam, B, in position to make an electrical connection with the disks on the axes of the frame, C. These cups are connected by a spirally coiled wire with the binding posts that receive the battery wires. The beams, A, B, are connected by rods, so that when it is desired to adjust the instrument, the parts will maintain their proper relation.

Upon one of the axes of the frame, C, there is an index that moves in front of the scale of degrees. Upon the other axis there is a small mirror, D, for receiving a beam of light and projecting it on a screen. By this arrangement a very long index is secured without additional weight.

The instrument as represented in the engraving is adjusted for the equator. In New York the axis of the wheel-supporting frame would have to be adjusted at an angle of 40° 41' with the horizon.

The instrument shown in the engraving should, when the axis of the frame, C, is adjusted equatorially, indicate 15° motion per hour in any latitude.

The arrangement of the wheel, the commutator, and connections is substantially the same in this instrument as in the one previously described.

BURSTING OF FLY-WHEELS BY GYROSCOPIC ACTION.

The theory of the bursting of fly-wheels, which has been accepted in the majority of cases, is that the centrifugal force due to a high velocity overcomes the cohesion of the particles of the material of which the wheel is composed.

Of course this explanation is entirely inadequate when applied to a wheel whose strength is sufficient to resist any tendency to fly to pieces from purely centrifugal force under the conditions of its use; but of the fact that such wheels burst no evidence is needed, and some cause other than centrifugal force must be assigned for the bursting.

Supposing the fly-wheel to be perfectly balanced and without defects in material or design, it may be driven without danger at any velocity usually considered within the limit of safety, so long as it continues to rotate in a plane at right angles to its geometrical axis. And it may be moved in the plane of its rotation or at right angles to it, that is, in the direction of the length of the shaft, without creating any more internal disturbance than would result from moving it in the same way while at rest. But when a force tending to produce rotation at right angles to the plane of the wheel's rotation is applied, the effect will be

THE GYROSCOPE.

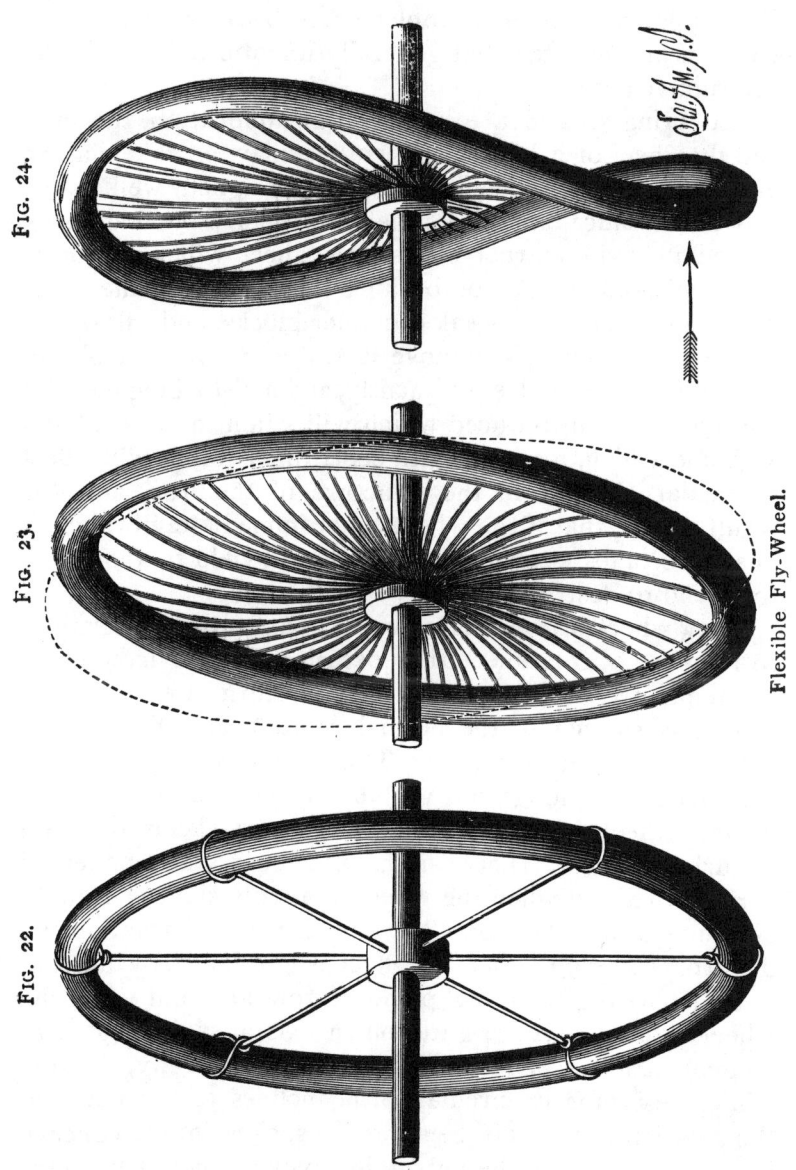

Fig. 24.
Fig. 23.
Fig. 22.

Flexible Fly-Wheel.

vastly different, and the result will be a tendency to rotate about a new axis between the other two, and the centrifugal strain upon the wheel is supplemented by a twisting strain, which is an important but generally unnoticed factor in the destructive action.

To bring this idea to a practical application, the shaft and fly-wheel of a high-speed engine may be taken as an example. Let the wheel be correctly designed, well made, and well balanced, and if its shaft is properly lined and supported in rigid journal boxes, the wheel will perform its office without danger of bursting; but support the same wheel and shaft upon weak plummer blocks, and allow one or both of its journals to move laterally at every stroke of the engine, or even less frequently, and a disturbing element will have been introduced which will strain the wheel laterally, and which, together with centrifugal force, will effect molecular changes in the structure of the iron, and the result will be that if the wheel is not immediately broken it finally becomes weakened, so that it will yield to the forces that tend to destroy it.

Any wheel whose axis is swung in a plane at right angles to its plane of rotation, either occasionally and irregularly or frequently and regularly, tends to turn laterally on an axis between that of the normal rotation and that of the extraneous disturbing force. This tendency exists in ordinary wheels, although not visible The engraving shows a flexible wheel, which clearly exhibits the effects of these disturbing forces. The rim is of rubber, the spokes of spring wire, and when the wheel is revolved very rapidly and moved in a plane parallel with its plane of rotation, no disturbance results, and no effect is produced by moving it at right angles to its plane of rotation; but when the wheel is turned even slightly on an axis at right angles to its geometrical axis by swinging the shaft laterally, the rim, while preserving its circular form, inclines to the plane of the rotation of its shaft, bending the spokes into a concave form on one side of the hub and convex on the other, showing the effects of the disturbing force on the figure of the wheel, as in Fig. 23.

THE GYROSCOPE.

When the disturbing force is rhythmical, lateral vibrations and wave motions are set up in the rim, which are out of all proportion to the extraneous force applied.

From this experiment it is evident that the lateral swinging of the shaft of a fly-wheel (for instance when its journal boxes are loose, or when the frame of the machine of which the fly-wheel forms a part is yielding) tends to weaken the wheel even when the lateral movement is slight; and where it is great, as when the shaft is broken, the twisting effect is correspondingly great, and the wheel or its support must yield.

No rotating machines are more subject to bursting than grindstones, and generally no rotating bodies of equal weight are mounted upon such small shafts or on such weak supports. The suspended ones are especially liable to the destructive action above described, as their frames are generally far too weak.

Fig. 24 illustrates the effect of a lateral blow on the rim of a fly-wheel. Of course the effect is much exaggerated in the flexible wheel, but it shows the form taken by the rim under a blow, the blow producing a much greater effect on the wheel while in motion than when at rest.

CHAPTER IV.

FALLING BODIES—INCLINED PLANE—THE PENDULUM.

" In a vacuum all bodies fall with equal rapidity." This is the first law of falling bodies. The well known guinea and feather experiment is a demonstration of this law. The

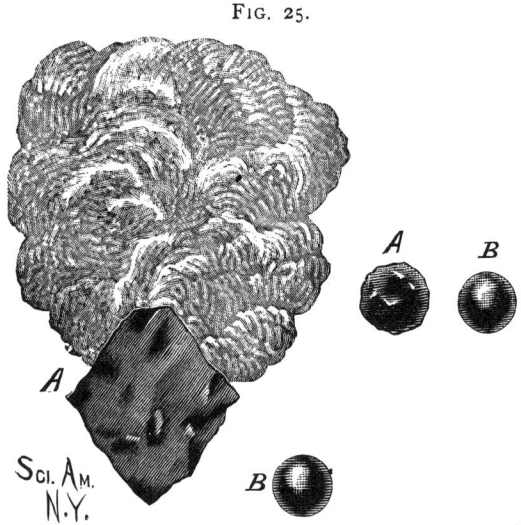

Fig. 25.

Effect of the Resistance of Air on Falling Bodies.

heavy body and the light one being dropped simultaneously in a tube deprived of air, reach the bottom at the same instant.

The converse of this experiment is illustrated in Fig. 25. In this case the retardation caused by the resistance of the air is clearly shown. A bunch of very loose cotton wool is attached to a small piece, A, of tin foil, and the cotton thus arranged is dropped simultaneously with the lead bullet, B. As would be expected, the bullet reaches the ground in about half the time required for the descent of the cotton.

FALLING BODIES—INCLINED PLANE—THE PENDULUM. 39

By rolling the cotton into a compact ball and inclosing it in the tinfoil, the surface exposed to the air will be very much diminished, and when the experiment is repeated with the cotton thus diminished in bulk, it is found that the two bodies fall with nearly equal rapidity.

The water hammer shown in Fig. 26 demonstrates that in a vacuum liquids fall like solids, without being broken up or divided. The water hammer consists of a glass tube half filled with water, which is boiled to expel the air, the tube being afterward sealed. When the tube is inverted, the water falls in a body, striking the opposite end of the tube, producing a sharp clink.

FIG. 26.

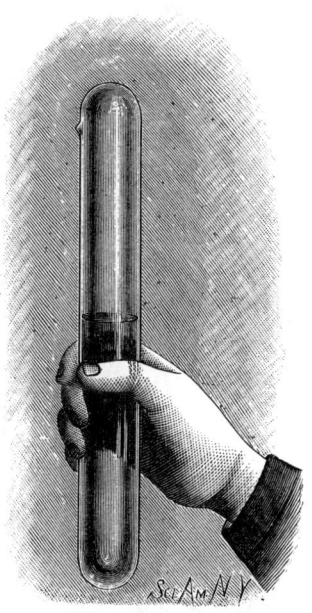

Water Hammer.

SWIFTEST DESCENT APPARATUS.

The descent of a falling body along an inclined plane is governed by the same law that controls the fall of free, unimpeded bodies, *i. e.*, " the spaces traversed are proportional to the squares of the times of descent." The law does not apply to the descent of a body along any curved path. A body descending a concave path will be accelerated most at the beginning of its fall. A body descending a convex path will start slowly, and will be increasingly accelerated as it approaches the end of its travel.

Three cases are here considered : First, that of a body rolling down an inclined plane; second, that of a body descending a concave circular curve; and third, that of a body descending a cycloidal curve. In the case of the inclined plane, if the body falls two feet in one second, it will fall eight feet in two seconds, eighteen feet in three seconds, and so on. In the case of the concave circular curve, the fall of the body will be accelerated rapidly at the start, and the body will reach the point of stopping quicker than the body on the inclined plane, although it travels over a longer distance. In the case of the cycloidal curve, the body acquires a high velocity at once, as its path at the beginning

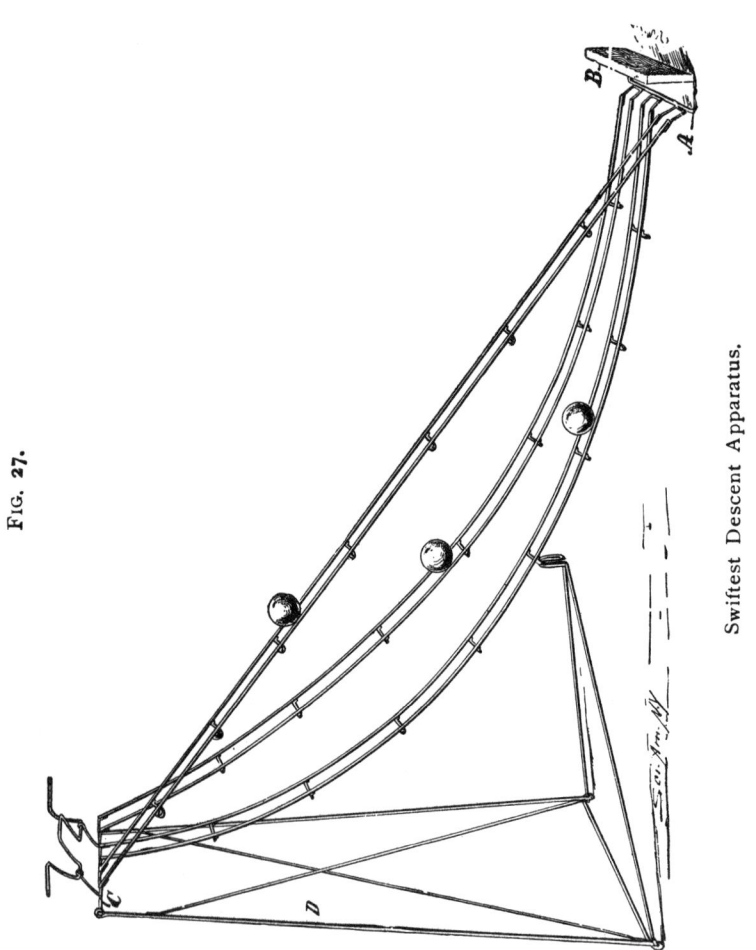

FIG. 27.

Swiftest Descent Apparatus.

is practically vertical. This curve has been called the curve of swiftest descent, as a falling body passes over it from the point of starting to the point of stopping in less time than upon any other path, excepting, of course, the vertical.

The cycloid has another property, in virtue of which it

FALLING BODIES—INCLINED PLANE—THE PENDULUM.

has been called the isochronal curve. A body will roll down this curve from any point in its length to the point of stopping in exactly the same time, no matter where it is started. For example, if it requires a second of time for a ball to roll from the upper to the lower end of the curve, it will also take one second for a ball to roll from the center of the curve to its lower end.

Apparatus for illustrating these principles is shown in Fig. 27. It does not differ much from the ordinary apparatus used for the same purpose. It is, however, made entirely of wire, and is arranged to fold, so that it occupies little space when not in use. The rails of the tracks are formed of one-eighth inch brass wire. These rails are connected by curved cross pieces having ends bent at right angles and soldered to the under surface of the rails. The lower ends of the rails are connected by angled wires with a cross bar, A, which is bent forward, then upward, to receive the board, B, forming the stop for the balls. The upper ends of the rails are connected by angled wires with a cross bar, C, which receives the loops of the wire leg, D. To the leg is jointed a brace which hooks over one of the cross pieces of the middle track.

To the upper cross bar are soldered wire eyes, supporting a wire bent so as to form three cranks for holding the balls, and releasing them all together. The rods of which the tracks are formed are about three feet long. The cycloid track is made first, the others being cut off to match. A method of laying out the cycloid curve is shown in Fig. 28. At the end of the base line, A D, draw the line, C D, perpendicular to A D. Describe a generating semicircle (in this case of nine inch radius) tangent to A D, at D. Through its center draw the line, E C, parallel to the base line. Divide the semicircle into any number of equal parts —six for example—and lay off on A D and E C distances equal to the radius C D $\times$ 3·1416, and divide A D and E C into six equal parts, C 1′, 1′, 2′, etc., equal to the divisions of the semicircle; draw chords, D 1′, D 2′, etc. From points 1′, 2′, 3′, etc., on the line, C E, with radii equal to that of the generating semicircle, describe arcs.

From points 1', 2', 3', 4', 5', on the line, D A, and with radii equal successively to the chords, D 1, D 2, D 3, D 4, D 5, describe arcs cutting the preceding, and the intersections will be the points of the curve required. Through these points the curve is drawn, and the wires for the cycloid track are bent so as to conform to this curve. The track, when completed, must sustain the same relation to a horizontal line as the curve in the diagram sustains to the base line, A D.

Another method of describing a cycloid is to fix a pencil in the edge of a disk and roll the disk on a level surface, without slipping, with a pencil in contact with a smooth board

FIG. 28.

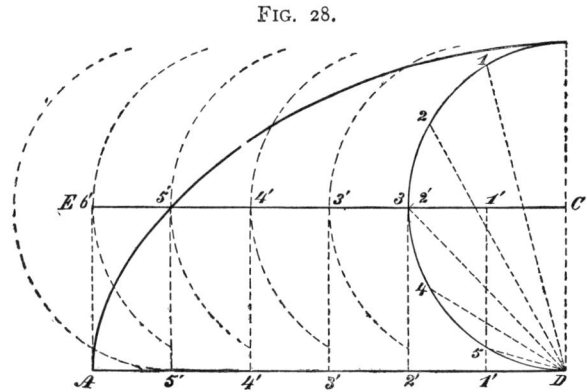

Method of Describing the Cycloid.

or a piece of paper, the curve being started with the pencil at the lowest point or in contact with the base line.

A ball is supported at the upper end of each track by the cranked wire, and when the three balls are liberated simultaneously by quickly turning up the cranked wire, it will be found that the ball on the cycloid reaches the point of stopping first, the ball on the circular curve coming next, the ball on the inclined plane being slowest of all.

If two cycloidal tracks be placed side by side, it will be found by trial that a ball started from the middle or at any point between the ends of one of the tracks will reach the point of stopping no sooner than the ball started at the top

FALLING BODIES—INCLINED PLANE—THE PENDULUM. 43

of the other track. In fact, if the tracks are accurately made, both balls, if started simultaneously, will reach the bottom at the same time.

DROPPED AND PROJECTED BALLS.

Although there is no shorter or quicker route for the descent of a falling body than that of a plumb line, it has been shown that a body projected horizontally with whatever force, and describing a long trajectory, will reach the earth in exactly the same time as another similar body simply dropped from the same height. There are many simple and ingenious devices for demonstrating this fact. If the experiment could be brought within convenient compass for observation, nothing would be better for the purpose than an ordinary gun, with powder as the propelling power, but this is of course out of the question. It is therefore necessary to resort to apparatus which may be used in an ordinary room, so that both projected and falling ball may be seen and heard. The apparatus is still a gun, but a very harmless and inexpensive one. It is a modified " Quaker gun," a well known toy used for shooting marbles.

Fig. 29 is a perspective view of the gun, showing it immediately after its discharge, and Fig. 30 is a longitudinal section showing the gun ready to be discharged. The gun consists of a wooden barrel chambered at the muzzle to receive the marble and provided with a rod attached to the breech piece, extending into the barrel and arranged to be propelled forward by a strong elastic rubber cord stretched over the breech piece, with its ends nailed to the sides of the gun barrel.

Two changes only are required to adapt the gun to scientific use. First, the notching of the rod passing through the barrel and the application of the trigger, D, for engaging the notches, and second, the support for the falling ball at the muzzle of the gun. The trigger, D, is merely a strip of sheet metal pivoted to the end of the barrel by an ordinary screw. In the muzzle of the gun at the under side is formed a slot, A, and in the end of the gun on opposite sides of the slot are inserted eyes, B. In these eyes is journaled a wire support, C, which holds the ball to be dropped, at one side of the muzzle and out of the path of the projected ball. The wire support, C, forms a lever, one end of which projects into slot in the barrel and is held by the ball in the muzzle. When the rod in the barrel is liberated by pulling the trigger, D, the ball in the muzzle is projected, thereby releasing the wire support, which immediately turns and allows the other ball to drop. It will be noticed that both balls reach the floor at exactly the same time, without regard to the amount of force applied to the projected ball.

FIG. 30.

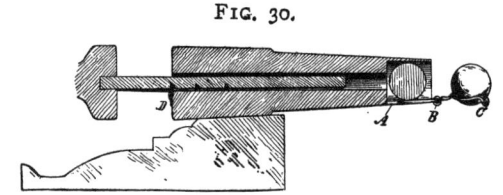

Longitudinal Section of Gun.

FIG. 29.

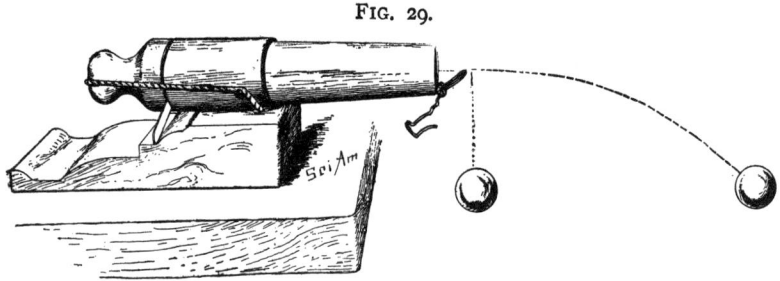

Dropped and Projected Balls.

The falling ball is impelled by the force of gravity only. The projected ball is acted upon by two independent forces—the force of gravity, which draws it toward the earth, and the projecting force, which tends to move it in a horizontal line. The projecting force is concerned only in carrying the ball horizontally forward, and does not in any way interfere with the action of gravity, but gravity brings the ball gradually nearer the earth, until it finally strikes. The gun in this experiment should, of course, be fired over a level plane.

FALLING BODIES—INCLINED PLANE—THE PENDULUM. 45.

THE PENDULUM.

A simple pendulum, which is a purely theoretical thing, is defined as a heavy particle suspended by a thread having no weight. The nearest possible approach to a simple pendulum is a heavy body suspended by a slender thread, as shown at A in Fig. 31, and although this is known as a compound or physical pendulum, its action corresponds very nearly with that of the simple pendulum. In the present

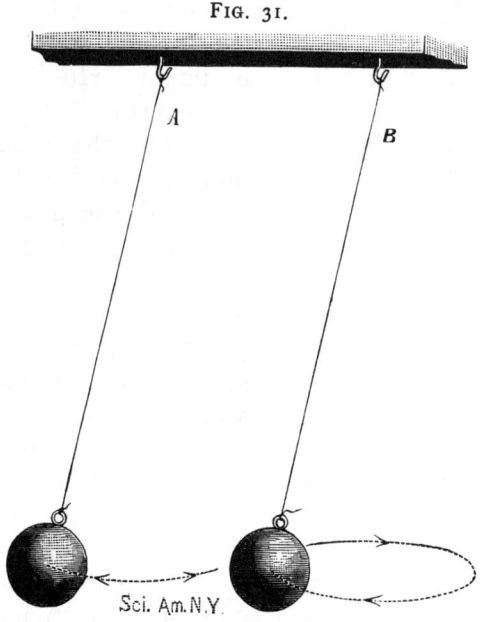

Oscillating and Conical Pendulums.

case the pendulum consists of a heavy bullet or lead ball suspended by a fine silk thread. This pendulum, to beat seconds in the latitude of New York, must be 39·1012 inches long. That is the distance between the point of suspension and the center of oscillation of the weight. This length varies in different places; e. g., at Hammerfest, in Norway, it is 39·1948, and at St. Thomas, one of the West India islands, 39·0207.

A seconds pendulum is one that requires one second for a single swing, or two seconds for a complete to-and-fro excursion. The distance through which the suspended weight travels in one swing is the amplitude of the pendulum. Galileo's discovery of the law of the pendulum in 1582 is a matter of common knowledge. He observed the regularity of the swinging of a lamp suspended from the roof of the cathedral of Pisa, and noticed that, whatever the arc of vibration, the time of vibration remained the same. He also determined the law of the lengths of pendulums by experiment. He found that, as the length of the pendulum increased, the time of vibration increased, not in proportion to the length, but in proportion to its square root. For example, while in New York it requires a pendulum 39·1012 inches long to beat seconds, the length for two seconds would be 156·4048 in. The length of a pendulum for any required time is found by multiplying the length of a seconds pendulum in inches by the square of the time the pendulum is to measure. In the above example, 39·1012 inches is the length of the seconds pendulum. Two seconds is the time to be measured. $2^2 = 4$. Therefore 39·1012 × 4 = 156·4048, the length of the two seconds pendulum. It is found that, barring the resistance of the air, all materials act alike when used for the weight of a pendulum. This is one proof of the uniformity of the action of gravitation on all substances.

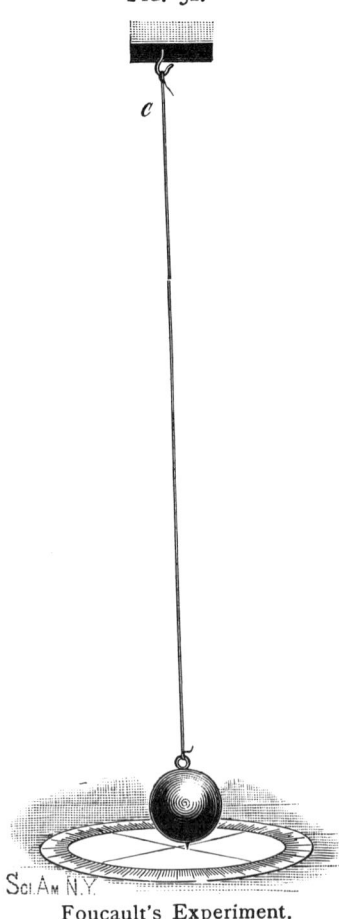

FIG. 32.

Foucault's Experiment.

In Fig. 31, at B, is shown a conical pendulum. It differs

FALLING BODIES—INCLINED PLANE—THE PENDULUM. 47

from the pendulum A only in the manner in which it is used; whereas the pendulum A is made to swing to and fro in a vertical plane, the pendulum B is started in a circle, as indicated by the dotted line. It is found by comparison that the pendulum B completes its circular travel in the same time that pendulum A requires to complete one to-and-fro vibration. The conical pendulum derives its name from the figure it cuts in the air.

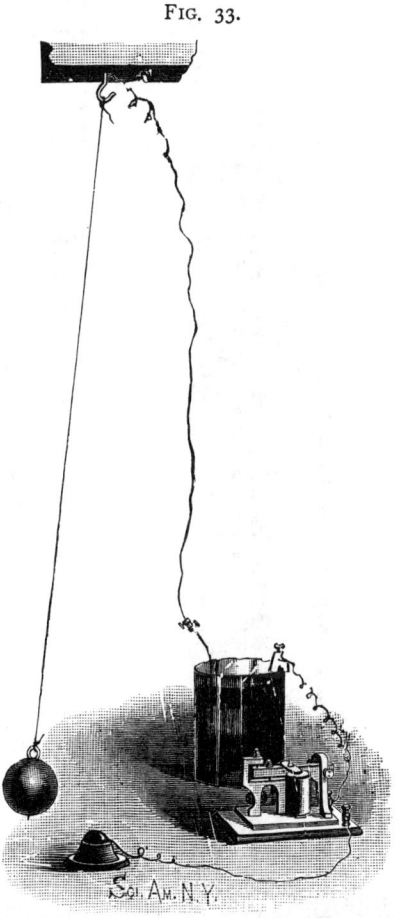

Fig. 33.

Pendulum with Audible Beats.

The pendulum has been used to determine the figure of the earth, also to show the earth's rotation. Foucault's celebrated experiment at the Pantheon at Paris consisted in vibrating a pendulum having a period of several seconds over the face of a horizontal scale. While the pendulum preserved the plane of its oscillation, the scale indicated a slow rotation. This experiment may be repeated easily on a small scale in the manner illustrated in Fig. 32. The ball, which must be a heavy one, is suspended by a very fine wire of considerable length, say from forty to fifty feet. It must be started very carefully to secure the desired result. To start it, a fine wire is tied around the equator of the ball. To this wire is attached a stout thread, by means of which the ball is drawn one side and held there until the pendulum is perfectly quiescent. The pendulum is then released by burning the thread.

In the course of a few minutes there will appear to be a slight change of its plane of vibration. The case is like that of the gyroscope already described. The plane of vibration remains really constant, but the rotation of the earth causes an apparent twisting of the plane. If the experiment be performed in the United States, and the plane of vibration be north and south at first, the northern limit will soon swing toward the right, as viewed from the south.

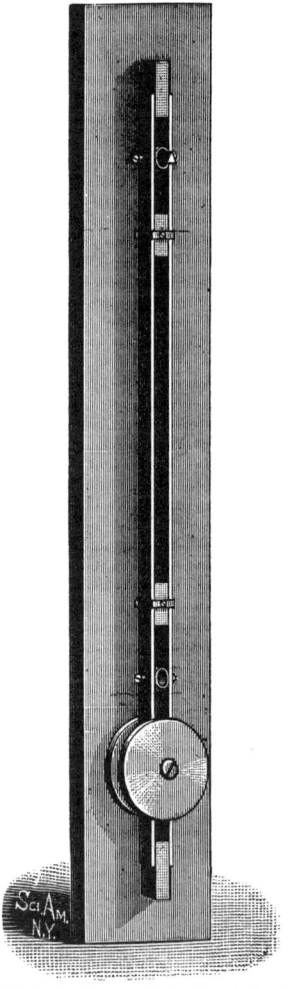

FIG. 34.

Kater's Reversible Pendulum.

A pendulum capable of producing audible beats is often desirable. Fig. 33 shows a simple, well known arrangement for producing audible beats by the aid of a telegraph sounder. The ball, in this case, is suspended by a fine wire. The under side of the ball is provided with a platinum point. A mercury globule is held by an iron cup in the path of the platinum point, and the pendulum, mercury, and sounder are in the battery circuit. By this arrangement an electrical contact is made for each swing of the pendulum, and the sounder is made to click each time the circuit is closed.

By means of Kater's reversible pendulum, the length of a simple pendulum having the same time of oscillation as the compound pendulum may be accurately determined.

In Fig. 34 is shown a slightly modified form of this pendulum, in which the rod is formed of two parallel bars of wood, separated by blocks at the ends and provided with two swiveled cylindric rings, be-

FALLING BODIES—INCLINED PLANE—THE PENDULUM. 49

tween which are placed two adjustable lead weights, held in place by crossbars secured to the weights by screws, and extending over the edges of the wooden bars. Below the lower swiveled ring are clamped lead weights, one upon either side of the bar, with a screw extending through one weight into the other. These weights are cheaply made by casting lead in small blacking box covers.

This pendulum is suspended upon a knife edge projecting from a suitable support, and the weights between the bars are adjusted until the time of vibration is the same for either position of the pendulum, it being reversed and oscillated first upon one of its rings as a center, then upon the other, until the desired adjustment is secured. Then the distance between the bearing surfaces of the rings will be the length of a simple pendulum which would vibrate in the same time as the compound pendulum.

MEASUREMENT OF TIME BY THE PENDULUM.

The application of the pendulum to the measurement of time dates from 1658. In that year Huyghens applied it to clocks. Singularly enough, this has proved to be the only practical use of any importance to which the pendulum could be adapted. The fact that millions of clocks have been made which depend on the pendulum for regulation proves the great value of Huyghens' invention.

A simple model, showing the application of the pendulum to clocks, is illustrated in Fig. 35. It is readily made, and serves to show how the pendulum acts in the regulation of a clock, and is useful for measuring seconds in experimental work. The frame is made entirely of hard wood. The three parallel plates are connected by wooden studs. The wooden arbor of the scape wheel is provided with steel wire pivots, the outer one being prolonged beyond the front plate to receive the second hand. The scape wheel consists of a disk of wood about three inches in diameter, provided with a circular row of steel pins, uniformly spaced and projecting from the face of the disk parallel with the arbor. With a disk of the size given thirty pins will be sufficient, with a larger disk sixty pins may be used.

50   EXPERIMENTAL SCIENCE.

Above the scape wheel arbor there is a wooden roller furnished with steel wire pivots. In the roller is inserted a steel wire forming the escapement or crutch, the ends of the wire being bent inward to form pallets which engage the scape wheel pins in alternation. The rubbing surfaces of the pallets are flattened and polished and the ends are beveled. In the roller is inserted a wire which extends down-

FIG. 35.

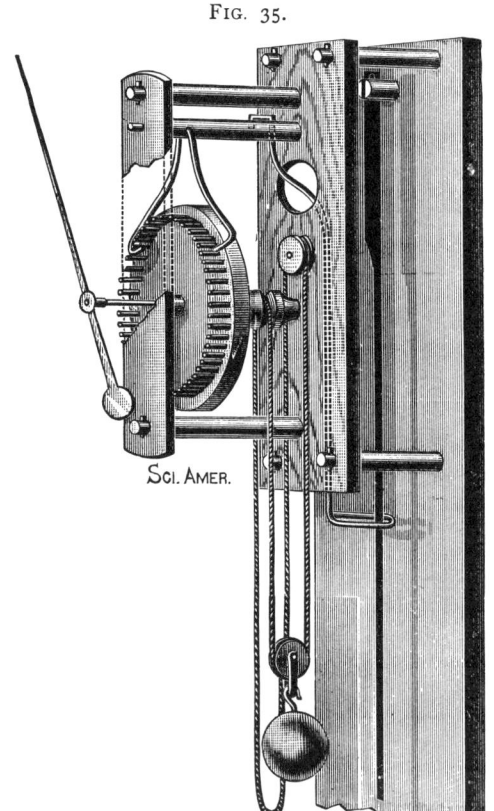

Application of the Pendulum to Clocks.

ward obliquely through a hole in the middle plate, and is finally bent into an oblong loop extending rearward. In a split stud in the back piece is inserted the flattened upper end of the pendulum rod. A small rivet passes through the upper extremity of the rod, and prevents it from slipping through the split stud. The rod passes through the

FALLING BODIES—INCLINED PLANE—THE PENDULUM. 51

oblong loop above referred to, and is provided on its lower end with an adjustable weight of 1½ to 2 pounds.

The scape wheel arbor is provided with a circumferential V-shaped groove forming a very small pulley for receiving the driving cord. Upon the middle plate above the arbor is fixed a circular block having a deep V-shaped circumferential groove for receiving and holding the endless driving cord, which passes round the arbor and grooved block as shown, and also passes around the pulley block attached to the weight. It is necessary to have the V-shaped grooves very deep and very narrow to enable them to pinch the driving cord. To insure uniformity in the action of the cord and weight, it is advisable to place in the second loop of the cord a pulley and connect with it a very light weight. When the driving weight has nearly run down, the cord may be pulled upward over the grooved block and fastened. The pendulum rod is made very thin and flexible at the upper end by hammering. The rod is made of wire of sufficient diameter to prevent springing by the action of the escapement, and the pendulum bob is adjustable. The distance between the center of the bob and the split stud is 39.1012 inches.

The motion of the pendulum is a result of the downward pull of gravity and the restraint of the pendulum rod. It is forced by gravity to move until the lowest point of its arc is reached, when the momentum acquired carries it forward and upward, in opposition to the earth's attraction, until its momentum is overcome by gravity, when it stops and is again drawn down by gravity, causing it to return to the lowest part of its arc and repeat the movement just described, but in the opposite direction. But for friction of the air and of its parts, the pendulum would swing on indefinitely without the propelling power.

The isochronism of the pendulum is perfect only when its amplitude of vibration remains the same, or when it is arranged to move in a cycloidal path. It is impossible to maintain constantly the same amplitude of vibration, and it is difficult to cause the pendulum to describe a true cycloid. A very close approximation to isochronism is secured by suspending the pendulum by means of a flat spring as above described and by limiting its swing to a very small arc.

The motion of a cycloid pendulum is very well illustrated by the cycloidal track and the ball shown in Fig. 36. The track is formed of steel bars smoothly finished, and the ball is of steel, hardened, ground, and polished, one of the kind used for ball bearings.

The period of oscillation of the ball rolling on the cycloid track is the same for all amplitudes. This may be readily proved by comparing two like instruments with the balls oscillating at different amplitudes.

A torsion pendulum is one that depends for its action upon the twisting and untwisting of an elastic suspension. The simplest pendulum of this class is the toy known as the

FIG. 36.

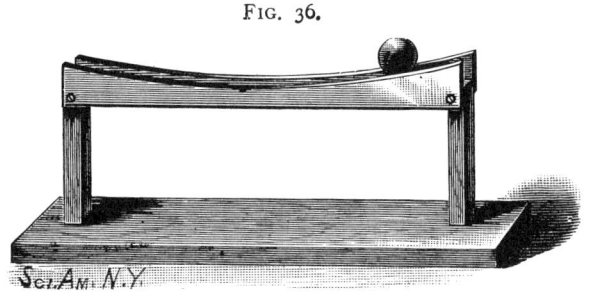

Cycloid Curve.

return ball. It consists of a wooden ball attached to the end of an elastic rubber cord. By grasping the free end of the cord and swinging the ball so as to cause it to roll in a circular path on the floor, the cord will be rapidly twisted. If, after twisting, the cord be fastened to a support, as shown in Fig. 37, it will be found that the ball will rotate rapidly by the untwisting of the cord. The momentum of the ball acquired during the untwisting will again twist the cord, but in the opposite direction. This pendulum will run more than an hour with a single winding. The period of such a pendulum, taken at random from a pile of return balls, was $1\frac{1}{2}$ minutes, the rubber cord when not extended being about a foot long.

By means of apparatus similar to that shown in Fig. 38,

FALLING BODIES—INCLINED PLANE—THE PENDULUM. 53

Coulomb determined the laws of the torsion of wires. The wire by which the weight is suspended is firmly secured to the hook, and the weight is provided with an index. The angle through which the index is turned from the position of rest is the angle of torsion. After turning the weight and releasing it, the elasticity of the wire returns it to the point of rest and the momentum of the weight carries it forward, twisting the wire in the opposite direction, until the weight reaches a point where the momentum of the weight is overbalanced by the resistance of the wire, when the wire again untwists, turning the weight in the opposite direction. These oscillations continue until the force originally applied is exhausted in friction. The oscillations within certain limits are very nearly equal.

A torsion pendulum, with a bifilar suspension, is shown in Fig. 39. The wheel is formed of a disk of metal, with a series of split lead balls pinched down upon its edge. The wheel weighs 1½ pounds. Its diameter is four inches. It has a double loop at the center for receiving the parallel suspending wires, which are ⅜ inch apart and 5 feet long.

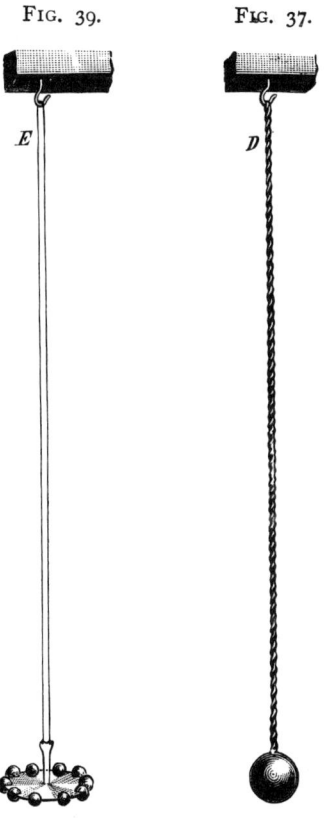

FIG. 39.   FIG. 37.

Torsion Pendulums.

No. 30 spring brass wire was used in this experiment. The period of the pendulum was five minutes.

The torsion pendulum has been successfully applied to clocks. Either of two results may be secured by its use. The time of running may be prolonged in proportion as the period of the torsion pendulum is longer than that of an oscillating one, or the number of gear wheels required in the clock may be greatly reduced. Ordinary clocks constructed on this principle run a year with a single winding. Clocks have been made on this plan which would run for one hundred years.

In the same year that Huyghens applied the oscillating pendulum to the clock, Hooke applied the spiral spring to the watch balance, thereby causing it to act as a pendulum.

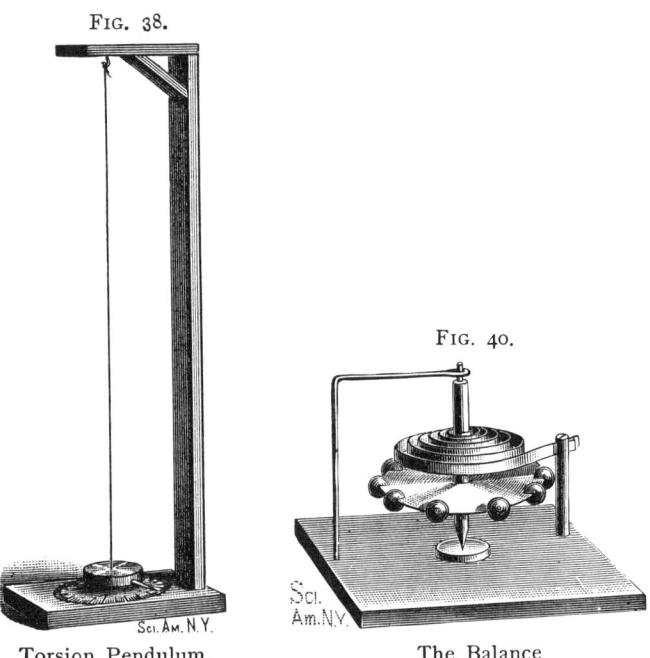

Fig. 38.

Fig. 40.

Torsion Pendulum. The Balance

The principle of Hooke's invention is illustrated by Fig. 40. The apparatus here shown has a vibratory period of one second. The staff rests at the bottom in a small porcelain saucer and turns at the top in a wire loop secured to the base board. The disk on the staff is loaded at its periphery with lead balls. A large watch main spring or music-box spring is attached to the staff and to a fixed standard. The oscillation may be quickened by using a stiffer spring or by removing some of the balls.

FALLING BODIES—INCLINED PLANE—THE PENDULUM. 55

In Fig. 41 is represented a model of a pendulum of recent invention which has been applied to clocks with some success.

Two cross bars are supported from the base by two wires. In the lower cross bar and in the base is journaled a wire having a hook at the upper end. This vertical wire carries a curved arm, to which is attached a thread having at its extremity a small weight, such as a button. The propelling power in this model consists of an elastic rubber band placed on the hook on the vertical rod, and received in a hook on the little crank shaft in the upper bar. The rubber band is twisted by turning the crank, and the crank is prevented from retrograde movement by the wire catch at the side of the bar.

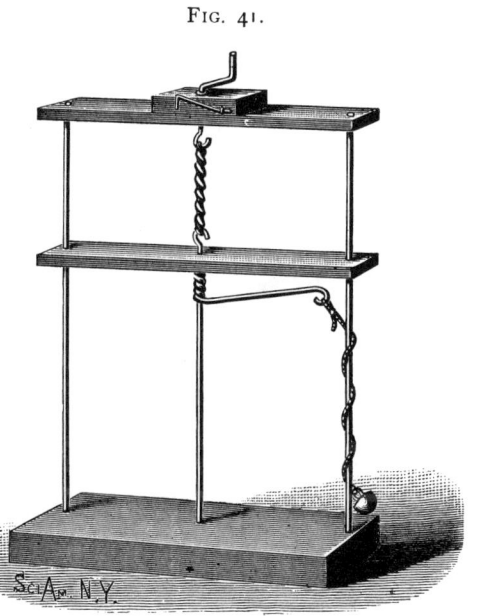

Fig. 41.

Flying Pendulum.

As the arm is carried around by the power stored in the rubber band, the weight on the thread is thrown outward by centrifugal force. When it reaches one of the side rods, it wraps the thread several times around the rod, thus holding the arm until the thread is unwound by the action of the weight, when the arm describes another half revolution and the operation just described is repeated.

CHAPTER V.

MOLECULAR ACTIONS.

Cohesion and adhesion are forces which hold together molecules or ultimate particles. Cohesion unites molecules of the same nature. It is exerted strongly in solids, to a less degree in liquids, and very little in gases.

Heat causes the mutual repulsion of molecules, and thus diminishes the force of cohesion. Solids, when strongly heated, expand, liquefy, and finally pass into a gaseous state, if not chemically changed at the temperature reached, e. g., wood, leather, etc. The tenacity, hardness, and ductility of bodies is due to cohesion.

The force of cohesion in liquids may be demonstrated by suspending a disk by a delicate filament of elastic rubber, noting the extension of the rubber, then placing the disk in contact with a body of water, as shown in Fig. 42, finally drawing upon the rubber until the disk separates from the water. It is found that a considerable extension of the rubber is required to detach the disk. By a more delicate experiment, in which the disk is suspended from a scale beam, the force of cohesion may be accurately measured. It is found by this experiment that

FIG. 42.

A Demonstration of Cohesion.

the material of the disk has no influence on the result, but that the weight required to detach the disk varies with the nature of the liquid. The fact that the disk retains a film of water after separation from the body of water shows that the force of cohesion of the water is less than the force of its adhesion to the disk.

In solids cohesion is often manifested in different degrees in different parts of the same body. The body is then under strain. Examples of bodies in this condition are to be found among iron castings and in unannealed glass ware.

Prince Rupert's drops, or Dutch tears, show in a striking manner how a body under sufficient internal strain may contain within itself the elements of destruction. These drops have a long, oval form, tapering at one end to a point, which is more or less curved. They are made by dropping melted glass into water, thus suddenly cooling the glass and putting it under great strain.

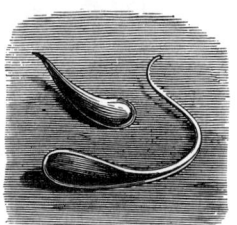

FIG. 43.

Prince Rupert's Drops.

The larger part of the drop may be struck with a hammer without breaking; but on breaking off the point, thus relieving the strain at one place, the glass instantly flies into pieces. So complete is the destruction, that the fragments are often like fine sand.

The Bologna flask is of the same nature as the Prince Rupert's drops. It is an unannealed glass flask, having a very thick bottom, which is under great strain. The flask will receive a hard blow without breaking, and a lead bullet may be dropped into it without producing any effect, but on dropping into it a quartz crystal, or in some other way slightly scratching the inner surface of the flask at the bottom, the flask at once goes to pieces. The action may be compared to the destruction of a superstructure of masonry by weakening or destroying the keystone of the arch which supports it.

A common example of action of this kind is met with in lamp chimneys, which break without any apparent cause. Engineers often find glass water-gauge tubes which will

58   EXPERIMENTAL SCIENCE.

readily stand steam pressure, but which, when scratched even imperceptibly on the inner surfaces, will break.

Adhesion is the term applied to the attraction between the surfaces of two bodies. In the experiment illustrated by Fig. 42 the water adheres to the disk, and the force of adhesion in this case is superior to the force of cohesion as manifested by the molecules of the water. If the moistening of the disk by the water is prevented by lycopodium dis-

Fig. 44.

Bologna Flask.

tributed on the surface of the water, there can be no adhesion.

Two pieces of plate glass pressed firmly together adhere strongly. This experiment succeeds in a vacuum, showing that atmospheric pressure plays no part in holding the glasses in contact.

In the arts, examples of adhesion are found in glues, cements, and solders.

MOLECULAR ACTIONS.

SURFACE TENSION.

The surface tension of liquids is manifested in various ways, notably in the formation of drops, as in rain, each drop becoming a perfect sphere Water sprinkled upon a surface it does not wet, for example, a dusty surface, or upon a surface covered with lycopodium, assumes spheroidal forms, as shown in Fig. 45.

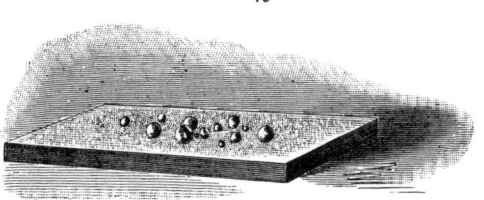

FIG. 45.

Surface Tension exhibited in Water Drops.

A pretty illustration of cohesion and surface tension is shown in Fig. 46. A few drops of olive oil are placed in a suitable vessel, and into the vessel is carefully poured a mixture of alcohol and water having the same specific gravity as the oil. The oil will be detached from the bottom of the vessel, and will, in consequence of the cohesion of its particles, assume a spherical form. Another method of performing this experiment is to introduce the oil into the center of the body of dilute alcohol by means of a pipette. By careful manipulation a large globule of oil may be introduced in this way.

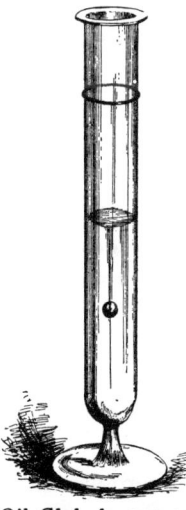

FIG. 46.

Oil Globule suspended in Equilibrium.

Liquids in large masses assume the form of the vessel in which they are contained, in consequence of the superior force of gravity.

From what has been said, as well as from what follows, it will be seen that liquids act as though they were inclosed in a tense superficial film. A glass tube pressed endwise into a body of mercury (Fig. 47) produces a deep depression before breaking the surface of the liquid. When a glass tube is presented in a similar way to the surface of water (Fig. 48), the effect is reversed, the water attaching itself to the surface of the glass with such force as to spread and lift the water in the immediate vicinity of the wall of the tube. In tubes of large diameter, the height to which water is lifted is slight, but in capillary tubes the height is considerable.

Fig. 49 shows the effect of the size of the tube on the height to which the liquid is raised by capillarity. The smaller the area of the upper end of the liquid column, the greater the concavity, and, as a consequence, the greater the strength of the surface film in comparison with the weight of the column raised.

When two glass plates are arranged at a slight angle with reference to each other, with their edges in contact, as

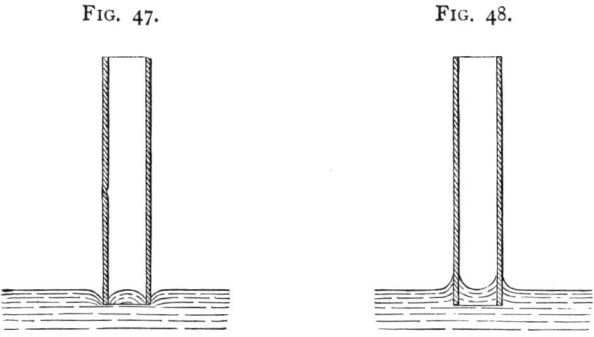

FIG. 47.　　　　FIG. 48.

shown in Fig. 50, the liquid exhibits the phenomenon shown by the tubes of different diameter, but to a less degree, owing to the contact of the edge of the surface film of the liquid with proportionately a smaller surface. When two glass plates are presented in a similar manner to the surface of a liquid which does not wet them, such as mercury or water covered with lycopodium, the effect is the opposite of that just described (Fig. 51). Capillary elevation and depression are more clearly shown by the experiment illustrated in Fig. 52. Two ½ inch glass tubes terminating in capillary tubes are bent into U shape and mounted upon a support. Into the larger end of one of the tubes is poured mercury, which flows into the smaller branch, but does not reach the level of the mercury in the larger branch.

The upper surface of the mercury in each branch of the tube is convex. When water is poured into the larger branch of the other tube, it rises in the capillary tube above its source, and its upper surface in each branch is concave.

FIG. 49.

A curious example of the effect of surface tension is shown in Fig. 53. The smaller end of a tapering tube is plunged several times into a vessel of water and withdrawn. Whenever it is drawn out of the water, the contraction of the water drop adhering to the lower end of the tapering tube forces the column higher within the tube, until at length a point is reached when equilibrium is established, the contractile force of the drop being balanced by the weight of the column of water contained by

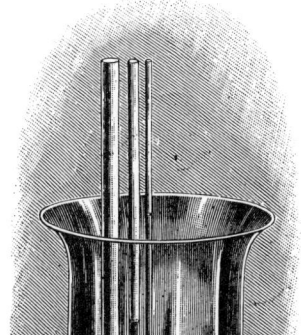

the tube and by the upward pull of the film at the upper surface of the water.

In Figs. 54 and 55 are illustrated experiments showing the force of capillary attraction and adhesion. In Fig. 54 is shown a ¾ inch tube open at one end and terminating in a capillary tubulure at the other end. By allowing the tube to sink for two or three inches in water, with the larger end downward, then placing a minute drop of water in the capillary end of the tube, the tube may be raised two or three inches, carrying with it the column of water contained by it. If the capillary end of the tube be closed by a small drop of water, and the larger end be plunged into water, as in Fig. 55, air will be retained in

FIG. 50.

the tube, and, as a consequence, the water cannot enter. An experiment showing a phase of capillarity is illustrated by Fig. 56. This experiment was originally intended for illustrating upon the screen tapestry and other designs formed of small squares, in colors; but it has another practical application, which is capable of considerable expansion. For projection, a piece of brass wire cloth, of any desired mesh, say from 12 to 20 to the inch, is mounted in a metallic frame to adapt it to the slide holder of the lantern, and the wire cloth is coated lightly with lacquer and allowed to dry.

FIG. 51.

The slide thus prepared is placed in the lantern and focused. The required design may now be traced by means of a small camel's hair brush, colored inks or aqueous solutions of aniline dyes being used. The small squares of the wire cloth are filled with the colored liquid, and show as colored squares upon the screen. Different colors may be placed in juxtaposition

FIG. 52.

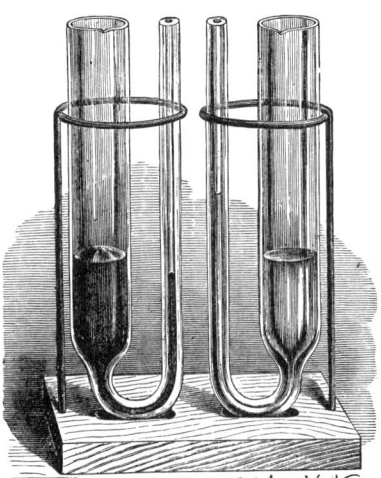

Capillary Elevation and Depression.

FIG. 53.

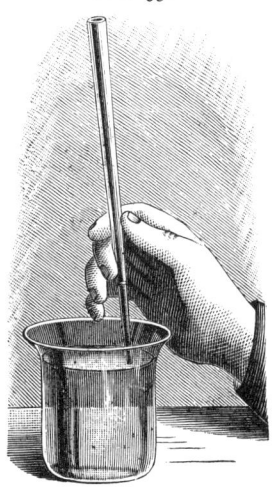

Effect of Surface Tension.

MOLECULAR ACTIONS. 63

without liability to mixing, and a design traced without special care will appear regular, as the rectangular apertures of the wire cloth control the different parts of the design.

FIG. 54. FIG. 55.

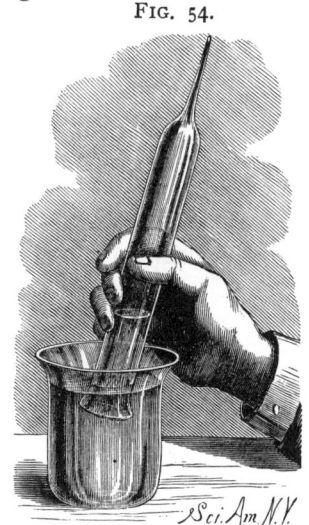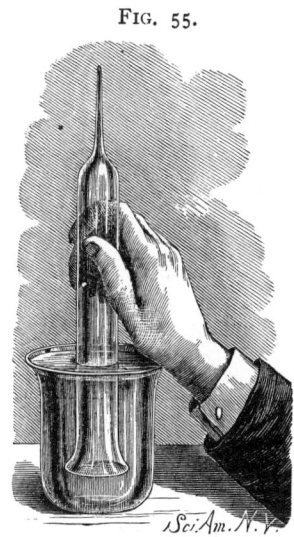

The colored liquid squares are retained in the meshes of the wire cloth by capillarity. A damp sponge will remove the color, so that the experiment may be repeated as often

FIG. 56.

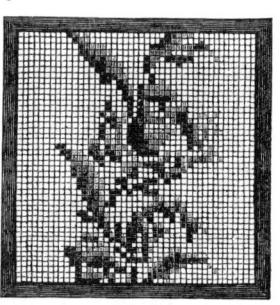

Method of Producing Designs on Wire Cloth.

as desired. In this experiment the colored squares have the appearance of gems. These designs may be made permanent by employing solutions of colored gelatine; but in this case the squares are so small that they are not very effective without magnification. Really elegant designs may be produced in this way for lamp shades, window and fire screens, signs, etc. The mesh of the wire cloth should be quite coarse, say 10 to the inch. The wire cloth is supported a short distance from a design drawn on paper, and the different colors are introduced into the meshes by means of an ordinary writing pen. The gelatine solution should not be very thick, and it must be kept warm. Ordinary transparent gelatine may be colored for this purpose by adding aniline. Colored lacquers answer admirably for filling the squares. The beauty of this kind of work and the simplicity of the method by which it is produced recommend it for many purposes.

ABSORPTION OF GASES.

The behavior of gases under certain conditions is of peculiar interest to the student of physics, since it involves actions which cannot be seen and which require purely mental effort for their comprehension. There are simple ways of demonstrating that certain actions do occur, but the exact mode of their occurrence is left to reason or conjecture.

In some of the following experiments molecular action proceeds with astonishing rapidity. One of the best examples of this rapid action is the absorption of gases by charcoal.

To illustrate absorption according to the usual method, a piece of recently heated charcoal is floated upon mercury and a test tube filled with carbonic acid gas or ammonia gas is inverted over it and quickly plunged into the mercury, Fig. 57. The absorption begins immediately and quickly forms a partial vacuum, which causes the mercury to rise in the tube.

When a quantity of mercury is not available, the experiment may be performed very satisfactorily in the manner illustrated by Fig. 58. A glass tube, closed at one end by a cork in which is inserted a short piece of smaller tube, is plunged open end downward into a tumbler partly filled with water. To a flask or bottle is fitted a cork in which is inserted a small glass tube, and the two small tubes are connected by a short piece of flexible rubber tubing. The flask is filled with carbonic acid gas,* and corked. One or two small pieces of fine charcoal are heated strongly in a closed vessel, such as a covered crucible, or upon the top of a stove. The cork of the flask is removed, and the charcoal is dropped

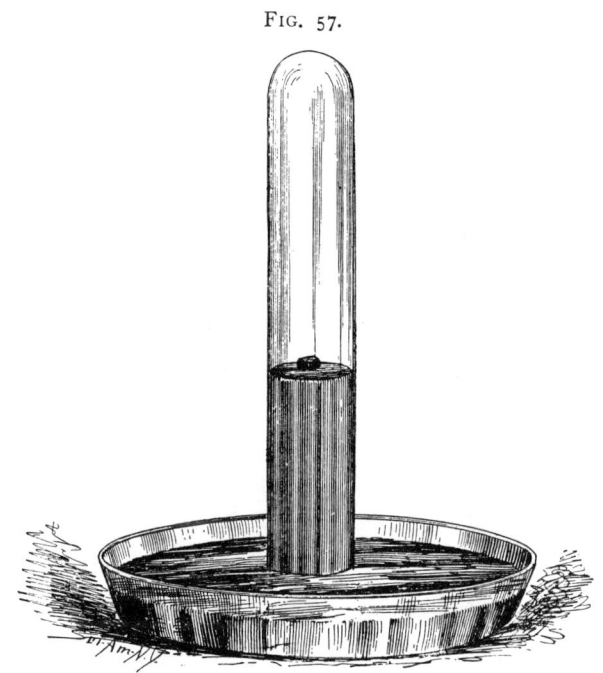

Fig. 57.

Absorption of Gases by Charcoal.

in and the cork replaced. If there are no leaks, the absorption of the gas by the charcoal will be immediately shown by the rise of the water in the tube in the tumbler. The coal will absorb 35 times its bulk of the gas. In the case of ammonia the volume of gas absorbed reaches 90 times the bulk of the charcoal. As the gases which are most easily

* Carbonic acid gas for this and subsequent experiments may be readily prepared by dissolving a small quantity of carbonate of soda (say 1 oz.) in water, in a tall glass or earthen vessel, then slowly adding a few drops of sulphuric acid. The gas will quickly fill the vessel to overflowing. The carbonic acid gas being much heavier than air, may be readily poured into the flask.

condensed to a liquid state are those which are absorbed with the greatest facility, it is fair to presume that the gases absorbed by the charcoal are in a liquid state. The well known purifying property of charcoal and other porous substances is referred to their absorptive power.

THE DIFFUSION OF GASES.

The tendency of gases to mix or diffuse one into the other is very strong. A simple experiment exemplifying

FIG. 58.

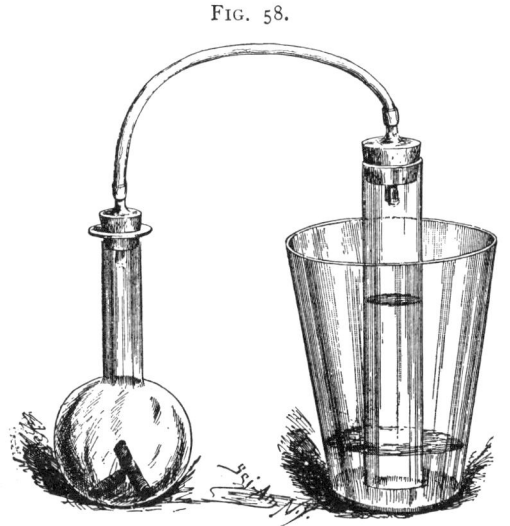

Absorption of Carbonic Acid Gas by Charcoal.

this tendency is illustrated by Figs. 59 and 60. A clean, dry porous cell, such as is used in galvanic batteries, is closed by a cork in which is inserted a small glass tube. A piece of barometer tube six or eight inches long is connected by rubber tubing with the tube of the porous cell. The end of the barometer tube is plunged into water and the porous cell is introduced into a vessel* filled with hydrogen or illuminating gas. The gas enters the porous cell so much more

* An ordinary fish globe answers admirably as a gas-containing vessel for this and similar experiments. It is readily filled with illuminating gas by placing it for a minute in an inverted position over a burner through which gas is flowing.

MOLECULAR ACTIONS.

rapidly than the air can escape through the pores of the cell that a pressure is created which causes the air to escape through the tube and bubble up through the water.

When the porous cell is removed from the glass globe, the reverse of what has been described occurs, the gas passing outward with much greater rapidity than the air can pass in, thereby producing a partial vacuum, which causes

FIG. 59.

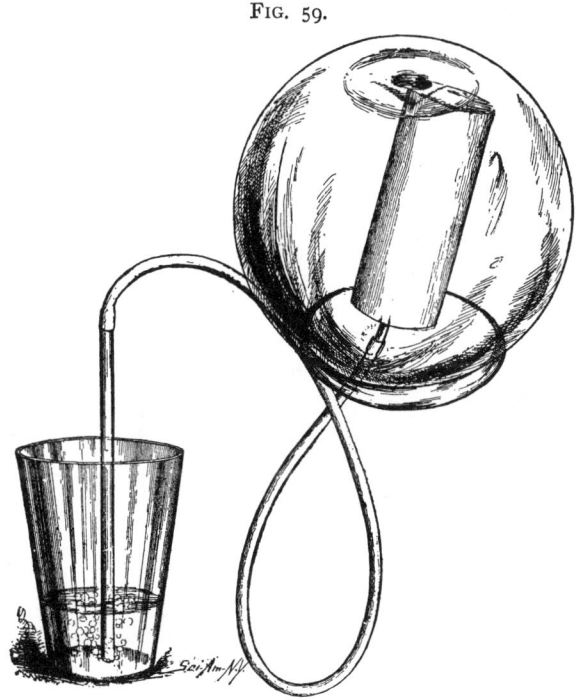

The Diffusion of Gases—Endosmose.

the water to rise to a in the glass tube, Fig. 60. These are examples respectively of endosmose and exosmose. In these experiments it is of vital importance to have tight joints, as the slightest leak will insure failure. The corks should fit tightly, and where they are not to be removed, they should be carefully sealed.

These experiments may be tried on a large scale by employing a porous Turkish water cooler instead of the porous cell, and using a larger and longer glass tube. A large bell glass or glass shade may serve as the gas-containing vessel. The action may be made more distinctly visible by coloring the water.

A convenient and inexpensive way of showing the same phenomena on a small scale is illustrated by Fig. 61. An ordinary clay tobacco pipe answers for the porous vessel. A short, centrally apertured cork is fitted to the bowl of the pipe, a glass tube, of about one-eighth inch internal diameter, is fitted to the bore of the cork, and the cork is carefully sealed. By connecting the stem of the pipe with

FIG. 60.

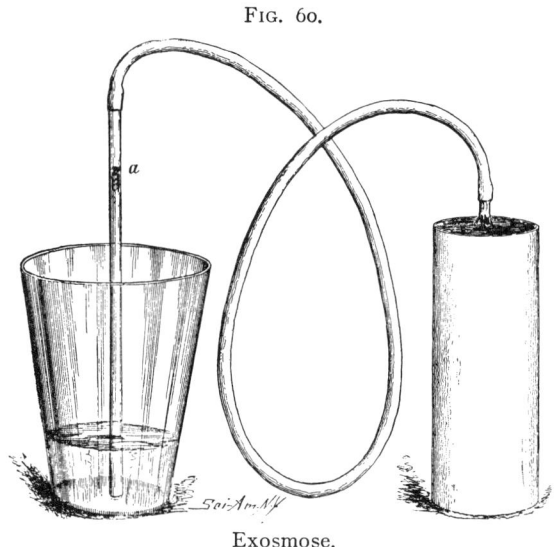

Exosmose.

a gas jet or hydrogen generator, by means of a flexible tube, and inserting the glass tube a short distance into water, the gas will bubble up through the water. After shutting off the gas at the burner, or by doubling or pinching the rubber tube, the water will immediately rise in the glass tube—showing that in the exchange of gas and air through the pores of the clay, the outward movement of the gas has been much more rapid than the inward movement of the air, thereby producing a partial vacuum, which causes the water to rise.

By breaking off the stem of the pipe near the bowl, the pipe and glass tube may be plunged in a deep glass jar, when the experiment may be proceeded with as follows:

A little water, say one-half inch in depth, is poured into the jar, after which the jar is filled with carbonic acid gas. Illuminating gas or hydrogen is allowed to flow through the pipe while it is removed from the jar, so as to drive out all the air and fill the pipe with gas. The gas is now shut off and the pipe is immediately placed in the jar, with the glass tube plunged in the water. The effect is the same as in the case of the air and gas, *i. e.*, the carbonic acid gas goes in and the hydrogen gas goes out; and when equilibrium is established, the pipe will contain some carbonic acid. This may be proved by removing the pipe from the jar and plunging the glass tube into some clear lime water, then allowing the gas to flow only long enough to force out the contents of the pipe. The presence of the carbonic acid is indicated by the milky appearance of the lime water, which is due to the formation of carbonate of lime.

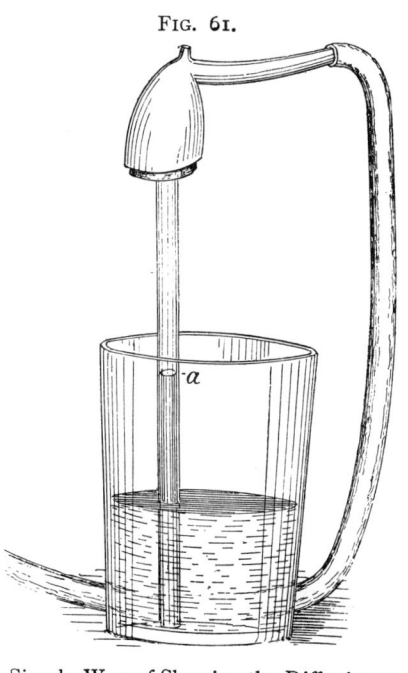

Fig. 61.

Simple Way of Showing the Diffusion of Gases.

There is sufficient carbonic acid in the exhalations of the lungs to show an action which is the reverse of that observed in connection with illuminating gas. When the pipe is blown through, and the end of the stem is quickly and completely stopped, one or two bubbles will escape from the glass tube, showing that the inward movement of the air through the pores of the clay is more energetic than the outward movement of the carbonic acid.

The diffusion of gases may be shown by the well known experiments illustrated by Figs. 62 and 63. A medium sized fish globe, a very small fish globe which will pass into the larger one, and a piece of bladder are the requisites for this experiment.

FIG. 62.

Pressure by Endosmose.

The small globe is filled with carbonic acid gas, and the bladder, previously moistened, is placed loosely over the mouth of the jar and tied so as to render the connection between the bladder and the globe air tight. A good way to insure a tight joint is to stretch a wide rubber band around the neck of the globe before applying the membrane. The large fish globe is filled with hydrogen or illuminating gas, and the small globe is placed under it as shown in Fig. 62. As the hydrogen passes inward through the membrane much more rapidly than the carbonic acid passes outward, the membrane is distended outwardly. It requires a little time to produce a visible effect. If the smaller globe is filled with hydrogen, and the large one with

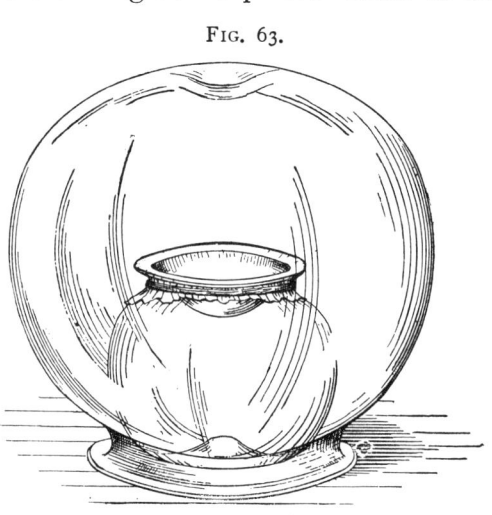

FIG. 63.

Partial Vacuum by Exosmose.

carbonic acid, the membrane will be distended inward, as shown in Fig. 63. In this latter case the experiment may be performed with the least trouble by placing the large globe with its mouth upward, and closing it by means of a plate of glass.

Endosmose proceeds from the rarer toward the denser gas. The law governing the diffusion of gases, according to Graham, is that *the force of diffusion is inversely as the square roots of the densities of the gases.*

When two miscible liquids are separated by a porous partition, they diffuse one into the other. A simple endosmometer for showing this action is shown in Fig. 64. It consists of a small funnel having its mouth closed by a piece of bladder held in place by a wide rubber band stretched around the rim of the funnel. The funnel thus prepared is immersed in water, for example, and is filled to the level of the water with sirup of sugar. The water passes through the bladder into the funnel and the sirup passes out. The rise of the liquid in the funnel indicates that the water enters more rapidly than the sirup escapes. The presence of the sirup in the water may be detected by taste. That the water passes through the membrane into the funnel may be proved by adding to the water a small quantity of sulphate of iron, and after the experiment has proceeded for a time, adding some tannin to the contents of the funnel. If sulphate of iron is present in the funnel, the sirup will turn dark upon the addition of the tannin.

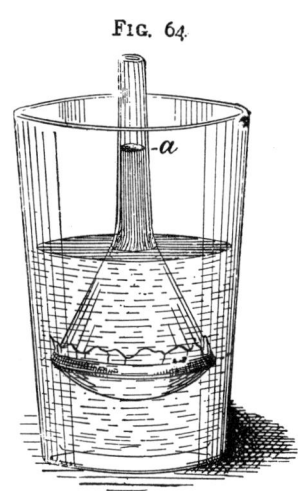

Fig. 64.

Endosmometer.

If the neck of the funnel proves to be too short, a glass tube may be connected with it by means of a short piece of rubber tubing.

CHAPTER VI.

LIQUIDS—PRESSURES EXERTED BY LIQUIDS.

Liquids are distinguished from solids by the great mobility of their molecules. The adhesion between the molecules of liquids produces more or less resistance to their free motion. This property, which is known as viscosity, is inherent in all liquids, some exhibiting extreme mobility,

FIG. 65.

Demonstration of Pascal's Law.

others having great viscosity. Ether is an example of a mobile liquid, and an example of a viscous one is found in glycerine.

Liquids are compressible to a very small degree only. They are, as we have already noticed (Chapter I), porous

PRESSURES EXERTED BY LIQUIDS. 73

and impenetrable, and, in consequence of their compressibility, they are elastic.

Pascal enunciated the following law of the pressures of liquids: " Pressure exerted anywhere upon a mass of liquid is transmitted undiminished in all directions, and acts with the same force on all equal surfaces, and in a direction at right angles to those surfaces."

To demonstrate this principle, the apparatus shown in Fig. 65 has been devised.

A hollow metallic globe is provided with openings at the top and bottom and upon four or more of its sides. Around these openings there are collars, over which are stretched and tied diaphragms of rather thick but elastic rubber, the upper diaphragm being omitted until the globe is filled with water. The globe being placed upon a suitable support, pressure is applied to the upper diaphragm, when it is found that the pressure is transmitted through the medium of the water not only to the diaphragm at the bottom of the globe, but in an equal degree to the diaphragms upon the sides of the globe, thus showing that the pressure is exerted by the water equally in all directions, and at right angles to the surfaces with which it is in contact. This is a simple illustration of Pascal's law.

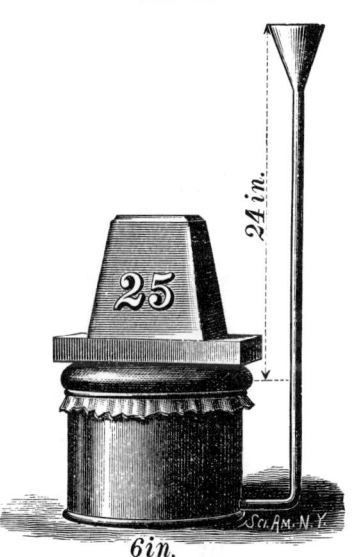

FIG. 66.

Pascal's Experiment.

Probably there is not a more striking example of the effects of hydrostatic pressure than that presented in Pascal's experiment, in which he burst a stout cask by inserting in it a tube about 30 feet high, and filling both the cask and tube with water. This experiment, in a modified form, is illustrated by Fig. 66. A tin cup of 6 inches diameter, and having a wired edge, is furnished with a leather or rubber cover, tied over the top of the cup so that it may have a motion of a half inch or more. In the side of the cup is inserted a tube which extends upward above the top of the cup 24 inches, and is furnished at its upper end with a funnel. The diameter of the tube is of no consequence; the result will be the same whether it is small or large. The cup is filled with water by submerging it with the tube in a horizontal position, with the tube uppermost, and alternately pressing in the flexible covering and then drawing it outward. This operation soon drives out the air and fills the cup with water. The cup is placed with the pipe in a ver-

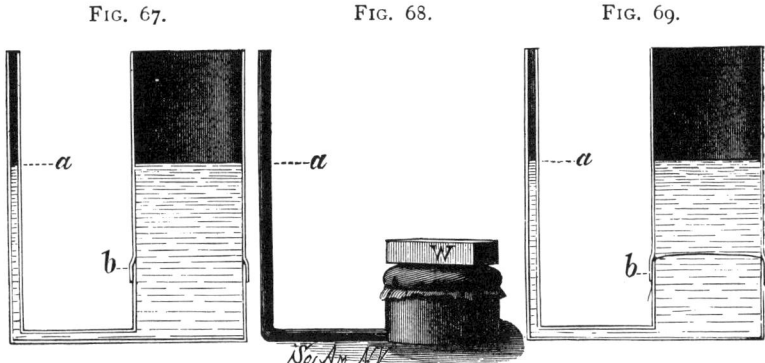

FIG. 67.   FIG. 68.   FIG. 69.

Equilibrium in Communicating Vessels.

tical position, and a board is laid over the flexible cover and pressed to expel all of the water above the rim of the cup.

Now, by placing a twenty-five pound weight upon the board and pouring water into the tube, the weight will be lifted and sustained. This experiment shows that a great pressure may be produced by a small column of water. In this case the cup, with its flexible cover, represents the large cylinder and piston of a hydraulic press, the tube stands for the pump cylinder, the small water column in the tube for the piston, and the weight of the column for the power applied. By increasing the height of the water column, the pressure will be correspondingly increased.

Fig. 67 shows two communicating vessels of different diameter. The larger one is divided at a point, b, near its base, and reunited by means of a packed joint. When water is poured into one of these vessels, it rises to the same level in both. By removing the upper portion of the larger vessel and tying a flexible cover over the lower part, it is found that a column of water in the smaller vessel extending to the point, *a*, will be exactly counterbalanced by a certain weight placed on the flexible cover, as in Fig. 68. The weight required will be exactly that of a column of water of the diameter of the larger vessel and equal in height to the distance between the flexible cover and the level of the smaller column, *a*. This may be shown by removing the weight, replacing the upper part of the larger vessel, as in Fig. 69, and filling it with water up to the level, *a*. The weight of water required in the larger vessel to thus lift the smaller column to the point, *a*, will be found to be the same as that of the weight removed.

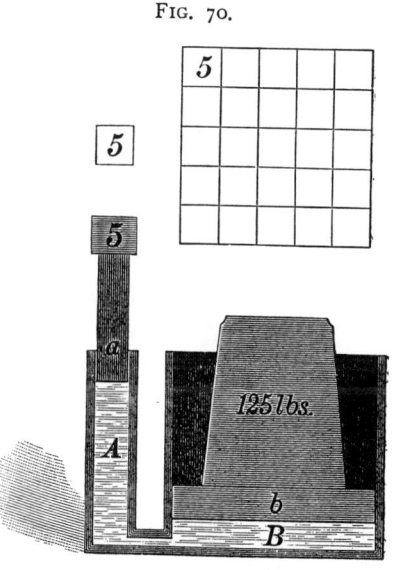

Fig. 70.

Principle of Hydraulic Press.

It seems puzzling that no variation in the size or form of the upper portion of the larger vessel can make any difference in the results, provided the same water level is maintained; but it must be remembered that the whole question is simply one of pressure per square inch. The weight will as readily balance a large column as a small one, the vertical height being the same in each case.

The enormous pressure developed in a hydraulic press is a subject of wonder, even to those who perfectly understand the principle involved in its operation. Men regard with interest anything that furnishes an exhibition of power, and it is difficult to avoid thinking that in the hydraulic press power is actually created in some mysterious way.

However, nothing of this kind happens. A hydraulic press is simply a power converter, in which a certain pressure per square inch, acting on a small area, is able to produce the same pressure per square inch on a large area, thereby multiplying the pressure. The sum total of all the power utilized in the press is exactly equal to the sum total of all the power applied to the press, less friction.

In Fig. 70 is illustrated a hypothetical hydraulic press, above which is given a diagram showing the relative areas upon which pressure is exerted. To the two communicating vessels, A, B, with square cross sections, are fitted the pistons, *a*, *b*. The piston, *a*, is one inch square, and consequently has an area of one square inch. The piston, *b*, is 5 inches square, and consequently has an area of 25 square inches. If the spaces below the pistons be filled with water, it will be found that, in consequence of the equal distribution of pressure throughout the confined body of water, a weight placed on the piston, *a*, will balance a weight twenty-five times as great placed upon the piston, *b*; that, for example, a downward pressure of five pounds upon the piston, *a*, will, through the medium of the water, cause a pressure of five pounds to be exerted on every square inch of surface touched by the water, and that the movable piston, *b*, having twenty-five times the area of the piston, *a*, and receiving on each square inch of its surface a pressure of five pounds, will be forced upward with a pressure of one hundred and twenty-five pounds.

A press of this description would have no practical value, inasmuch as a movement of the piston, *a*, through the space of five inches would lift the piston, *b*, only one-fifth of an inch. To lift the piston, *b*, five inches would necessitate a piston, *a*, having a length of one hundred and twenty-five inches (over ten feet).

To obviate this difficulty, the pump piston of a hydraulic press is of a reasonable length, and valves are provided by means of which the short piston, by acting repeatedly, will accomplish the same results as would in the other case require a very long piston.

In Figs. 71 and 72 is shown a very simple and easily constructed hydraulic press, which has considerable utility. It is made of pipe fittings, valves, rods and bolts, that are all procurable almost anywhere.

To the baseboard is secured a flange, into which is screwed a short piece, A, of gas pipe. On the upper end of the pipe is screwed a coupling, into which is inserted a bushing from which the internal thread has been removed. In the bushing and in the pipe, A, is inserted a rod of cold rolled iron,

FIG. 71.

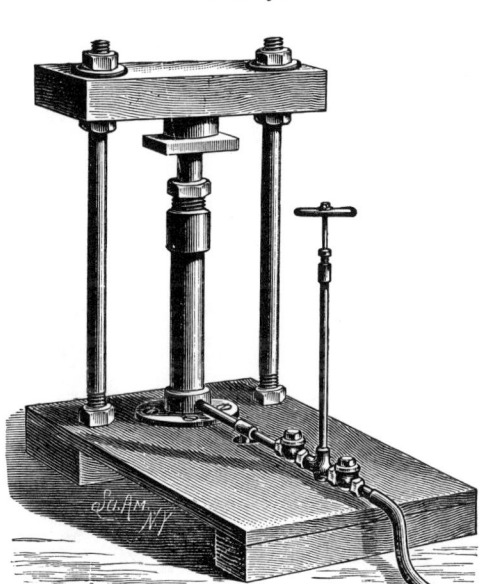

Simple Hydraulic Press.

a bar of brass, or a short section of shafting, and the space in the coupling around the rod is filled with hemp packing, which may be compressed, if required from time to time, by screwing in the bushing. The flange at the bottom of the pipe, A, is connected with the pump, B, by the pipe, C, in which is inserted a discharge, as shown. The pump cylinder is inserted in a crosstee, to opposite sides of which are attached ordinary check valves. The tee is fastened to the base by a plugged piece of pipe, extending through the base and provided with a nut, which clamps the base tightly. The barrel of the pump is in all respects like the press barrel, except in size. The piston consists of a ¼ inch brass rod, to the upper end of which is attached a tee handle.

A heavy bar of wood is supported over the pipe, A, by bolts extending through the base and through a re-enforc-

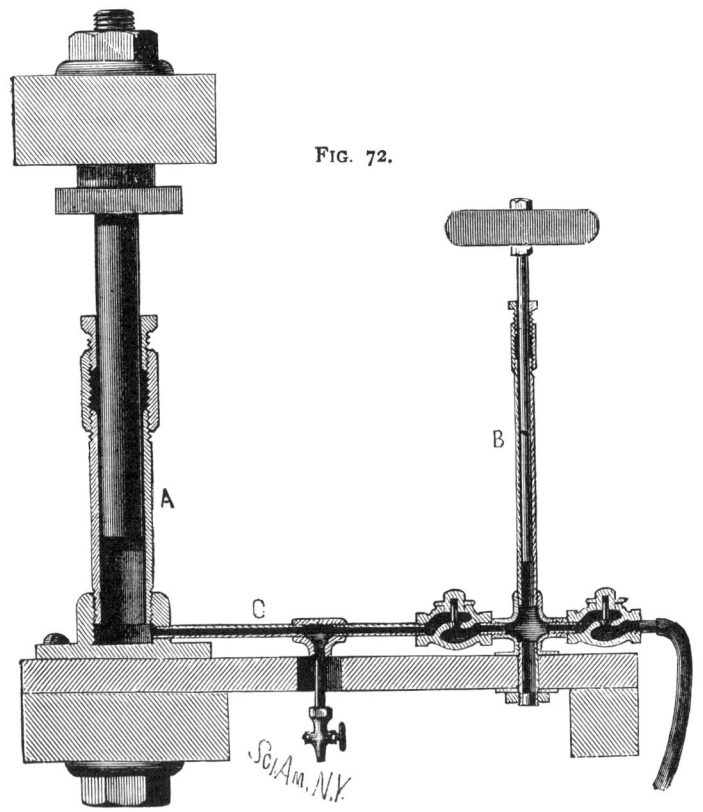

FIG. 72.

Sectional View of Simple Hydraulic Press.

ing bar under the base. The check valves both open toward the cylinder, A, and the outer one is provided with a rubber suction pipe. Water is drawn into the pump by lifting the piston and forced into the press barrel by the descent of the piston. The proportion of the pressure attained, to the power applied, will be as the area of the large piston to the area of the small one. With pistons of respectively 2 inches and ¼ inch diameter, a pressure of 3,000 pounds may be produced easily. If it is desired to create a greater pressure, the barrel, A, may be made of hydraulic tubing, and a lever may be applied to the pump piston, or the diameter of the barrel, A, and its piston may be increased.

LATERAL PRESSURES.

In some experiments already described it was shown that hydrostatic pressure is equally distributed on all sides of the containing vessel. Fig. 73 illustrates an experiment

FIG. 73.

Reactionary Apparatus.

in which are shown the effects of removing pressure from a portion of one side of the vessel, thus allowing the pressure to act upon the opposite side of the vessel in such a manner as to cause it to move. This experiment is arranged to show this action in two ways, one so as to propel the vessel forward, the other so as to cause it to turn.

The apparatus consists of a tall tin can—such as is used by fancy bakers for wafers or fine crackers—mounted upon a wooden float provided with a lead ballast to keep it in an upright position. In one side of the can at the bottom is inserted a short tube, a, and in diametrically opposite sides of the can, also at the bottom, are inserted longer tubes, b, which reach over the wooden block and have their ends

80   EXPERIMENTAL SCIENCE.

turned in opposite directions. All of the tubes are stopped, and the float is placed in a large vessel of water, when the can is filled with water and the stopper of the tube, *a*, is withdrawn, thereby allowing water to escape from the can, and by reaction drive the can backward.

When the straight tube, *a*, remains closed, and the bent tubes, *b*, are opened, the reaction of the issuing streams results in the rotary movement of the apparatus. The

FIG. 74.

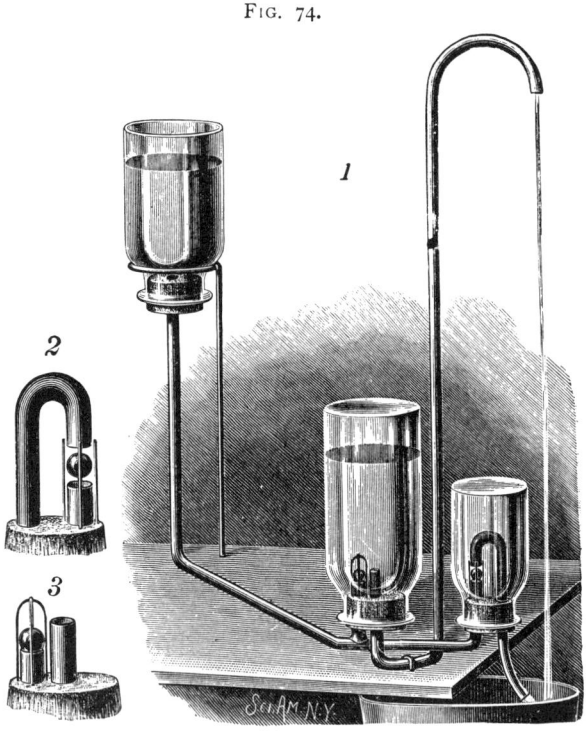

Hydraulic Ram.

apparatus arranged in this way illustrates the principle of Barker's mill.

The hydraulic ram, a simple form of which is illustrated in Fig. 74, depends for its action on the momentum of the water column and upon the elasticity of air. The reservoir in the present case consists of an inverted glass bottle having no bottom, and provided with a perforated stopper in

PRESSURES EXERTED BY LIQUIDS. 81

which is inserted one end of a tube, preferably lead, on account of the facility with which it may be cut and bent. The other end of the tube is branched, one branch extending through a stopper inserted in an inverted bottle which serves as an air chamber. The other branch of the tube extends to the overflow valve. In the stopper of the air chamber is inserted a second tube, which is bent upward and curved over, forming the riser.

The smaller bottle, which serves as a valve chamber, is provided with a stopper which receives the branch of the supply tube and an overflow tube. The arrangement of these tubes is shown in detail at 2, the curved tube being the overflow, the straight one the inlet. To the inlet and overflow tubes is fitted a valve consisting of a metal ball or a marble. The fitting is accomplished by simply driving the ball against the end of each tube, so as to form valve seats. Four wires are inserted in the stopper around the inlet tube to prevent the escape of the valve. The distance which should separate these tubes, as well as the weight of the ball valve, is determined by experiment.

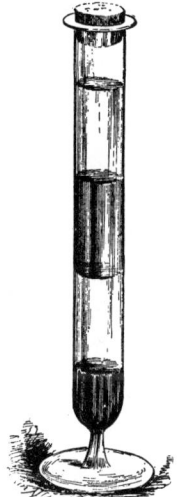

FIG. 75.

Vial of Four Liquids.

In the air chamber above the branch of the supply tube is confined a ball valve by a cage formed of wires inserted in the stopper, as shown at 3. This valve is fitted in the manner already described.

The discharge tube extends above the level of the reservoir. The reservoir and the tubes are supported by wire loops and standards inserted in a base board.

Water flows from the reservoir through the valve chamber and out at the overflow. When the velocity of the flow is sufficient to carry the valve in the valve chamber up against the end of the curved overflow tube, the overflow is immediately checked, and the momentum acquired by the water causes it to continue to flow for an instant into the air chamber, compressing the air in the chamber, and causing the water to rise in the discharge tube. As soon as equilibrium is established, the valve in the air chamber closes and the valve in the valve chamber falls away from its seat on the overflow tube, allowing the water to discharge again, and so on, this intermittent action continuing so long as there is water in the reservoir. The water discharged by the riser is only a fraction of that flowing out of the reservoir.

We have already noticed (Fig. 66) that a liquid will assume the same level in communicating vessels. The size and form of the vessels is immaterial. The smaller one may be inclined, curved, or bent in any form and the larger one may have any capacity, still the result will be the same.

FIG. 76.

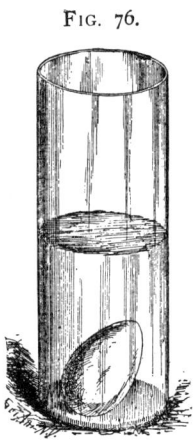

Egg in Fresh Water.

FIG. 77.

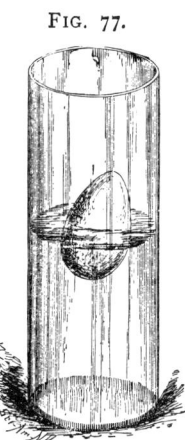

Egg Buoyed up by Salt Water.

FIG. 78.

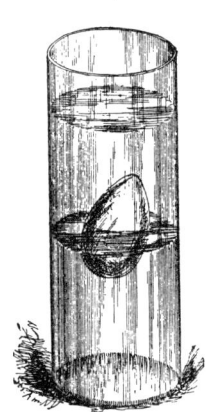

Egg in Equilibrium between two Liquids of Different Densities.

When, however, the vessels contain liquids of different densities, the level will be no longer the same. In such case the lighter liquid will stand higher.

When several liquids of different densities which do not mix are contained in the same vessel, there will be stable equilibrium only when the liquids are arranged in the order of their densities, the heavier liquid being, of course, at the bottom. This is illustrated by the "vial of four liquids," shown in Fig. 75. A test tube with a foot makes a convenient receptacle for the liquids. In the bottom of the tube is

PRESSURES EXERTED BY LIQUIDS. 83

placed mercury. The second liquid in order is a saturated solution of carbonate of potash in water. The third is alcohol, colored with a little aniline red to mark the division of the liquids more clearly. The fourth is kerosene oil. When these liquids are shaken up, they mix mechanically, but when the tube is at rest the liquids quickly arrange themselves in their original order.

The experiment illustrated in Figs. 76, 77, and 78 shows the effects of liquids of different densities. Two pint tumblers or similar vessels are necessary for this experiment. Half fill one with water and the other with strong brine. Into the water drop an egg. It goes to the bottom (Fig. 76). An egg dropped into the brine floats (Fig. 77). By carefully pouring the brine through a long funnel or through a funnel with an attached tube, which will reach to the bottom of the tumbler containing the pure water, the water and the egg will be lifted, and the egg will float in equilibrium at the middle of the tumbler.

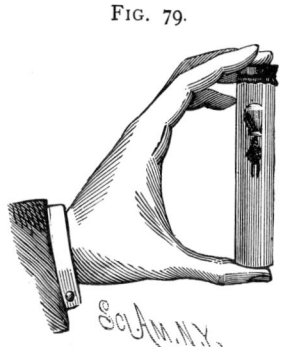

FIG. 79.

The Cartesian Diver.

The first experiment shows that the egg is a little more dense than pure water, the second that brine is more dense than the egg, and the third that the egg can be supported in equilibrium between two liquids of different densities.

The hydrostatic toy known as the Cartesian diver illustrates the several conditions of floating, immersion, and suspension in equilibrium. In a tall, slim glass tube, closed at the bottom and filled with water, is placed a porcelain or glass figure having a glass bulb attached to its head. The glass bulb has a small hole in the bottom, and is filled partly with water and partly with air, the proportion of air and water being such as to just allow the bulb to float. The top of the tube is closed by a piece of flexible rubber tied over its mouth. The pressure of the fingers upon the rubber communicates pressure through the water to the air contained by the bulb, causing the air to occupy less space and increasing the weight of the bulb in proportion to the amount of water forced in. As the weight of the bulb increases the diver descends, and when the finger is removed from the elastic cover of the tube, the air by its own elasticity regains its normal volume, and the bulb, becoming lighter, rises to the top of the jar.

CHAPTER VII.

GASES.

Gases are elastic fluids in which the molecular force of repulsion is superior to the force of attraction. Expansion, the most characteristic property of gases, is due to this force. The limit of the expansive force of a gas is unknown. If there were no opposing causes, it would appear that the particles of a gas might separate indefinitely.

The expansive force of the atmosphere is opposed by the earth's attraction; the air is thus in a state of equilibrium.

The expansibility of air is shown by inclosing a small quantity of it at atmospheric pressure in an elastic rubber balloon,* and placing the balloon in the receiver of an air pump, then removing the atmospheric pressure from the exterior of the balloon by exhausting the receiver. The air in the balloon will expand, distending it as shown in Fig. 80.

FIG. 80.

Dilatation of Balloon in a Vacuum.

In former experiments illustrating the diffusion of gases, it was shown that carbonic acid gas was very much heavier than air, by pouring the gas from one vessel to another, thus to a great extent displacing the air in the receiving vessel, in the same manner as it would be displaced by the pouring in of a liquid. In the case of pure hydrogen or illuminating gas, the order

---

* The small inflatable balloons applied to the toy squawkers, and which may be bought in any toy store for three cents, answer perfectly for this experiment.

of things was reversed; *i. e.*, to fill the vessel it was necessary to invert it, so that the air might be displaced by the rising of the gas, which is so much lighter than air.

To show visibly that one gas is heavier than air and the other lighter, a pair of balances may be pressed into service. If the balances are not at hand, a pair may readily

FIG. 81.

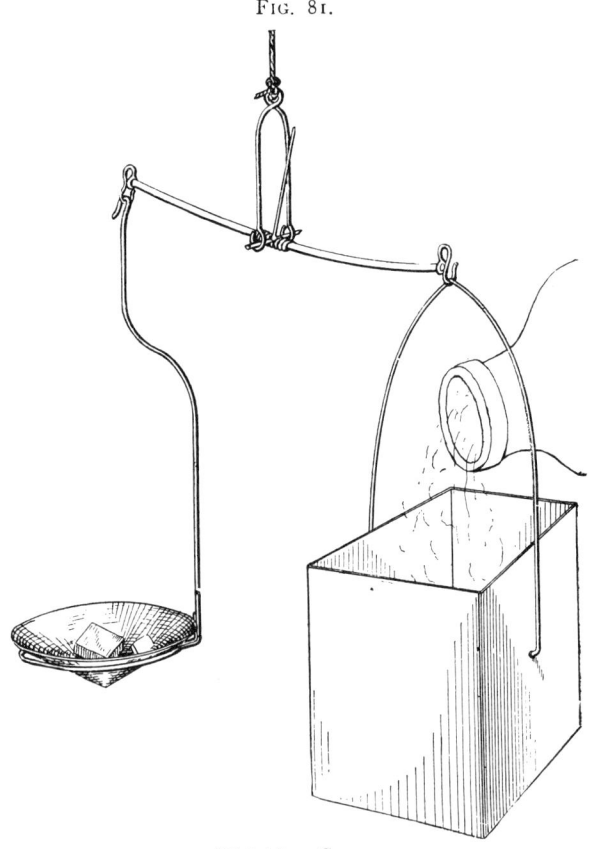

Weighing Gases.

be made of wire, as shown in the engraving. All the pivots should be made V-shaped, to reduce the friction to a minimum. The pivot of the beam should be a little higher than the bearing surface of the hooks at the ends of the beam. The conical scale pan may be made of paper, by radially slitting a disk, overlapping the edges, and sticking them together. The paper box for receiving the gas is five inches in each of its dimensions, and is suspended from the scale beam by a wire stirrup, so that it may be reversed. After bringing the scale to equilibrium in air by placing some small weights in the pan, the air contained by the box may be displaced by pouring in carbonic acid gas. The box will immediately descend, showing that carbonic acid gas is

FIG. 82.

Gas Wheel.

heavier than air. Allowing the weights in the pan to remain the same, the paper box is inverted, when the carbonic acid falls out, and air takes its place. The balance beam again becomes horizontal. Now, by opening a jar of hydrogen under the box, the air is again displaced, this time, however, by the rising of the inflowing gas. When the greater portion of the air is replaced by hydrogen, the box rises, showing by its buoyancy that its contents are lighter than air. If the balance is allowed to remain for a time, the gas will be diffused, and the balance beam will return again to the horizontal position.

To determine the weight of air, a globe provided with a stop cock is completely exhausted and weighed. Air is then admitted and the globe is again weighed, when its weight will be greater than before. The difference between the

FIG. 83.

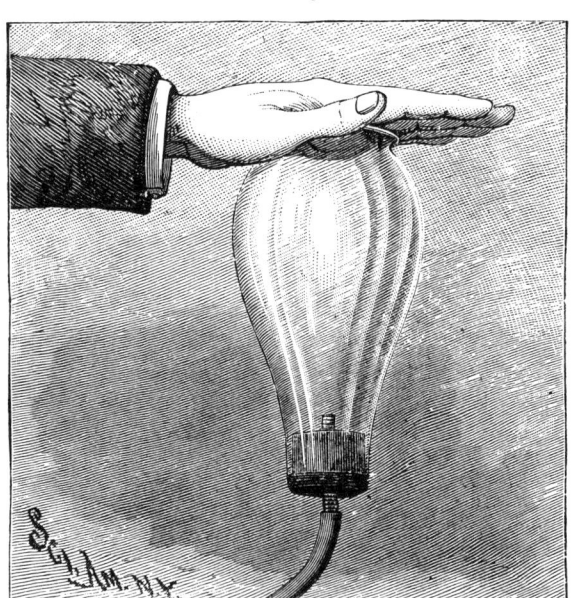

Hand Glass.

weight in the first and second cases will be the weight of the air contained by the globe.

One hundred cubic inches of dry air under an atmospheric pressure of 30 inches, and at the temperature of 60° Fahrenheit, weigh 31 grains. The same volume of carbonic acid under the same conditions weighs 47·23 grains, 100 cubic inches of hydrogen weigh 2·14 grains.

Air at the same pressure and at a temperature of 32° is about $\frac{1}{773}$ as heavy as water.

GASES.   89

In Fig. 82 is shown a very simple wheel, to be operated by gases. The wheel consists of a disk of light but stiff card board, mounted between two corks on a straight knitting needle, and provided around its periphery with buckets formed of squares of writing paper, attached to the periphery of the disk by two adjoining edges so as to form hollow cones, as shown. The knitting needle is journaled in wire or wooden standards, and lubricated so that it may turn freely. Carbonic acid gas may be generated in a

FIG. 84.

Rubber Forced Inward by Air Pressure.

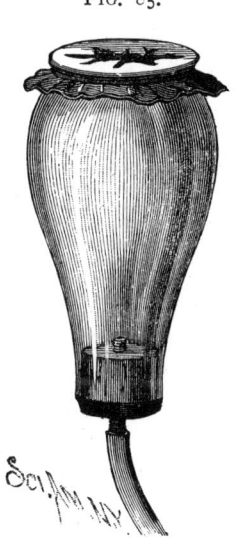

FIG. 85.

Crushing Force of the Atmosphere.

pitcher and poured upon the wheel in the manner illustrated. By making the wheel large enough and carefully balancing it, it may be turned by liberating hydrogen gas under the mouths of the buckets.

To exhibit some of the effects of atmospheric pressure, all that is required besides an air pump, or aspirator, is a large and heavy lamp chimney.

The lamp chimney needs no other preparation for use than the insertion of a five-sixteenths inch tube in the center of the cork and the thorough sealing of the cork with its tube in the smaller end of the chimney.

A very striking and instructive experiment consists in exhausting the air from the chimney by applying the suction tube of the pump to the tube at the closed end of the chimney, while the palm of the hand is applied to the large open

Fig. 86.

Weight Lifted by Air Pressure.

end of the chimney. As the air is exhausted from beneath the hand, the pressure of the atmosphere exerted on the hand drives the palm down into the chimney, as shown in Fig. 83, and as the exhaustion proceeds, the pressure becomes painful and difficult to endure.

It is easy under such circumstances to realize that the

GASES. 91

atmosphere has a very appreciable weight. The same fact may be illustrated by tying over the open end of the chimney a thin piece of elastic rubber, then exhausting the air from the chimney, allowing the external air to press the rubber down into the chimney, as shown in Fig. 84.

The disruptive power of atmospheric pressure is illustrated by the rupturing of a thin piece of bladder tied over the open end of the chimney, as shown in Fig. 85. When the air is exhausted from the chimney, the bladder, if thin enough, will burst with a loud report. If the bladder will not readily burst, the rupture may be started by puncturing it with the point of a knife.

In Fig. 86 is illustrated a similar experiment, in which the inwardly pressed diaphragm is made to raise a weight. A piece of rubber cloth is tied over the open end of the chimney, and a hook is fastened to its center by sewing. The cloth is heavily coated with rubber cement around the sewing of the hook. A weight is placed on the hook, and the air is exhausted as before. The upward pressure of the atmosphere raises the weight. This experiment illustrates the action of a form of vacuum brake now extensively in use; the weight representing the brake.

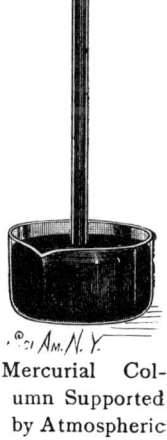

FIG. 87.

Mercurial Column Supported by Atmospheric Pressure.

THE BAROMETER.

The pressure of the atmosphere is plainly exhibited in the mercurial barometer, the simplest form of which is shown in Fig. 87. It consists of a glass tube about 36 inches in length, closed at one end and completely filled with mercury, the open end being plunged into a vessel of mercury. The column will stand at a height of about 30 inches above the level of the mercury in the vessel, showing that the pressure of the atmosphere under ordinary circumstances is equal to that of a column of mercury of about the height given. The weight of water being to that of mercury as 1 to 13·59, the height of a water column supported by the atmosphere would be about 34 feet.

The original mercurial column experiment of Torricelli was followed by an experiment by Pascal which proved conclusively that the support of the mercurial column was due to atmospheric pressure. It consisted in making simultaneous observations of two barometers, one situated at a high altitude, the other at a lower level. It was thus shown by the descent of the mercurial column, at a high elevation, that atmospheric pressure diminishes in proportion to the ascent.

AN INEXPENSIVE AIR PUMP.

The engraving illustrates an efficient air pump for both exhaustion and compression, which may be made from materials costing one dollar and fifty cents, and with the expenditure of not more than two or three hours' labor.

With this pump, the entire range of ordinary vacuum and plenum experiments may readily be performed by the aid of a few well known and inexpensive articles, such as lamp chimneys, fish globes, a tumbler or so, and pieces of sheet rubber, bladder, etc.

Fig. 88 illustrates the manner of using the pump. Figs. 89 to 92 inclusive are sectional views of the pump and its valves. Fig. 93 shows a form of valve for the compression pump, and Fig. 94 shows the application of a foot pedal to the pump. The materials required are as follows: A piece of so-called pure rubber tubing $1\frac{3}{4}$ inches external diameter, 1 inch internal diameter, and 9 inches long; a piece of pure rubber tubing 1 inch external diameter, $\frac{5}{8}$ inch internal diameter, and 5 inches long; a piece of heavy pure rubber tubing $\frac{5}{8}$ inch external diameter and 4 feet long; two wooden valve castings (shown in Fig. 90); a strip of the best oiled silk, $\frac{3}{8}$ inch wide and 8 or 10 inches long; and some stout thread.

The piece of one inch rubber tube is cut diagonally at an angle of about 30°, so as to divide it into two similar pieces. The wooden valve casing is pierced longitudinally with a one-sixteenth inch hole and transversely with a hole $\frac{1}{2}$ inch square, and thoroughly shellacked or soaked in melted paraffine to render it impervious to air. The longitudinal hole is cleared out, and the walls of the square transverse hole are smoothed. One of the walls of the square hole into

Fig 88

Testing Simple Air Pump.

which the one-sixteenth hole enters forms one valve seat, and the other forms the other valve seat. The valves each consist of two thicknesses of the oiled silk strip stretched loosely over the valve seat, and secured by the thread wound around the wooden valve casing. It will, of course,

94   EXPERIMENTAL SCIENCE.

be understood that when the valve casings are placed in
the 1 inch rubber tubing, and the 1 inch tubes are placed in
the ends of the larger tube, as shown in Fig. 89, the valves
must both be capable of opening in the same direction, so

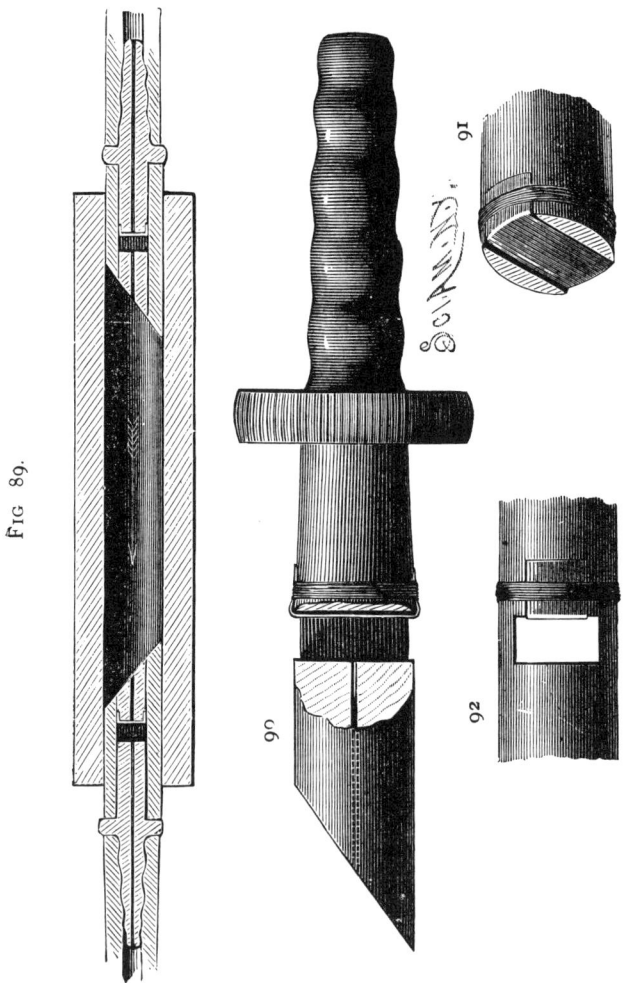

FIG. 89.—Longitudinal Section of Simple Air Pump. FIG. 90.—Valve Casing Partly in Section
FIG. 91.—Transverse Section showing Valve in Perspective. FIG. 92.—Plan View of Valve.

that the air may pass through the pump as indicated by the
arrow, entering by one valve and escaping by the other.

The pieces of rubber tube inclose the valve casings, so
that each valve has a little air-tight chamber of its own to

work in. The beveled ends of the rubber tube are arranged as shown in the engraving, and the inner ends of the wooden valve casings are beveled to correspond, so that when the large rubber tube is placed on the floor and

FIG. 93.

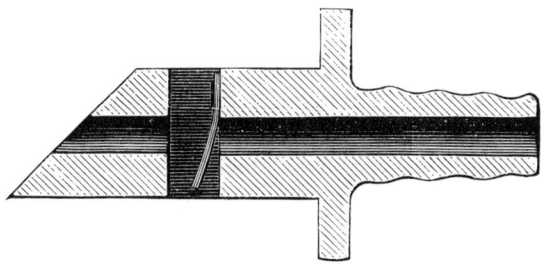

Valve for Compression Pump.

pressed by the foot, there will be very little air space left in the pump. The four-foot rubber tube is attached to one end of the pump for vacuum experiments, and to the opposite end for plenum experiments. To avoid any possibility

FIG. 94.

Treadle for Air Pump.

of the sticking of the valves, the valve seats are rubbed over with a very soft lead pencil, thus imparting to them a slight coating of plumbago, to which the oiled silk will not adhere. As an elastic rubber pump barrel of the kind described requires considerable pressure of the foot to insure the successful operation of the pump, it is advisable to construct a treadle like that shown in Fig. 94. It consists of two short boards hinged together, the lower one having a shallow groove for the reception of the middle part of the pump. The edges of the upper board are beveled at about the same angle as the ends of $1\frac{1}{4}$ inch rubber tube. The width of the hinged boards should be somewhat less than the length of the chamber in the pump. A mark is made on the side of the larger tube at one end to indicate the top, the proper position for the pump being that shown in Fig. 88.

The pressure of the foot on the side of the pump barrel expels the air through the discharge valve, and when the barrel is released, its own elasticity causes it to expand, and while regaining its normal shape it draws the air from any vessel communicating with the suction valve.

A vacuum sufficient for most of the ordinary experimental work may be produced by means of this pump in a short time. A gauge may be improvised by attaching the suction pipe to a piece of barometer tube about 30 inches long, and dipping the end of the tube in mercury, using a yard measure as a scale, as shown in Fig. 88. The pump will be found to compare favorably with piston pumps.

When it is desired to construct a pump of this kind for compressing air or for a low vacuum, the elastic tube forming the pump barrel may be larger and thinner, and the hole through the wooden valve casing may be made larger, as shown in Fig. 93, and the oiled silk valve may be replaced by a simple rubber flap valve, held in place by a single tack.

The fish globe forms the receiver of the air pump. It is closed by the soft rubber disk, which is supported by the wooden disk, the rubber being secured to the wood by four common screws passing through the rubber into the wood, about midway between the center and circumference of the rubber. Both the board and the rubber are apertured to receive a five-sixteenths brass tube, provided with a fixed collar at the top of the wood, and with a screw collar at the inner end which is turned down upon the rubber, clamping it to the wood, and at the same time making an air-tight joint around the tube.

The suction tube of the pump is applied to the small brass tube, and the soft rubber disk is pressed down upon the mouth of the globe, when the operation of producing a vacuum is begun. After a few strokes of the pump, the cover will be retained on the globe by atmospheric pressure, and will need no further holding by the hand.

A great deal of experimental and practical work may be

FIG. 95.   FIG. 96.

Water Boiling in Vacuo.   Bell in Vacuo.

done with the simple air pump described in the foregoing pages. The apparatus required for the vacuum experiments costs less than the pump. It consists of a fish globe 6 in. in diameter, a disk of thick, soft rubber large enough to cover the fish globe, a plain disk of wood as large as the rubber, two 3 in. pieces of five-sixteenths inch brass tubing, a lamp chimney with a flange on the lower end, a cork fitting the small end of the chimney, a thin piece of bladder, a thin piece of very elastic rubber, a small bell, a tumbler, a small rubber balloon, some sealing wax, some stout thread, and a piece of small wire.

98   EXPERIMENTAL SCIENCE.

The fact that water boils at a temperature below 212° when the atmospheric pressure is removed, is exhibited by placing a tumbler of hot, but not boiling, water in the receiver, as shown in Fig. 95, then exhausting the air from the receiver.

The bell suspended in the receiver by a light elastic rubber band stretched across a wire fork, whose shank is inserted in the tube of the receiver cover, as shown in Fig. 96, may be distinctly heard when rung in the receiver before exhaustion, but after exhausting the receiver, the bell will be heard feebly, if at all, thus showing that the air when rarefied is a poor sound conductor.

FIG. 97.

Destruction of Life by Removal of Air.

The inability of rarefied air to support life is shown by the experiment illustrated by Fig. 97. A mouse in the receiver soon dies when the air is exhausted.

A device for use in connection with the simple air pump for desiccating and for removing air from microscope mounts is shown in Fig. 98. It consists of an ordinary fruit jar having soldered in its cover a short tube, which is adapted to receive the suction tube of the air pump. The objects to be treated are placed in the jar, the cover put on and made tight, and the suction pipe of the pump is applied.

These are mostly well-known vacuum experiments, adapted to the simplified apparatus. There are, of course, many others that may be performed with equal facility by means of this air pump.

With the pump arranged for compression, a large number of experiments of a different character may be performed. A reservoir will be needed, like that shown in Fig. 99. It consists of a piece of ordinary leader, such as may be procured from any tinman. It should be 3 or 4 in. in diameter and 3 or 4 feet long. Heads are soldered on the ends, and all the seams are made air tight by soldering. A five-sixteenths inch tube is inserted in one end, and another in the

FIG. 98.

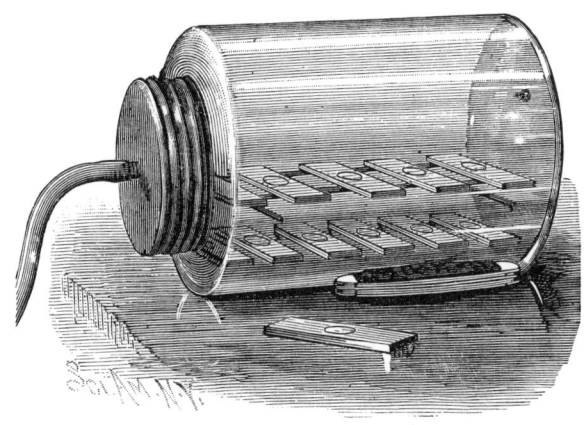

Withdrawing Air from Microscope Slides.

side. The discharge end of the pump is connected with one of the tubes of the reservoir, and a rubber tube, having at one end a one-sixtcenth inch nozzle of metal or glass, is connected with the other tube of the reservoir. The air

FIG. 99.

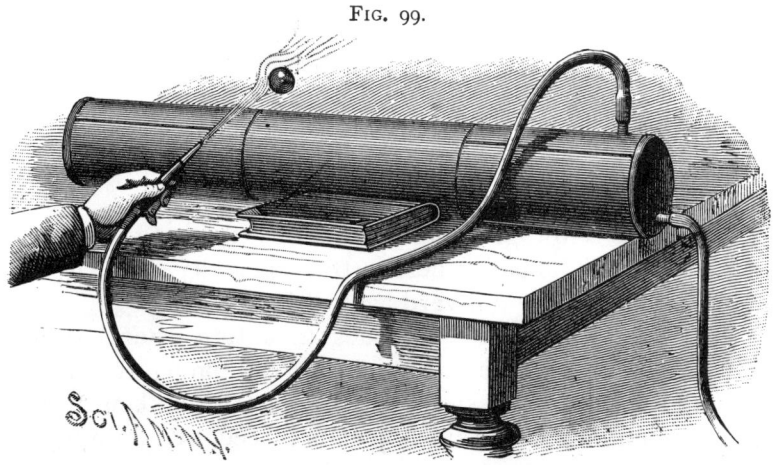

Compressed Air Reservoir and Ball Experiment.

may be confined in the reservoir by doubling the discharge tube or applying to it an ordinary pinch cock. A light ball of cork may be supported in the air jet while the nozzle is held in an inclined position, as shown in Fig. 99.

By connecting the discharge pipe of the reservoir with a spool, in the manner shown in Fig. 100, the familiar experiment of sustaining a card, together with an attached weight, by blowing down on the card may be performed.

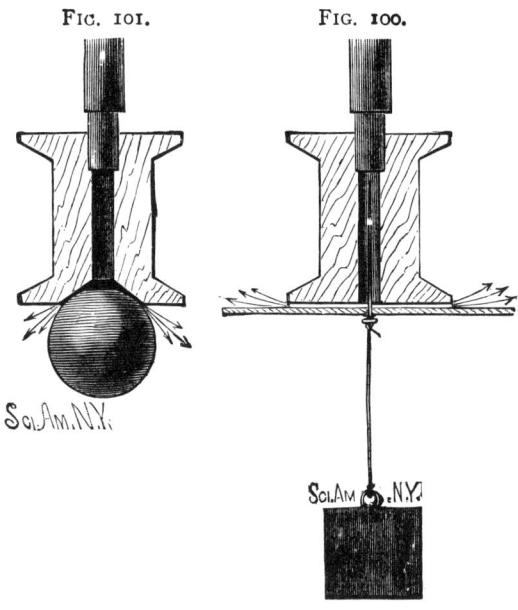

FIG. 101. FIG. 100.

Ball Experiment. Card Experiment.

A pin passing through the card into the central aperture of the spool prevents the card from slipping.

Fig. 101 shows a simple way of exhibiting the ball experiment. The ball is held in the concavity of the spool by blowing forcibly outward against it.

In these cases the air issues in a thin sheet, which adheres to and carries away the air adjoining the upper surface of the object supported, thereby producing a partial vacuum into which the object is forced by atmospheric pressure.

GASES.

In Fig. 102 is shown an atomizer which may be used in connection with the reservoir and air compressor for atomizing liquids for various purposes. In the present case it is represented as an atomizing petroleum burner. A burner of this kind yields a very intense heat, and produces a flame 2 or 3 ft. long. The oil in the vertical tube adheres to the air forced through the horizontal tube and is carried

FIG. 102.

Atomizing Petroleum Burner.

forward with the air in the form of fine spray, which readily burns as it is ejected from the nozzle. The vacuum formed in the vertical tube is supplied by oil forced up by atmospheric pressure.

ASPIRATORS FOR LABORATORY USE.

Wherever a head of water of ten feet or more is available, an aspirator is by far the most convenient instrument for producing a vacuum for filtration and fractional distillation. It is also adapted to a wide range of physical experiments.

Besides the advantage of convenience and compactness, the aspirator has the further advantage over piston air pumps in the matter of cost. It may be had at prices varying from $1.50 to $4 or $5.

Two kinds are in general use—one of glass, known as Bunsen's filter pump, and shown in Figs. 103 and 104; the other of brass, shown in Figs. 105, 106, and 107.

The glass aspirator can be purchased of almost any dealer in druggists' sundries or chemical glassware. Any expert glass blower can make it in a short time.

This instrument consists of an elongated bulb terminating in a crooked tube at the bottom and having a tapering nozzle

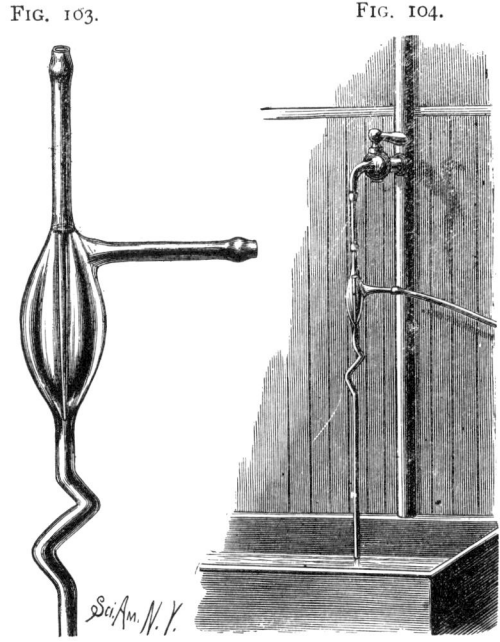

Bunsen Filter Pump.

inserted in the top and welded. The lower end of the nozzle is located directly opposite and near the crooked discharge tube. A side tube is connected with the bulb at a point near the junction of the nozzle and bulb.

This aspirator is used in the manner indicated in Fig. 104, *i. e.*, the upward extension of the nozzle is connected with a tap by a short piece of rubber tubing, and the side tube is connected by a piece of rubber tubing with the vessel to be exhausted. When the water is allowed to flow through the aspirator, it leaps across the space between the nozzle and discharge tube and carries with it by adhesion the air from the bulb, which is continually replaced by air from the vessel being exhausted.

It is necessary to securely fasten the ends of the rubber tube connected with the tap, or the water pressure may force it off, thus causing the breaking of the instrument. To secure the best effects with this pump, it is necessary to connect a vertical tube 25 to 30 feet long with the discharge end of the pump.

The metallic aspirator shown in Figs. 105, 106, and 107 is of course free from all danger of being broken in use, and it has other qualities which render it superior to the glass instrument, one of which is a much higher efficiency, another is its ability to retain the vacuum should the flow of water be accidentally or purposely discontinued. It can be screwed directly on the water tap, and needs no additional pipe to cause it to work up to its full capacity; and where a head of water is not available, it may be inserted in a siphon having a vertical height of ten feet or more.

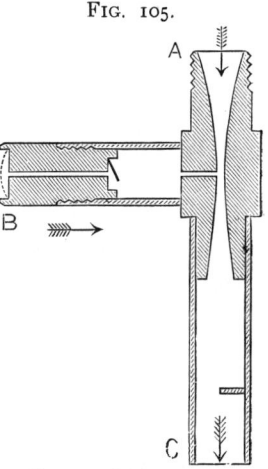

FIG. 105.

Chapman's Aspirator.

This instrument is known as the Chapman aspirator. Like all instruments of its class, it is based on the principle of the Giffard injector. The construction of the aspirator is shown in section in Fig. 105. The water enters at A, as indicated by the arrow. The air enters at B, and both air and water are discharged at C. The water in going through the contracted passage forms a vacuum at the narrower part into which the air enters. The starting of the instrument is facilitated by a diaphragm which half closes the discharge tube. The water is prevented from entering the air pipe by a small check valve shown in the interior of the lateral tube. Much of the efficiency of this instrument is due to the accuracy with which the contracted passage is formed. A slight change in the shape of this passage seriously affects the results.

The vacuum produced by this aspirator is equal to that of the mercurial barometer, less the tension of aqueous vapor. That is to say, when the barometer is at 30 inches, the vacuum produced by the aspirator will be about $29\frac{1}{2}$ inches. Such a vacuum can be produced by water under a pressure of five and one-half pounds.

In Fig. 106 is shown the aspirator applied to a Geissler

FIG. 106.

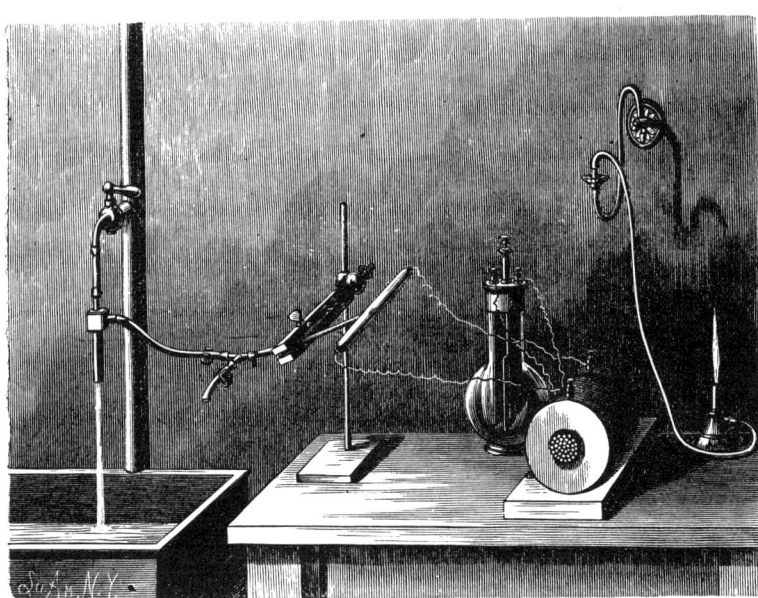

Exhausting Geissler Tube.

tube. It quickly exhausts an 8 inch tube, so that the discharge of an induction coil will readily pass through. By placing a tee in the connecting pipe, the Geissler tube can be filled with different gases. Each will exhibit its peculiar color as the spark passes. The vacuum is not high enough for a perfected Geissler tube, but it is sufficient for the greater part of vacuum experiments. The aspirator can be arranged to produce a continuous blast sufficient for the

GASES. 105

operation of a blowpipe, and for other uses requiring a moderate amount of air or gas under pressure.

The method of accomplishing this is illustrated in Fig. 107. The instrument is arranged to discharge into a bottle or other vessel having an overflow, and the air for the blast is taken out through the angled tube inserted in the stopper of the bottle. The amount of air pressure is regulated by the water pressure and the height of the overflow pipe.

For many vacuum experiments a plate provided with a

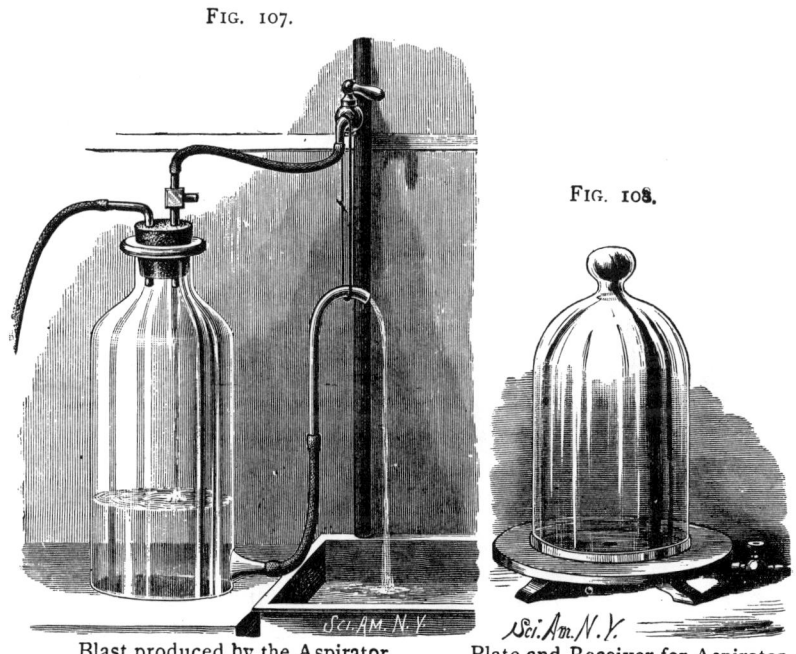

Fig. 107.

Fig. 108.

Blast produced by the Aspirator.   Plate and Receiver for Aspirator.

central aperture, and having a tube extending from the aperture to the edge of the plate, will be found useful. The tube is provided with a suitable valve, which closes communication with the aspirator, and which also serves to admit air, when required, to the receiver fitted to the plate. This plate and various accessories are like the plate and accessories of a piston air pump. Communication is established between the tube of the plate and the aspirator by means of a pure rubber tube, which is practically air tight.

MOUTH VACUUM APPARATUS.

Although the vacuum apparatus already described is very simple, it is quite practicable to perform many experiments of this class by using the mouth as an air pump, thus dispensing almost entirely with mechanism. The operation of producing a partial vacuum is facilitated by employing a valve such as is shown in the left hand figure of Fig. 109. This valve consists of a thick tube of hard wood, having a bore of about $\frac{1}{16}$ inch. One end of the tube is corrugated to receive a rubber pipe, and over the other end is tied a valve of elastic rubber. By connecting this valve with a stopped glass tube by means of a flexible rubber pipe and a jet tube in the manner shown, and then sucking the air through the valve, a partial vacuum may be quickly formed in the tube. The vacuum will be retained by the valve, so that when the valve is disconnected from the jet tube, while the latter is immersed in water, the pressure of the external air will cause the water to enter the glass tube through the jet in the form of a fountain. It is obvious that many of the foregoing experiments may be tried in a similar way.

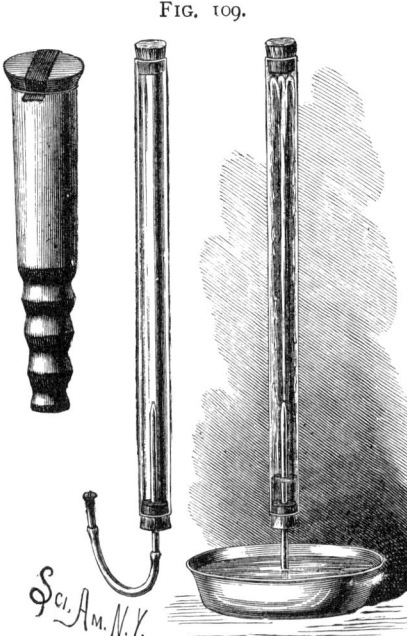

FIG. 109.

Mouth Vacuum Apparatus.

ANCIENT INVENTIONS OPERATED BY AIR PRESSURE.

More than two thousand years ago, Hero (or Heron), a philosopher and mathematician of Alexandria, invented the fountain shown in the annexed engraving. This device,

FIG. 110.

Hero's Fountain.

because of its antiquity, as well as its simplicity and completeness, is very interesting and instructive.

As represented in the engraving, it may be classed with toys, or at most regarded as only an apparatus for illustrating a scientific principle; but it is more than this. It is the progenitor of a number of modern inventions for raising water and producing air pressure.

The curious feature of the apparatus is that it apparently causes the water to rise above its own level by its own pressure, but such is not the case. Its action is due to the transference of the pressure of one column of water to another column of water at a higher level, through the medium of a column of confined air. It is as truly a case of the application of external power as it would be if a steam air compressor were applied.

The water to be elevated is contained by the upper bulb, which communicates at its lower side with the fountain nozzle, and at its upper side with the downwardly curved tube connecting with the top of the lower bulb. A tube connecting with the lower side of the lower bulb extends upward to the level of the upper bulb, and terminates in a flaring cup.

The upper bulb having been filled with water and the lower bulb with air, the fountain is started by pouring a small quantity of water into the cup, which by flowing downward through the tube connected with the cup exerts a pressure on the air contained by the lower bulb. This pressure is equal to the weight of the column of water in the tube. The air pressure thus created is transferred to the top of the upper bulb by the air column rising from the lower bulb through the tube connecting the two bulbs, so that the pressure of the water column descending from the cup, less a very small allowance for friction, is effective in forcing the water out of the upper bulb through the fountain nozzle.

The proper inclination of the apparatus directs the water jet so that the water falls into the cup and replaces the water used in creating the air pressure in the lower bulb.

When the lower bulb is filled with water, and the water has been entirely discharged from the upper bulb, the action of the apparatus ceases; but it may be again started by inverting the fountain, allowing the water in the air bulb to run into the upper or water bulb, then righting it and again pouring a little water into the cup.

This device was employed during the last century for elevating water in the mines of Hungary.

In Fig. 111 is shown an interesting modification of Hero's fountain. The apparatus is made of glass, to illustrate the principle on which it operates. It consists of a volute coil of tubing connected at its center with a hollow shaft that communicates with a hollow journal box, from which a standpipe rises. When this coil is turned in the direction indicated by the arrow, water and air assume in the coiled tube positions relative to each other as shown in the engraving; the water being arranged in a series of curved columns on one side of the center of the wheel, the air being correspondingly disposed on the opposite side of the center. The height to which the water will be raised by this machine is equal to the sum of the heights above their upwardly curved lower ends of all the curved columns of water contained by the coil. It will be noticed that the pressure of one curved column of water in the coil is communicated to the next through the intervening air, which weighs practically nothing.

This machine was invented by Wirtz, of Zurich, in 1746.

"In 1784 a machine of this kind was made at Archangelsky that raised a hogshead of water in a minute to an elevation of 74 feet, and through a pipe 760 feet long."

INERTIA OF AIR.

Although air is a light and extremely mobile fluid, it has sufficient inertia to permit of the flight of birds, the operation of windmills, and the propulsion of sailing vessels. The aerial top shown in Fig. 112 is dependent upon the inertia of the air. This top is simply a metallic screw wheel, adapted to be revolved by means of a string in the same manner as an ordinary top.

With the application of a sufficient amount of force, this top will rise to a height of 150 to 200 feet. It can hardly be

FIG. 111.

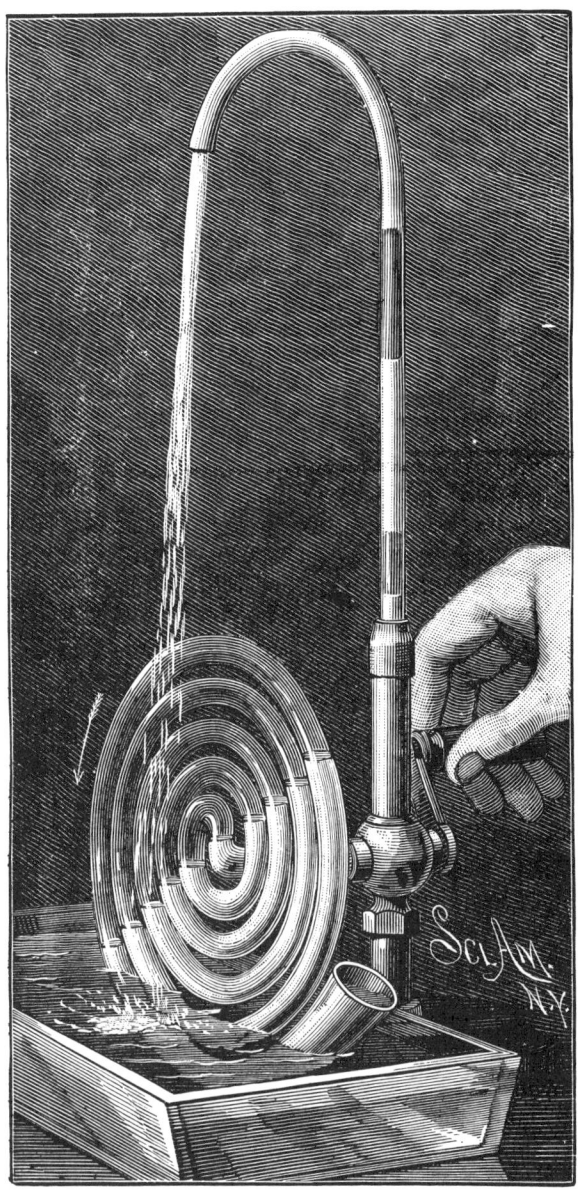

Wirtz's Pump.

called a flying machine, as it does not carry its own motive power. In the next illustration, however, is shown a flying machine which in one sense carries its own power, that is, stored power.

It consists of a light frame furnished at one end with a slender rattan bow inclosed in a little bag of tissue paper, which forms a sort of rudder when the fly-fly ascends, and opens like an umbrella when it descends, forming a parachute, which greatly retards the fall. In the crosspiece of the opposite end is journaled a little shaft formed of a wire having on its inner end a loop receiving a number of rubber bands, which are fastened to the opposite end of the frame. To the outer end of the little shaft is secured a piece of cork, in which are inserted two feathers inclined at an angle with the plane of the shaft's rotation, and oppositely arranged with respect to each other.

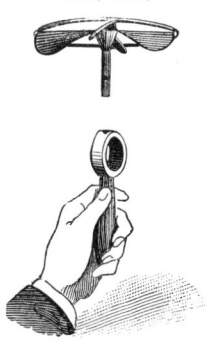

Fig. 112.

Aerial Top.

By turning the propeller wheel thus formed, the rubber bands are twisted, and sufficient power is stored in them to turn the propeller wheel in the direction opposite to that required for winding, and thus propel the device through the air.

Another device still more nearly approaching the ideal flying machine is shown in the annexed cut, Fig. 114 being a perspective view of the entire bird, and Fig. 114a an enlarged perspective view of the working parts. It is known as Penaud's mechanical bird.

Fig. 113.

The Fly-Fly.

It is a pretty toy, imitating the flight of a bird very well indeed. It soars for a few seconds, and then requires rewinding. Two Y-shaped standards secured to the rod forming the backbone of the apparatus support at their upper ends two wires, upon which are pivoted two wings formed of light silk. The wings are provided with light stays, and are connected at their inner corners with the backbone by threads. In the Y-shaped standards is journaled a wire crank shaft carrying at its forward end a transverse wire forming a sort of balance, and serving also as a key for winding. The inner end of the crank shaft is provided with a loop to which are attached rubber bands which are also secured to a post near the rear end of the apparatus. Two connecting rods placed on the crank are pivotally connected with the shorter arms of the levers of the wings. The rear end of the backbone is provided with a rudder.

FIGS. 114 AND 114*a*.

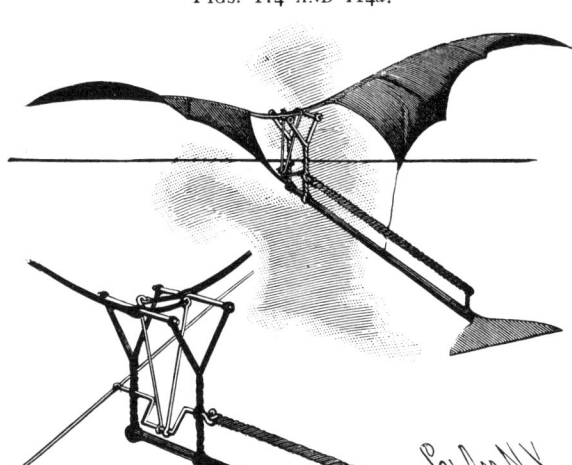

Mechanical Bird.

The rubber bands are twisted by turning the shaft by means of the cross wire. When the shaft is released, it is turned by the rubber bands in a reverse direction, causing the crank to oscillate the wings, which beat the air in a natural manner, and propel the device forward. The principle of the inclined plane is involved here, but the plane, instead of being rotated, as in all the cases mentioned above, is reciprocated.

The toy boomerang, which is, in some respects, similar to the regular article, cannot perform all the feats with which the more pretentious implement is credited; but it can be projected, and made to return over nearly the same path.

The toy boomerang is made of a piece of tough cardboard cut on a parabolic curve as shown in the engraving, one arm of the boomerang being a little longer than the other. When laid on an inclined surface, as shown in the engraving, and snapped by a pencil held firmly in one hand and drawn back and released by the fingers of the other hand, the boomerang is set in rapid rotation by the blow, and is at the same time projected, the first part of the trajectory being practically in the continuation of the plane in which the boomerang is started; but when the momentum which carries it forward is exhausted, the boomerang still revolves, and maintains its plane of rotation, so that when it begins to fall, instead of describing the same trajectory as ordinary projectiles, it makes a circuit to one side and comes back toward the point of starting. The flatness or curvature of the boomerang and the form of its edges, as well as the position in which it is placed for starting, and the speed and manner of starting, all have an effect in determining the outward as well as the return course of the projectile.

FIG. 115.

Boomerang.

VORTEX MOTION.

Every one has noticed the symmetrical wreaths of smoke and steam occasionally projected high into the air on a still day by a locomotive; similar rings may often be noticed after the firing of a gun. It is not uncommon to see a smoker forming such wreaths with his mouth. These rings are simply whirling masses of air revolving upon axes curved in annular form, the smoke serving to mark the projected and whirling body of air, thus distinguishing it from the surrounding atmosphere. The whirls would exist without the smoke, but they would, of course, be invisible.

FIG. 116.

Vortex Rings.

All the apparatus needed for producing vortex rings at will is an ordinary pasteboard hat box, having a circular hole of 4 or 5 inches diameter in the cover. Two pads of blotting paper are prepared, each consisting of six or eight pieces. Upon one pad is poured a small quantity of muriatic acid and upon the other a similar quantity of strong aqua ammonia. These pads are placed in the box and immediately a white cloud is formed, which consists of particles of chloride of ammonium so minute as to float in the air.

By smartly tapping opposite sides of the box, a puff of air is sent through the circular opening of the cover, carrying with it some of the chloride of ammonium. The friction of the air against the edges of the cover retards the outer portion of the projected air column, while the inner portion passes freely through, thus imparting a rotary motion to the body of air adjoining the edge of the cover, the axis of revolution being annular. After the ring is detached, the central portion of the air column continues to pass through it, thus maintaining the rotary motion.

When two rings are projected in succession in such a manner as to cause one to collide with the other, they behave much like elastic solid bodies. By making the aperture in the box cover elliptical, the rings will acquire a vibratory motion.

By fastening the box cover loosely at the corners, the box may be turned upon its side and rings may be projected horizontally.

It is obvious that smoke may be used in this experiment in lieu of the chloride of ammonium.

CHAPTER VIII.

SOUND.

The student of acoustics need not go beyond the realm of toys for much of his experimental apparatus. The various toy musical instruments are capable of illustrating many of the phenomena of sound very satisfactorily, if not quite as well as some of the more pretentious apparatus.

Sound is a sensation of the ear, and is produced by sonorous vibrations of the air. It may be in the nature of a mere noise, due to irregular vibrations, like the noise of a wagon on the street, or it may be a sharp crack or explosion, like the cracking of a whip or like the sound produced by the collision of solid bodies. The clappers, or bones, with which all boys are familiar, are an example of a class of toys which create sound by concussion, and the succession of sounds produced by the clappers are irregular, and clearly distinct from musical sounds. A succession of such sounds, although occurring with considerable frequency and perfect regularity, will not become musical until made with sufficient rapidity to bring them within the perception of the ear as a practically continuous sound. The rattle, or cricket, produces a regular but unmusical sound.

Fig. 117.

Clappers.

The wooden springs of the cricket snap from one ratchet tooth to another, as the body of the cricket is rapidly swung around, making a series of regular taps, which, taken all together, make a terrific noise, having none of the characteristics of musical sounds. That a musical sound may be made by a series of taps is illustrated by the buzz, a toy consisting of a disk of tin having notched edges and provided with two holes on diametrically opposite sides of the center, and furnished with an endless cord passing through the holes. The disk is rotated by pulling in opposite directions on the twisted endless cord, allowing the disk to twist the cord in the reverse direction, then again pulling the cord, and so on.

Fig. 118.

The Cricket, or Rattle.

If, while the disk is revolving rapidly, its periphery is brought into light contact with the edge of a piece of paper, the successive taps of the teeth of the disk upon the paper produce a shrill musical sound, which varies in pitch according to the speed of the disk. Such a disk mounted on a shaft and revolved rapidly is known as Savart's wheel.*

Fig. 119.

The Buzz.

It is ascertained by these experiments that regular vibrations of sufficient frequency produce musical sounds, and that concussions, irregular vibrations, and regular vibrations having a

*See chapter on experiments with the scientific top.

slow rate, produce only noises. It has been determined that the lowest note appreciable by the ear is produced by sixteen complete vibrations per second, and the highest by 24,000 complete vibrations per second.

VIBRATING RODS.

The zylophone and metallophone are examples of musical instruments employing free vibrating rods supported at their nodes. The zylophone consists of a series of wooden rods of different lengths, bored transversely at their nodes, or points of least vibration, and strung together on cords. The instrument may either be suspended by the cords or

FIG. 120.

The Zylophone.

laid upon loosely twisted cords situated at the nodes. By passing the small spherical wooden mallet accompanying the instrument over the wooden rods, very agreeable liquid musical tones are produced by the vibration of the rods, and when the rods are struck by the mallet they yield tones which are very pure, but not prolonged.

The cheaper forms of zylophone are tuned by slitting the rods transversely at their centers on the under side, by means of a saw, to a depth required to give them the flexibility necessary to the production of the desired tones. The rods are divided by the nodes into three vibrating parts, the parts between the nodal points and the ends being nearly one-half of the distance between the nodes.

FIG. 121.

The Metallophone.

The metallophone is similar in form to the zylophone, but, as its name suggests, the vibrating bars are made of metal—

FIG. 122.

Music Box.

hardened steel. The bars rest at their nodes on soft woolen cords, secured to the upper edges of a resonator forming the support of the entire series of bars. The resonator is tapered both as to width and depth, and serves to greatly increase the volume of sound, although it does not act as a perfect resonator for each bar.

When a bar is struck, its downward movement produces an air wave which moves downward, strikes the bottom of the resonator, and is reflected upward in time to re-enforce the outwardly moving air wave produced by the upward bending of the bar.

The metallophone yields sweet tones which are quite different in quality from those produced by the vibration of wooden bars.

The music box furnishes an example of the class of instruments in which musical sounds are produced by the vibra-

FIG. 123.

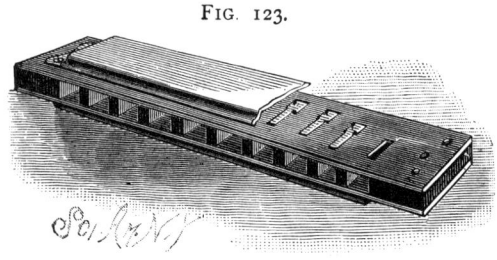

Mouth Organ, or Harmonica.

tion of free reeds or tongues rigidly held at one end and free to vibrate at the other end. The tongues of the music box are made by slitting the edge of a steel plate, forming a comb, which is arranged with its teeth projecting into the paths of the pins of the cylinder, which are distributed around and along the cylinder in the order necessary to secure the required succession of tones. The engagement of one of the pins of the cylinder with one of the tongues raises the tongue, which, when liberated, yields the note due to its position in the comb.

The tongues are tuned by filing or scraping them at their free or fixed ends, or by loading them at their free ends. In this instrument the sonorous vibrations are produced by the tongue, which itself has the desired pitch.

SOUND.

REEDS.

In reed instruments the sounds emitted by the reeds are greatly strengthened by resonance. The mouth organ or harmonica is a familiar example of a simple reed instrument without accurately adjusted resonators.

FIG. 124.

The Bugle.

When reeds are employed in connection with resonating pipes, as in the case of the reed pipes of an organ, the pipe synchronizes with the reed, and re-enforces the sound. When the reed is very stiff, it commands the vibrations of the air column, and when it is very flexible, it is controlled by the air column.

The horn is a reed instrument in which the lips act as reeds, and the tapering tube serves as a resonator.

LONGITUDINAL VIBRATION OF RODS.

The foregoing are examples of the transverse vibration of rods. The annexed figures illustrate apparatus in which the longitudinal vibration of rods is shown.

By grasping a steel rod at the center between the thumb and finger, each of its two ends being free, and striking it upon the end with a hammer, the rod can be made to yield a sound of very high pitch. By holding one end firmly in a vise, and skillfully rubbing the rod, by pulling it

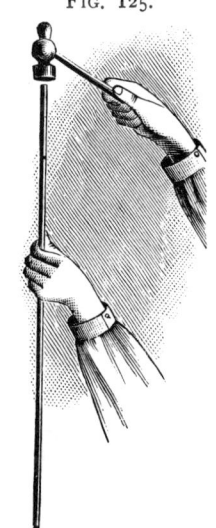

FIG. 125.

Longitudinal Vibration of a Steel Rod.

between the fingers with a cloth or piece of leather covered with powdered resin, a note an octave lower will be emitted.

Marloye's harp, shown in Fig. 126, depends upon the longitudinal vibration of rods. This instrument consists of a number of pin rods of different lengths inserted in a sounding box or solid block of wood, and tuned by cutting them off at such lengths as to cause them to yield the notes of the diatonic scale. The instrument is played by rubbing the rods lengthwise by the thumb and finger covered with powdered resin. The sounds produced by the instrument resemble those of a flute.

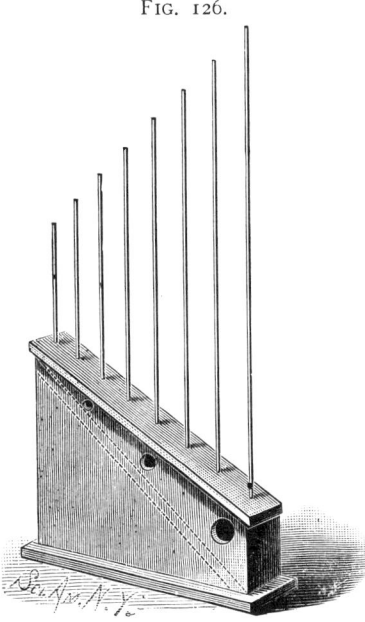

FIG. 126.

Marloye's Harp.

PIPES.

The ancient Pandean pipes present an example of an instrument formed of a series of stopped pipes of different lengths. These pipes

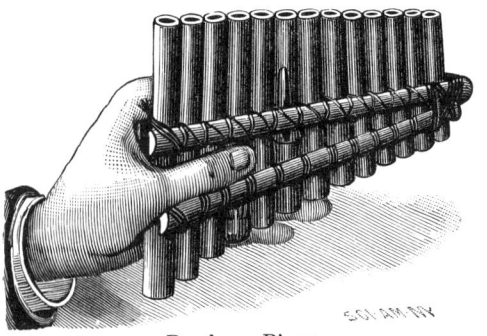

FIG. 127.

Pandean Pipes.

are tuned by moving the corks by which their lower ends are stopped, and the air is agitated by blowing across the end of the tubes. The flageolet is an open pipe in which the air is set in vibration by blowing a thin sheet of air through the slit of the mouthpiece against the thin edge of the opposite side of the embouchure. The rate of the fluttering produced by the air striking upon the thin edge is determined by the length of the pipe of the instrument, the length being varied to produce the different notes, by open-

FIG. 128.

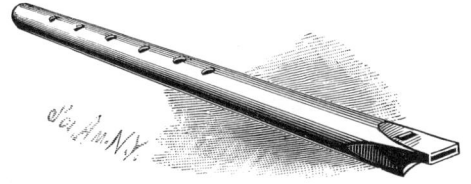

Flageolet.

ing or closing the finger holes. By comparing the flageolet with the Pandean pipes, it is found that for a given note the open flageolet pipe must be about twice as long as the Pan pipe. When all the finger holes of the flageolet are closed, it is then a simple open pipe, like an organ pipe, and, if compared with the Pan pipe yielding the same note, it is found to be just twice as long as the closed pipe. If, while

FIG. 129.

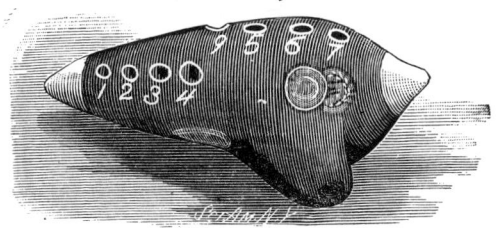

Ocorina.

the holes are closed, the open end of the flageolet pipe be stopped, the instrument will yield a note an octave lower if the blowing be very gentle. These experiments show that the note produced by a stopped pipe is an octave below the note yielded by an open pipe of the same length, and the same as that obtained from an open pipe of double the length.

The ocorina is a curious modern instrument, of much

124   EXPERIMENTAL SCIENCE.

the same nature as the flageolet. It is, however, a stopped pipe, and shows how tones are modified by form and material, the latter being clay. It produces a mellow tone, something like that of a flute.

STRINGED INSTRUMENTS.

The zither, now made in the form of an inexpensive and really serviceable toy, originated in the Tyrol. It consists of a trapezoidal sounding board, provided with bridges, and having 24 wire strings.

Its tones are harp-like, and with it a proficient player can produce agreeable music. Much of the nature of the

FIG. 130.

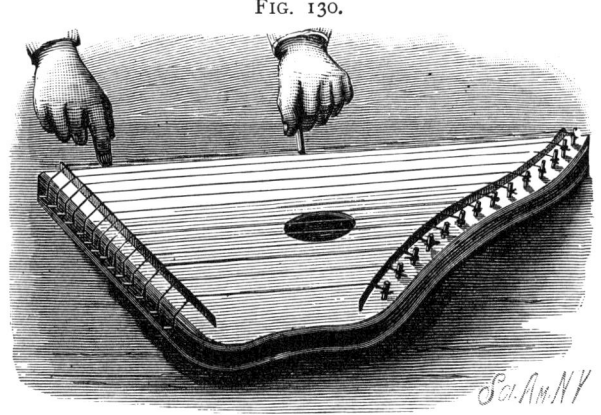

Zither.

vibration of strings may be exhibited by means of this instrument. On damping one of the strings by placing the finger or a pencil lightly against its center, and vibrating the string, at the same time removing the pencil, the string will yield a note which is an octave higher than its fundamental note.

By examining the string closely, it will be ascertained that at the center there is apparently no vibration, while between the center and the ends it vibrates. The place of least vibration at the center of the string is the node, and between the node and the ends of the strings are the venters. It will thus be seen that the string is practically divided into two equal vibrating segments, each of which produces a note an octave higher than that of the open string. That the note is an octave higher than the fundamental note may be determined by comparing it with the note of the string which is an octave above in the scale of the zither.

By damping the string at the end of one-fourth of its length, the remaining portion of the string divides itself into three ventral segments, with two nodes between.

The division of the string into nodes and venters occurs whenever the string is vibrated, and all of the notes other than the fundamental are known as harmonics, and impart to the sound of the string its quality.

By tuning the first two strings in unison, the vibration of one string by sympathy with the other string may be shown.

CONDUCTION OF SOUND.

The string telephone, although not a musical instrument, nor even a sound producer, exhibits an interesting feature in the conduction of sounds. It consists of two short tubes or mouthpieces, each covered at one end with a taut parchment diaphragm, the two diaphragms being connected with a stout thread. By stretching the thread so as to render it taut, a conversation may be carried on over quite a long distance, by talking in one instrument and listening at the other. The vibration of one diaphragm, due to the impact of sound waves, is transmitted to the other diaphragm by the thread.

FIG. 131.

String Telephone.

In the toys illustrated we have a representative of the Savart's wheel in the buzz; of the pipe organ in the Pan pipes, the flageolet, and the mouth organ; of band instruments in the bugle; and of the piano, harp, and other stringed instruments in the zither.

HARMONIC VIBRATIONS.

Impulses which, occurring singly or at irregular intervals, are incapable of producing any noticeable effects may, when made regularly, under favorable circumstances, yield astonishing results. The rattling of church windows by air waves generated by a particular pipe of the organ, a bridge strained or broken by the regular tramp of soldiers or by the trotting of horses, the vibration of a six or eight story building by a wagon rumbling over the pavement, a factory vibrated to a dangerous degree by machinery contained within its walls, a mill shaken from foundation to roof by air waves generated by water falling over a dam, are all familiar examples of the power of regular or harmonic vibrations.

Harmonic vibrations result from regularly recurring impulses, which may be very slight indeed, but when the effects of the impulses are added one to another, the accumulation of power is sometimes very great.

To secure cumulative effects, the impulses must not only be regular in their occurrence, but the body receiving the impulses must be able to respond, its vibratory period must correspond with the period of the impulses, and, further than this, the impulses must bear a certain relation to a particular phase of the vibration, in order that they may act upon the vibrating body in such a way as to augment its motion rather than diminish it.

There are railroad bridges that vibrate alarmingly when crossed by locomotives running at a certain speed, the vibrations being caused by the comparatively slight lack of balance in the driving wheels and connecting rods. For this reason the speed is restricted on such bridges.

During the early tests of the East River bridge between New York and Brooklyn, it was found that the structure was so massive and its vibratory period so slow that it could not be injuriously affected by the marching of men or the trotting of horses; consequently, travel proceeds on this bridge as upon any highway.

A well known English physicist is reported to have said

that with suitable appliances he could break an iron girder by pelting it with pith balls. An experiment of this kind would certainly show in a striking manner the effects of very slight rhythmic impulses. As it is manifestly impracticable to perform such an experiment, an easier method of illustrating harmonic vibrations must be sought.

In the accompanying engravings, Fig. 132 shows how a bar of steel may be set in active vibration by drops of water. The bar is supported at nodal points upon angular pieces

FIG. 132

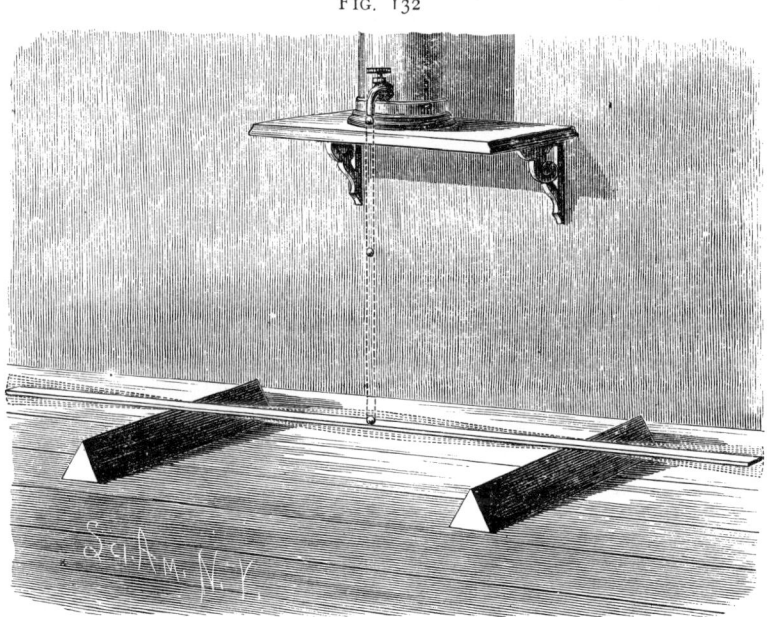

Harmonic Vibration.

of wood. Above the center of the bar is arranged a faucet, which communicates with the water supply. The bar is first vibrated by hand, and the faucet is adjusted so that the water drops in unison with the vibrations of the bar. The motion of the bar is then stopped, and the water is allowed to drop on it. The bar soon begins to vibrate, and in a short time the vibration acquires considerable amplitude. In Fig. 133 is shown an experiment in which the intermittent pull of an electro-magnet is made to accomplish the same thing. In this case the steel bar forms a part of the circuit. The magnet is provided with a light wooden spring-pressed arm, carrying a contact point and a conductor. This arm is arranged to follow the bar up and down through the upper half of its excursion, breaking the contact at the median position of the bar. The magnet becomes alternately magnetized and demagnetized, and the bar is alternately pulled down and released. The bar used in these experiments is $\frac{1}{4}$ inch thick, $1\frac{1}{4}$ inches wide, and 8 feet

FIG. 133.

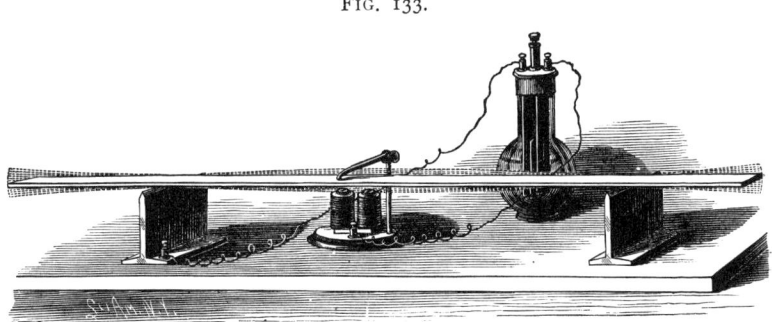

Vibration by Magnetic Impulse.

long. A much larger bar might be used. Without doubt, even an iron girder of great size and weight might be set in active vibration by the same means.

SIMPLE SOUND RECORDER.

In Fig. 134 is shown a simple device for recording sounds autographically.* The propelling of the smoked plate under the stylus is accomplished by simply inclining the support of the plate and allowing the plate to slide off quickly by its own gravity.

This apparatus consists of a wooden mouthpiece like that of a telephone, with a parchment diaphragm glued to its back, and provided with a tracing point, which is slightly inclined downward toward the guide for the plate.

This tracing point is a common sewing needle, having its pointed end bent downward. It is cemented at the eye end

* See also chapter on projection.

SOUND.   129

to the center of a diaphragm by a drop of sealing wax. The mouthpiece is attached to a base supporting the crosspiece upon which the smoked plate is placed.

A thin strip of wood fastened by two common pins—one at each end—serves as a guide for the smoked plate.

To prevent the tracing point from being deflected laterally by the moving glass, a needle is driven down into the baseboard in contact with the tracing point.

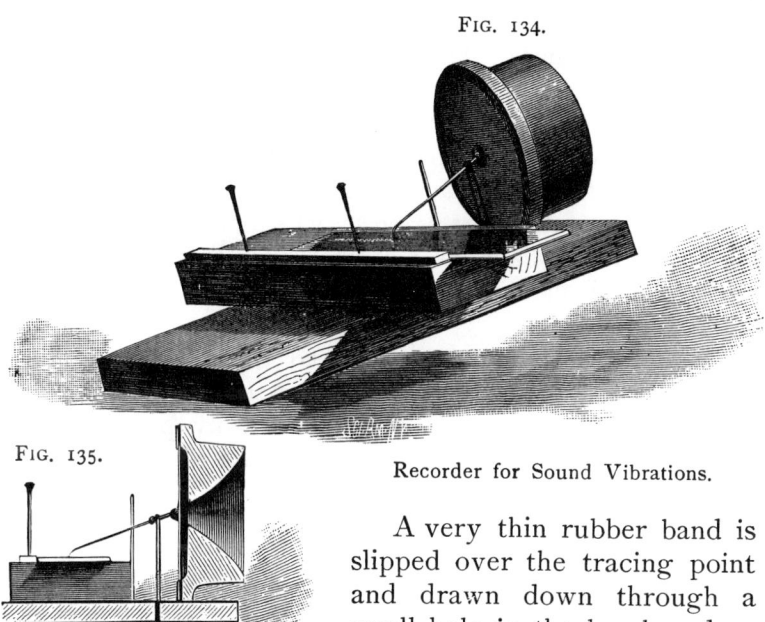

FIG. 134.

FIG. 135.

Recorder for Sound Vibrations.

A very thin rubber band is slipped over the tracing point and drawn down through a small hole in the baseboard, as shown in Fig. 135, until the necessary tension is secured for keeping the point in delicate but continuous contact with the smoked plate.

The best plates for the purpose of making the tracings are the microscope slide glasses with ground edges. They may be readily smoked over a gas jet turned down quite small, or over a candle or kerosene lamp. The flame in any case should be small and the film of smoke fine and very thin.

The smoked plate is placed on the support and against the guide and under the needle, and the instrument is inclined until the plate rests against the guide. Now the mouth is placed near the mouthipece, and a vowel is uttered, while the instrument is inclined sidewise at a sufficient angle to permit the glass to slide off quickly. Of course the glass should fall only a very short distance, and it is well to provide a soft surface for it to alight on.

If all this is done with the slightest regard for precision, a beautiful tracing will be secured, which will show the composite nature of each sound wave. The regularity and uniformity of the entire tracing is surprising, considering the comparatively crude means employed in producing it.

The beginning of the sinuous line is somewhat imperfect, owing to the slow initial movement of the plate in its descent, but the greater portion is perfect.

After having made one line, the pins holding the guide are moved forward, placing the guide in a new position, when the operation of tracing may be repeated with another vowel. Monosyllables and short words may be recorded. If the plate is made long enough, it will, of course, receive an entire sentence.

These tracings may be covered with a second microscopic glass plate to protect them, or they may be mounted as a microscopic object for a low power by putting a thin cover over them in the usual way. Used as lantern slides, they give fine results.

VIBRATING FLAMES.*

The most perfect exhibition of vibrating flames can be made only with expensive apparatus; but the student can get very satisfactory results by the employment of such things as are shown in Fig. 136. A candle, a rubber tube, an oblong mirror, and a piece of thread are the only requisites, excepting the support for the mirror—which in the present case consists of a pile of books—and a little paper funnel inserted in the end of the rubber tube and forming the mouthpiece.

The thread is tied around opposite ends of the oblong mirror, and the mirror supported by passing the thread through the upper book of the pile, which juts over to allow

* See also chapter on experiments with scientific top.

the mirror to swing freely without touching the books.
The mirror is made to vibrate in a horizontal plane by
giving it a twisting motion. One end of the rubber tube is
placed very near the base of the candle flame, and the other
end, which is provided with the paper mouthpiece, is placed
before the mouth and a sound is uttered which causes the
air contained by the rubber tube to vibrate and impart its
motion to the candle flame. The vibratory character of the
flame is not noticeable by direct observation, but on viewing the flame in the swinging mirror, separate images of

FIG. 136.

Simple Method of Producing and Viewing Vibrating Flames.

the flame will be seen. These images are combined in a
series which, with a certain degree of accuracy, represent
the sound waves by which the fluctuations of the flame are
produced.

To show that these images result from a vibrating flame,
it is only necessary to view the flame in the mirror. When
no sound is made in the mouthpiece, only a plain band of
light will be seen.

A somewhat more convenient arrangement of mirrors is
shown in Fig. 137. In a baseboard is inserted a wire, one-
eighth inch or more in diameter and about a foot long. On

this wire is placed an ordinary spool, and above the spool a thin apertured board (shown in the detailed view), the board being about 8 inches long and 6 inches wide. The board is perforated edgewise to receive the wire. In the upper edge of the board, half way between the center and end, is inserted a wire, upon which is placed a small spool, serving as

FIG. 137.

Rotating Mirror.

a crank by which to turn the board. Upon opposite sides of the board are placed mirrors of a size corresponding to that of the board, the mirrors being secured to the board by strips of paper or cloth pasted around the edges. The image of the flame is viewed in the mirrors as they are revolved.

SPEAKING FLAME.

The speaking flame apparatus shown in the annexed engravings is based on the principle of the annular burner often used in producing the oxyhydrogen light, the principal difference being in the diminished annular orifice. The construction of the burner is clearly shown in Fig. 138, the detached illustration being an enlarged sectional view of the end of the burner. Gas is taken through the central tube, and the flexible speaking tube is connected with the outer tube of the burner. When the apparatus is used for producing musical and articulate sounds, a resonator is attached, as shown in Fig. 138. In this figure the resonator is broken away to show its position relative to the burner.

By screwing the cap of the burner up or down, an adjustment may be secured which will cause the flame to reproduce any sounds uttered in the mouthpiece attached to the flexible speaking-tube. With a fine adjustment articulate speech or any note of the musical scale within the compass of the human voice may be reproduced by the flame.

The slight air waves which reach the burner through the flexible pipe act directly upon the base of the flame; this

FIG. 138.

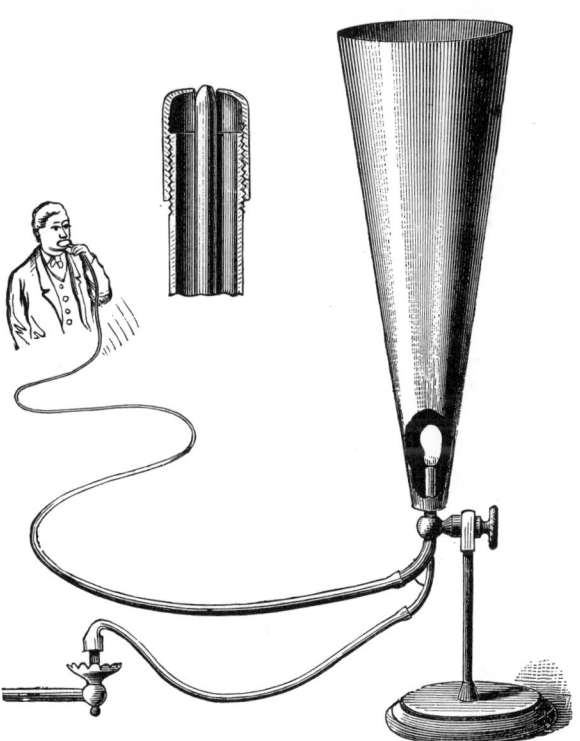

The Speaking Flame.

portion of the flame being more sensitive to disturbing influences than any other. This fact has been determined by experiments on sensitive flames, such as are described further on. By speaking in the mouthpiece while the gas is cut off from the burner, it is found that no sound proceeds from the burner, thus showing conclusively that the sounds are produced by the flame.

FIG. 139.

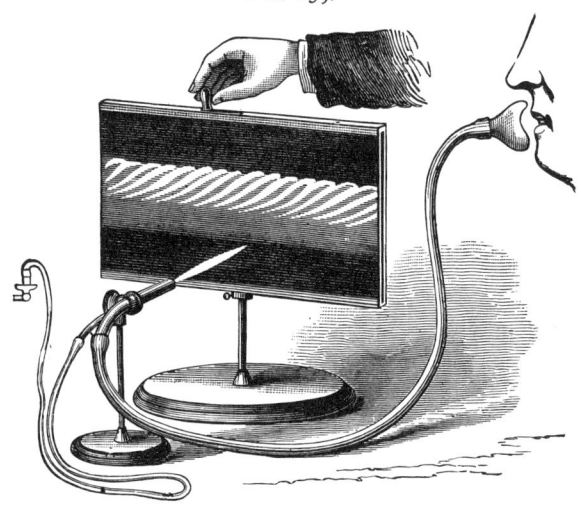

Vibrating Flame Apparatus.

FIG. 140.

Circular Mirror.

SOUND. 135

FIGS. 141 TO 144.

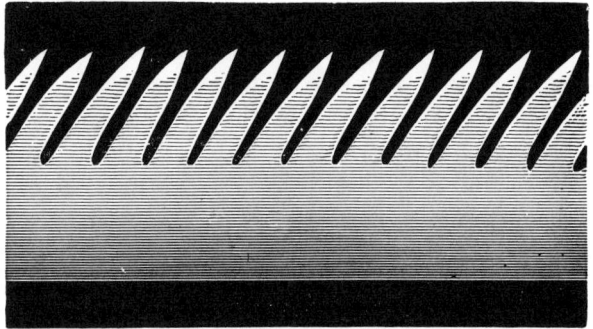

Manometric Flame.

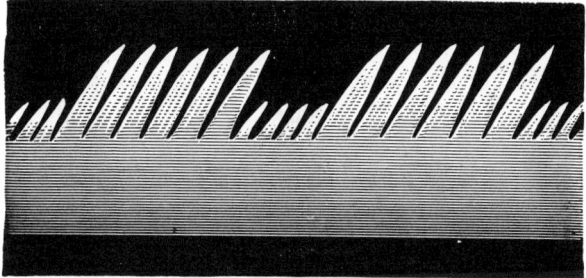

A Trill.

A Rope of Flame.

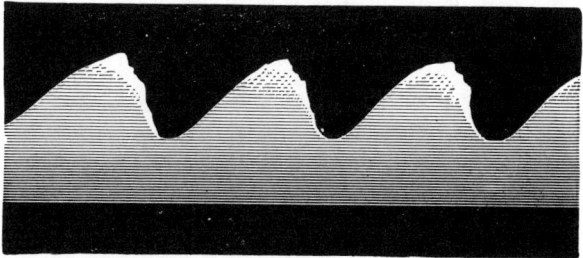

Waves.

With a continuous speaking-tube explosive sounds are liable to extinguish the flame, but this difficulty may be avoided by cutting a longitudinal slit, an inch or so in length, in the speaking-tube near the mouthpiece.

When sounds are uttered in the mouthpiece with sufficient intensity to cause the flame to respond audibly, the sound waves induce longitudinal vibrations of the flame, which produce sounds varying in pitch and intensity with those uttered in the mouthpiece.

In Fig. 139 is shown a method of analyzing the vibrating flame. By means of a revolving mirror an image of each separate flame may be seen. In fact, the results are identical with those secured by Koenig's manometric capsule.

A circular mirror mounted obliquely on a spindle, as shown in Fig. 140, so that it will wabble, is effective in analyzing these flames. The image in this case has a crown-like appearance.

In the experiment here shown a flute is employed as the source of sound.

In Figs. 141, 142, 143, and 144 are illustrated some of the flame images seen in the revolving mirror.

COMPOSITION OF VIBRATIONS.

The optical method of studying sonorous vibrations has the advantage over other methods in being of interest not only to the student of acoustics, but also to those who care only for beautiful effects and have no regard for the lessons they teach.

As incidental to scientific work, the effect of beautiful experiments on the latter class may be worth a little consideration, as it not infrequently happens that the mere onlooker is lured into the paths of science by such means.

Among physical experiments, none are more attractive or instructive than those connected with the subject of sound. The experiments of M. Lissajous are particularly interesting, but when the figures are produced by the apparatus employed by Lissajous, a costly set of instruments will be required.

In the annexed engraving are shown two pieces of apparatus for producing these figures; that shown in Fig. 145 being quite inexpensive, that shown in Fig. 146 being a little

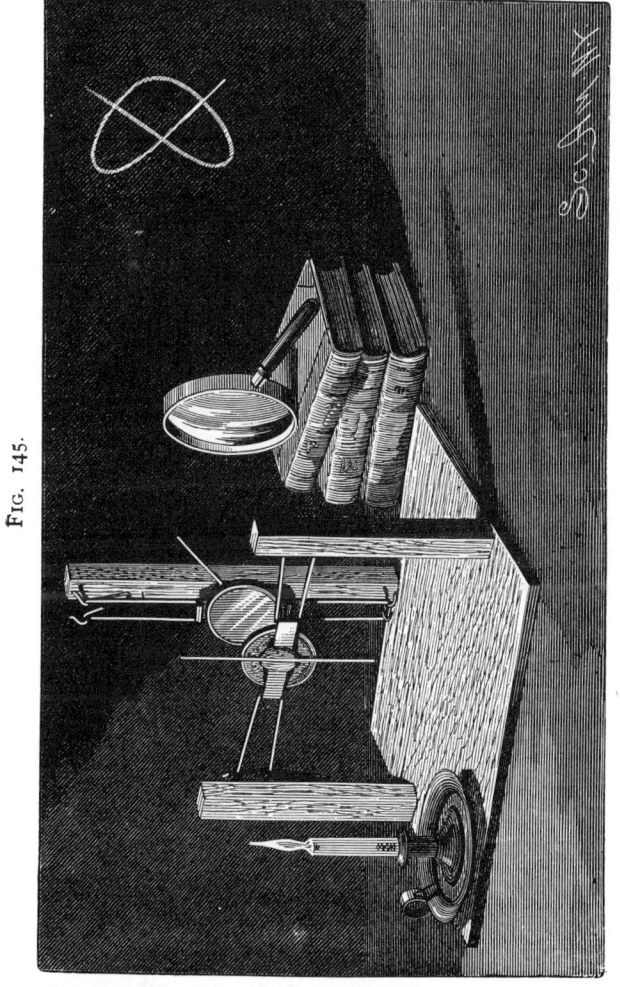

Fig. 145.

Simple Apparatus for Producing Lissajous' Figures.

more costly, and, at the same time, more efficient in its performance.

The device shown in Fig. 145 consists essentially of two plane mirrors, supported by torsional bands of rubber, one being supported so as to vibrate in a vertical plane, the other in a horizontal plane, the mirrors being arranged with respect to each other so that the light received by one mirror will be reflected upon the face of the other mirror, by which it will in turn be projected through a double convex hand glass of long focus, to be finally received on the wall or screen.

The mirrors employed in the construction of this instrument are the small, inexpensive circular pocket mirrors sold on the street corners. They are about $1\frac{1}{2}$ inches in diameter. To adapt them for use, a strip of tin, having its ends curled up to form hooks, is secured to the back of each mirror by means of sealing wax.

A baseboard provided with three standards supports the mirrors in the position of use. In one of the posts near the top are inserted two ordinary wire hooks, and near the bottom are inserted two similar hooks. Rubber bands received in these hooks are inserted in the hooked ends of the strip of tin attached to the back of the mirror. Several wire nails are driven into the face of the standard, for convenience in increasing or diminishing the tension of the rubber bands, the bands being drawn forward between the hooks and slipped over one or the other of the nails to increase the tension.

The mirror thus mounted on the vertical rubber bands will, when struck lightly, vibrate in a horizontal plane. To change the rate of vibration, a weight is attached to the back of the mirror by means of beeswax. In the present case the weight consists of a piece of wire about 6 inches long. By varying the position of the wire on the mirror, *i. e.*, by placing it at different angles with the rubber bands that support the mirror, the rate of vibration may be greatly varied.

The second mirror is mounted in substantially the same way, the only difference being that the rubber bands are arranged horizontally, and supported by two posts instead of one. This mirror vibrates in a vertical plane, and its rate of vibration is changed in the manner above described. A candle or other source of light is arranged so that the light from it will fall on one mirror and be reflected to the other mirror, which in turn will project it through the lens to the wall. When the mirrors are set in vibration, a figure of more or less complicated character will be produced upon the wall. If the two mirrors vibrate in unison, a straight line, or an ellipse, or a circle will be produced. If one mirror vibrates twice as fast as the other, the figure will have the form of figure 8. The figures may be varied to an almost unlimited extent by changing the tension of the rubber bands, and by shifting the wire weights. As the various figures which may be produced are illustrated in most works on physics and on sound, it will be unnecessary to illustrate them here.

The apparatus shown in Fig. 146 will now be understood with little explanation, as the principle on which it operates is the same as that of the more simple form. The mirrors are each supported by two parallel steel wires, which are really but parts of the same wire. The extremities of the wire are securely fastened in the T-shaped head of a bolt, which in the case of the horizontal wires extends through one of the posts, and receives a milled nut, by which the tension of the wires may be varied.

The wire at its mid-length passes around a small sheave in the other post, so that as the wire is tightened the tension of its two branches will be equalized. The vertical wires are supported in the same way by studs projecting from the central post—the lower stud being provided with a sheave for receiving the wire, the upper stud being mortised for receiving the tension screw.

The mirrors are attached by small clamps which embrace both wires, and the arms supporting the adjustable weights are pivoted to the clamps. The weights may be swung in the plane of the mirror, and they are made adjustable on their supporting arms.

The best illumination aside from sunlight is that of a small parallel beam from an oxyhydrogen or electric lantern. The apparatus may be coarsely adjusted by turning the weighted arms on their pivots, and a finer adjustment may be secured by increasing or diminishing the tension of the wires.

Fig. 146.

Apparatus for Compounding Rectangular Vibrations.

RE-ENFORCEMENT OF SOUND.

The re-enforcement of sounds by the vibration of confined masses of air may be readily investigated without apparatus, that is, such apparatus as is commonly employed in acoustical experiments. A very simple experiment illustrating the fact that a sound may be strengthened by a confined body of air is illustrated in Fig. 147. The only

FIG. 147.

Re-enforcement of Vocal Sounds.

requisite for tnis experiment is a paper tube 16 or 18 inches long and about 3 inches in diameter, or, in the absence of such a tube, a sheet of thick paper rolled into a tube will answer. This tube should be held with one end near the mouth, the opposite end being closed by the palm of the hand. By making a sound continuously with the voice, gradually rising in pitch, for example by singing O, with the voice rising from the lowest note it is capable of making, toward the highest note, a point will be found where the sound is largely increased. This increase of sound will occur at the same point in the scale each time the experiment is tried with the same tube, thus showing that the dimensions of the tube are in some way related to the re-enforced note, and to that only. It will also be noticed that the vibrations of the air in the resonant tube not only affect

FIG. 148.

Selective Power of a Resonant Vessel.

the auditory apparatus, but also have sufficient power to be plainly perceptible to the sense of touch, the vibrations being felt by the hand.

Another very simple experiment showing the same phenomenon in a different way is illustrated in Fig. 148. In this case the resonant vessel consists of a vase. Any vessel of substantially the same form may be used. The size is not very material, but by making several trials of different vessels a particular one will be found which will yield better results than others on account of being of the correct dimensions. The experiment consists in holding the vase obliquely in close proximity to the ear, then running the chromatic scale upon any instrument having sufficient range, preferably upon a piano or organ. Some note of the scale will sound much louder than any of the others. By tilting the vase slightly in one direction or the other, so as to cause the ear to partly close the mouth of the vase, the resonant qualities may possibly be improved, as the movement of the vase in this manner amounts to tuning the resonator.

In Fig. 149 is represented an experiment in which the mouth is employed as a resonator, and an ordinary tea bell as the source of the sound. The tuning is effected by moving the tongue back and forth, also by opening or closing the lips. By a few trials a position of the mouth will be found which will cause it to respond to the sound of the bell and act as an efficient resonator.

The familiar instrument shown in Fig. 150 is used in connection with the mouth as a resonator. In this example the reed of the Jew's harp is made to yield a variety of tones, dependent upon the adjustment of the mouth and the force of the breath. The fundamental note of the reed is the clearest and best, and always distinctly heard. The forced overtones are less satisfactory, but suffice for playing tunes that are recognizable.

The experiment with the bell, represented in Fig. 151, is very striking, and is easily performed. The bell is simply an old fashioned clock bell or gong fastened on the end of a small wooden handle by a common wood screw. The resonator is a paper tube of about two-thirds the diameter of the bell, provided with a movable portion or diaphragm, as shown at A. Although the bell may be set in vibration by rapping it with the knuckles or striking it with a large sized rubber eraser, it may be more satisfactorily sounded by drawing a well resined bow over its edge. The bell is held over the mouth of the paper tube, and the diaphragm is moved up or down in the tube until a position is reached in which the bell will yield a full tone, which is much louder than it is capable of giving when used without the resonator. The diaphragm is then fastened by means of sealing wax or glue.

To re-enforce one of the overtones of the bell, the opposite end of the tube is gradually shortened by paring off narrow strips from its edge until it responds to the high tone which the bell is capable of giving out when bowed in a particular way. Now, by causing the bell to vibrate strongly and placing it near opposite ends of the resonator in alternation, it will be found that the deeper cavity will

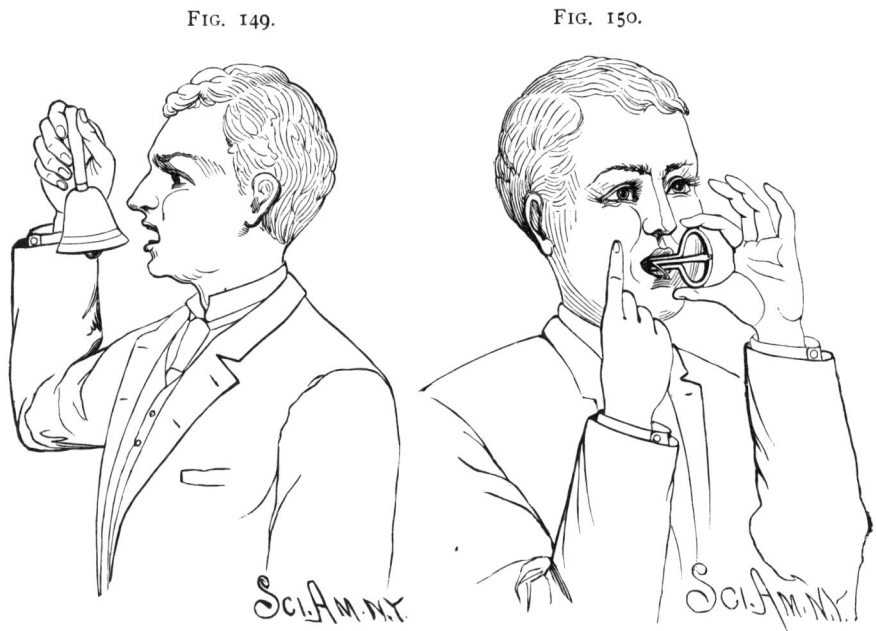

FIG. 149.   FIG. 150.

The Mouth used as a Resonator.   Experiment with the Jew's Harp

respond only to the grave note of the bell, while the shallower cavity will re-enforce only the overtone to which it is tuned. In this experiment it will be found a little more convenient to have separate resonators for the different tones.

In Fig. 152 is shown an experiment which is substantially the same as that just described in connection with the bell. In this case two tuning forks, A and C, are used as sound producers, and to each fork is adapted a resonator

FIG. 151.

Bell and Resonator.

FIG. 152.

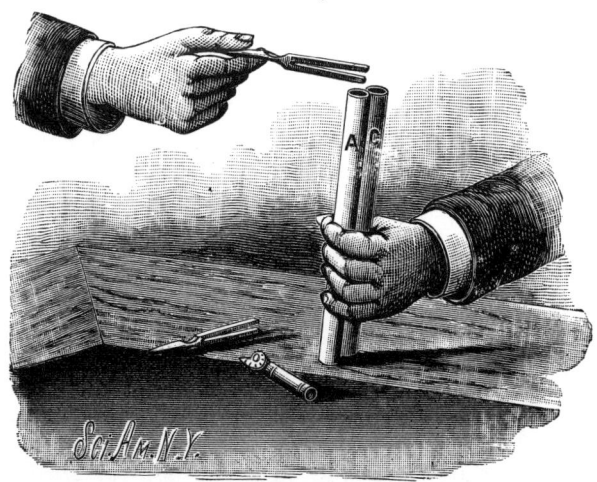

Tuning Forks and Resonant Tubes.

consisting of a paper tube about ⅜ inch in diameter and 8 or 10 inches long. Each tube is tuned to the fork in connection with which it is to be used by inserting a cork and moving it until the length of the inclosed air column is such as to respond to the fork. It will be found that the A resonator will respond only to the A fork, and the C resonator will re-enforce only the sound of the C fork.

In all these cases the resonant tube or cavity corresponds in depth to about one-quarter of a wave length of the particular sound which it is adapted to re-enforce. The wave proceeding from the sounding body strikes the bottom of the resonant chamber and is reflected back in time to proceed with the other half of the wave moving in the opposite direction, greatly augmenting its volume.

The combination of two series of sound waves may be made to produce silence if the relation of the two series be such that the air condensations of one series coincide with the rarefactions of the other series. This may be demonstrated by holding a tuning fork over its appropriate resonator and turning it until the plane of vibration of the fork is at an angle of 45° with the axis of the resonating tube. By placing the fork in the same position relative to the ear, the same phenomenon may be observed without the resonator.

MUSICAL FLAMES.

The experiments of Tyndall and others on sounding flames are so interesting and so easily repeated with very simple appliances, that the student of physics, particularly in the department of acoustics, should not fail to repeat them. The production of musical sounds by means of flames inclosed in resonant tubes is especially easy. One form of this experiment is illustrated by Fig. 153.

For the mere production of sounds, a metal tube will answer, but for the analysis of the flame by which the sound is produced, a glass tube will be required. This tube, whether of metal or glass, may be 40 inches long and one inch internal diameter. It should be supported in a fixed vertical position in a suitable support, a filter support, for example. In a lower arm of the support is placed a glass tube three-eighths inch in diameter, having its upper end drawn to a small circular aperture, which will allow sufficient gas to escape to form a pointed flame about 2½ inches

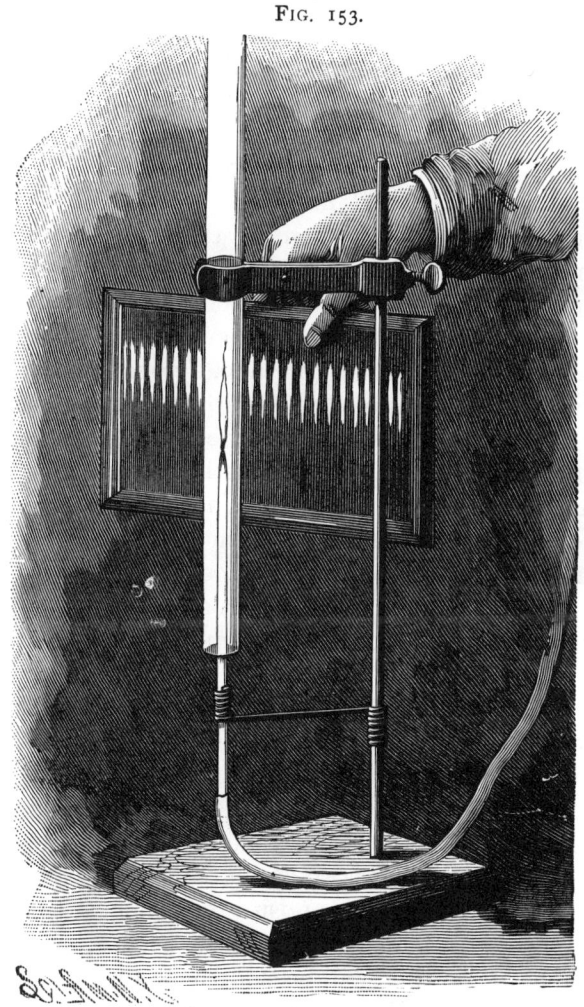

Fig. 153.

Production of Sounding Flames.

in height. The tube is drawn down by heating it near one end until it softens, by continually turning it in a gas flame, then quickly removing it from the flame, and drawing it out as far as possible. By making a nick with a fine file in one side of the tube, at a point where it is about one-sixteenth inch in diameter, the tube may be broken squarely. It may then be tried as a burner. If the flame yielded by gas at full pressure is less than two inches in length, the tube should be again broken off at a point where it is a little larger in diameter, and if the opening happens to be too large, it may be reduced by holding the extreme end of the tube in a gas flame until it partly fuses, when it will contract.

The small glass tube is connected with the gas supply, and the jet is lighted and inserted centrally in the larger tube, and moved slowly upward in the tube until a clear musical note is heard. If the flame is full size, the note will be the fundamental note of the tube. By turning off the gas so as to make the flame three-fourths to one inch high, and again inserting the burner in the tube, a point will be found between its former position and the lower end of the tube at which a tone of higher pitch will be heard. This is one of the harmonics. If the burner with the small flame be carried further upward into the tube, a point will be reached where both the fundamental and harmonic will be produced simultaneously. These tones are produced by rapidly recurring vibrations of the flame, which are rendered uniform by the vibratory period of the column of air contained in the tube.

There are two methods of analyzing these flames. One consists in simply shaking the head, or quickly rolling the eyes from side to side, thereby enabling the eye to receive the impressions of the successive flames in different positions on the retina. The other consists in viewing the image of the flame in a revolving or oscillating mirror. By holding a looking glass in the hand, opposite the flame, as shown in the engraving, and oscillating the glass, what appears to be a single flame in the tube will be shown in the mirror as a succession of flames of like form connected at their bases.

Another way of showing the periodic character of the flame consists in revolving a disk having alternating radial bands of black and white, in proximity to the tube, so that the disk is illuminated only by the light of the intermittent flame. When the disk attains a proper speed, the

SOUND. 149

intermittent illumination will cause it to appear stationary. This beautiful experiment is due to Toepler.

By employing a concave mirror instead of a plane one as described above, the image of the flame may be projected upon a screen.

A SIMPLE PHONOGRAPH.

This instrument, which is shown in perspective in Fig. 154, in section in Fig. 155, and in plan in Fig. 156, has

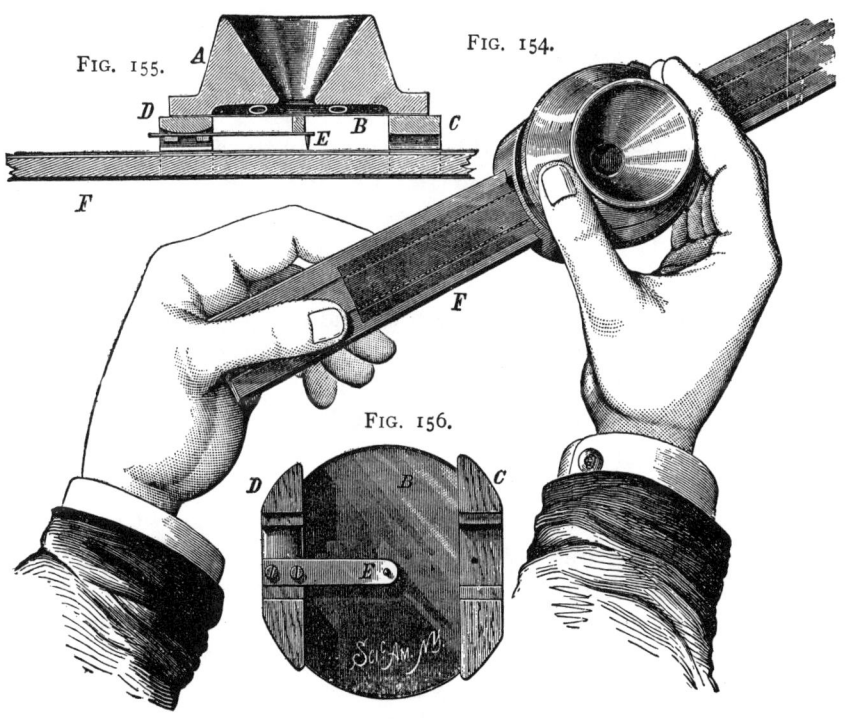

A Simple Phonograph.

a mouthpiece, A, to which is attached a thin ferrotype plate diaphragm, B, by means of a good quality of sealing wax or cement.

Upon the outer face of the diaphragm, and at opposite edges, there are guides, C D, for receiving the wooden strip, F. These guides present only a slight bearing surface to the strip. The guide, D, is rounded to receive the spring, E, which is secured to it by two screws, by which also the spring is adjusted so as to bear with more or less force on the small rubber block which rests upon the center of the diaphragm.

A needle, which is sharpened like a leather sewing needle or awl, is soldered to the spring, and is located directly opposite the center of the diaphragm. The guides, C D, are placed so that the median line of the strip, F, is at one side of the needle. This strip has four slight longitudinal grooves, two on each side, which are made with an ordinary carpenter's gauge. These grooves are located so that when the strip is moved through the guides, one or the other of them will pass over the needle. A piece of beeswax is rubbed over the sides of the strip to give it an adhesive coating for receiving the foil used in recording the sounds.

The foil, which should be rather heavy, must be cut into strips wide enough to extend beyond the grooves in the wooden strip. The foil is laid on the wooden strip and burnished down with the thumb nail, so that it will adhere. The strip thus prepared is placed in the guides, C D, and the needle is adjusted so that it indents the foil slightly as the stick is moved along.

By talking in the mouthpiece, and at the same time moving the strip along with a smooth, steady motion, the sounds are recorded on the foil. By passing the strip again through the guides, so that the needle traverses the same groove, and applying to the mouthpiece a paper funnel or resonator, the sounds or words spoken into the instrument will be reproduced. It is even possible to record the sounds on a plain strip of wood so that they may be reproduced. The engraving is about two-thirds the actual size of the instrument.

THE PERFECTED PHONOGRAPH.

Ten years ago a young man went into the office of the *Scientific American*, and placed before the editors a small, simple machine about which very few preliminary

151

Fig. 157.

Edison's New Phonograph.

remarks were offered. The visitor without any ceremony whatever turned the crank, and to the astonishment of all present the machine said: " Good morning. How do you do? How do you like the phonograph?" The machine thus spoke for itself, and made known the fact that it was the phonograph, an instrument about which much was said and written, although little was known.

It was the latest invention of Edison, and the editors and employes of the *Scientific American* formed the first public audience to which it addressed itself. The young man was Mr. Thomas A. Edison, even then a well known and successful inventor. The invention was novel, original, and apparently destined to find immediate application to hundreds of uses. Every one wanted to hear the wonderful talking machine, and at once a modified form of the original phonograph was brought out and shown everywhere, amusing thousands upon thousands; but it did not by any means fulfill the requirements of the inventor. It was scarcely more than a scientific curiosity or an amusing toy. Edison, however, recognized the fact that it contained the elements of a successful talking machine, and thoroughly believed it was destined to become far more useful than curious or amusing. He contended that it would be a faithful stenographer, reproducing not only the words of the speaker, but the quality and inflections of his voice; and that letters instead of being written would be talked. He believed that the words of great statesmen and divines would be handed down to future generations; that the voices of the world's prima donnas would be stored and preserved, so that, long after their decease, their songs could be heard. These and many other things were expected of the phonograph. It was, however, doomed to a period of silence. It remained a toy and nothing more for years. Finally it was made known to the public that the ideal phonograph had been constructed; that it was unmistakably a good talker; and that the machine, which most people believed to have reached its growth, had after all been refined and improved until it was capable of faithfully reproducing every word, syllable, vowel, consonant, aspirate and sounds of every kind.

During the dormancy of the phonograph, its inventor secured both world-wide fame and a colossal fortune by means of his electric light and other well known inventions. He has devoted much time to the phonograph, and has not only perfected the instrument itself, but has established a large factory provided with special tools for its manufacture, in which phonographs are to be turned out in great numbers.

The original instrument consists of three principal parts—the mouthpiece, into which speech is uttered; the spirally grooved cylinder, carrying a sheet of tin foil which receives the record of the movements of the diaphragm in the mouthpiece; and a second mouthpiece, by which the speech recorded on the cylinder is reproduced. In this instrument the shaft of the cylinder is provided with a thread of the same pitch as the spiral on the surface of the cylinder, so that the needle of the receiving mouthpiece is enabled to traverse the surface of the tin foil opposite the groove of the cylinder. By careful adjustment this instrument was made to reproduce familiar words and sentences, so that they would be recognized and understood by the listener; but in general, in the early phonographs, it was necessary that the listener should hear the sounds uttered into the receiving mouthpiece of the phonograph to positively understand the words uttered by the instrument.

In the later instruments, such as were exhibited throughout the country and the world, the same difficulty obtained, and perfection of articulation was sacrificed to volume of sound. This was necessary, as the instruments were exhibited before large audiences, where, it goes without saying, the instrument to be entertaining had to be heard. These instruments had each but one mouthpiece and one diaphragm, which answered the double purpose of receiving the sound and of giving it out again. Strangely enough, the recently improved phonograph is more like the original one than any of the others. It is provided with two mouthpieces, one for receiving and one for reproducing.

The new phonograph, which is shown in Fig. 157, is of about the size of an ordinary sewing machine. In its con-

154 EXPERIMENTAL SCIENCE.

Fig. 158.

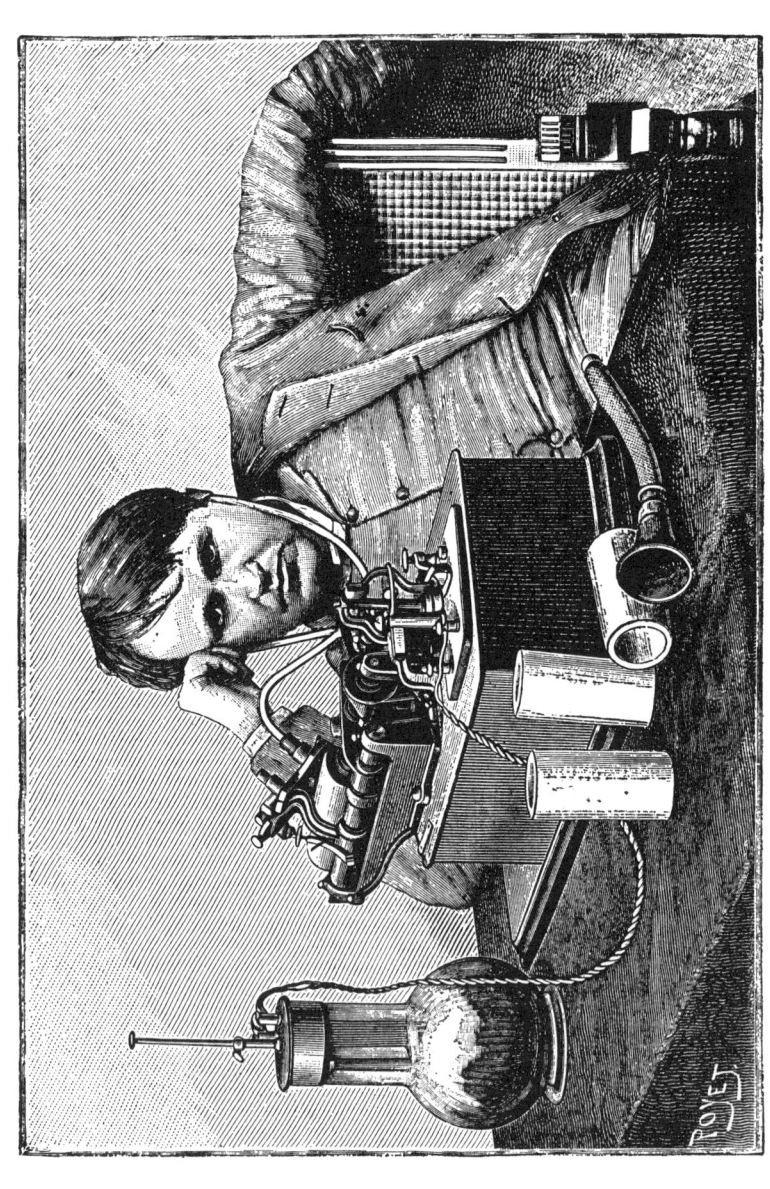

Edison Listening to the first Phonogram sent from England

struction, it is something like a very small engine lathe; the main spindle is threaded between its bearings, and is prolonged at one end to receive the hardened wax cylinder upon which the sound record is made. Behind the spindle and the cylinder is a rod upon which is arranged a slide, having at one end an arm adapted to engage the screw of the spindle, and at the opposite end an arm carrying a pivoted head, provided with two diaphragms, whose positions may be instantly interchanged when desirable. One of these diaphragms is turned into the position of use when it is desired to talk to the phonograph, and when the speech is to be reproduced, the other diaphragm takes its place. The glass diaphragm, which receives the speech and makes the impressions upon the cylinder, is shown in Fig. 159. The needle by which the impressions are made in the wax is attached to the center of the diaphragm, and pivotally connected to a spring arm attached to the side of the diaphragm cell. The device by which the speech is reproduced is shown in section in Fig. 160. The cell contains a delicate glass diaphragm, to the center of which is secured a stud connected with a small curved steel wire, one end of which is attached to the diaphragm cell. The spindle of the phonograph is rotated regularly by an electric motor in the base of the machine, which is driven by a current from one or two cells of battery. The motor is provided with a sensitive governor which causes it to maintain a very uniform speed. The arm which carries the diaphragms is provided with a turning tool for smoothing the wax cylinder preparatory to receiving the sound record.

The first operation in the use of the machine is to bring the turning tool into action and cause it to traverse the cylinder. The turning tool is then thrown out, the carriage bearing the diaphragms is returned to the position of starting, the receiving diaphragm is placed in the position of use, and as the wax cylinder revolves, the diaphragm is vibrated by the sound waves, thus moving the needle so as to cause it to cut into the wax cylinder and produce indentations which correspond to the movements of the diaphragm. After the record is made, the carriage is again returned to the point of starting, the receiving diaphragm is replaced by the reproducing diaphragm, and the carriage is again moved forward by the screw, as the cylinder revolves, causing the point of the reproducing diaphragm to traverse the path made by the recording needle. As the point of the curved wire attached to the diaphragm follows

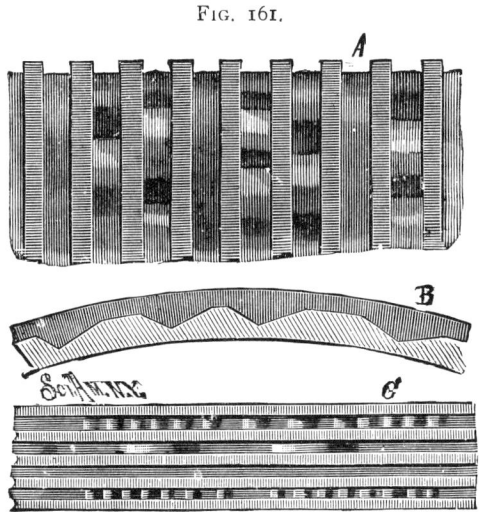

Phonographic Record Magnified.

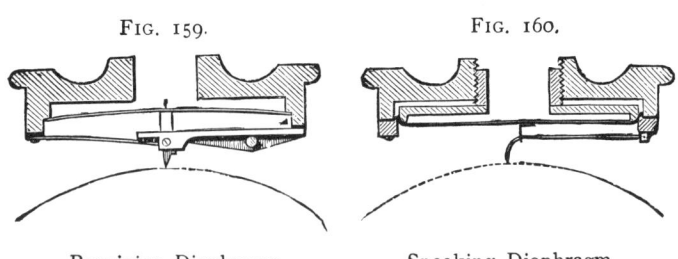

Receiving Diaphragm.   Speaking Diaphragm.

the indentations of the wax cylinder, the reproducing diaphragm is made to vibrate in a manner similar to that of the receiving diaphragm, thereby faithfully reproducing the sounds uttered into the receiving mouthpiece.

A crucial test of the capabilities of this machine was recently made in our presence, at Edison's laboratory, near

Llewellyn Park, Orange, N. J. A paragraph from the morning newspaper was read to the machine in our absence, and when upon our return to the instrument it was reproduced phonographically, every word was distinctly understood, although the names, localities, and the circumstances mentioned in the article were entirely new and strange to us. Another test of the perfection of the machine was the perfect reproduction of whistling and whispering, all the imperfections of tone, the half tones and modulations even, being faithfully reproduced. The perfect performance of the new instrument depends upon its mechanical perfection—upon the regularity of its speed, the susceptibility of the wax cylinder to the impressions of the needle, and to the delicacy of the speaking diaphragm. No attempt is made in this instrument to secure loud speaking—distinct articulation and perfect intonation have been the principal ends sought.

A highly magnified section of the phonograph cylinder, showing the indentations, is illustrated in Fig. 161; A representing a section of the face of the cylinder, B a transverse section of a portion of the cylindrical wax shell, and C showing a less magnified face view of a small portion of the cylinder.

The new phonograph is to be used for taking dictation for taking testimony in court, for reporting speeches, for the reproduction of vocal music, for teaching languages, for correspondence, for civil and military orders, for reading to the sick in hospitals, and for various other purposes too numerous to mention.

Imagine a lawyer dictating his brief to one of these little machines; he may talk as rapidiy as he chooses, every word and syllable will be caught upon the delicate wax cylinder, and after his brief is complete he may transfer the wax cylinder to the phonograph of a copyist, who may listen to the words of the phonograph and write out the manuscript. The instrument may be stopped and started at pleasure, and if any portion of the speech is not understood by the transcriber, it may be repeated as often as necessary.

In a similar manner a compositor may set his type directly from the dictation of the machine, without the necessity of "copy," as it is now known. Mr. Edison says that the whole of "Nicholas Nickleby" could be recorded upon four cylinders, each 4 inches in diameter and 8 inches long, so that one of these instruments in a private circle or in a hospital could be made to read a book to a number of persons. This is accomplished by means of a multiple earpiece.

The little wax cylinders upon which the record is made are provided with a rigid backing, and the cylinders are made in different lengths; the shortest—one inch long—having a capacity of 200 words, the next in size 400 words, and so on. These cylinders are very light, and a mailing case has been devised which will admit of mailing the cylinders as readily as letters are now mailed. The recipient of the cylinder will place it on his own phonograph and listen to the phonogram—in which he will not only get the sense of the words of the sender, but will recognize his expression, which will, of course, have much to do with the interpretation of the true meaning of the sender of the phonogram.

Fig. 158 is a life-like picture of Mr. Edison photographed while he listened to his first phonogram from abroad.

A very interesting and popular use of the phonograph will be the distribution of the songs of great singers, sermons and speeches, the words of great men and women, music of many parts, the voices of animals, etc., so that the owner of a phonograph may enjoy these things with little expense.

It may even be pressed into the detective service and used as an unimpeachable witness. It will have but one story to tell, and cross examination cannot confuse it.

REFLECTION AND CONCENTRATION OF SOUND.

The particular action of sound to be dealt with here is that of reflection, examples of which are presented in every echo; and whispering galleries are but the exhibition of the same thing, although more rare. A few of them have a world-wide reputation.

In his article on sound in the "Encyclopædia Metropolitana," Sir John Herschel mentions the abbey church of St.

Albans, where the tick of a watch may be heard from one end of the edifice to the other. In Gloucester Cathedral a gallery of octagonal form conveys a whisper 75 feet across the nave. In the whispering gallery of St. Paul's the faintest sound is conveyed from one side of the dome to the other, but is not heard at any intermediate point. The dome of the capitol at Washington is an excellent whispering gallery. These effects are due to an accidental arrangement of the walls.

Sails of ships are sometimes inflated by the wind so that they act as concentrating reflectors of sound. Arnott says that in coasting off Brazil he heard the bells of San Salvador from a distance of 110 miles, by standing before the mainsail, which happened at the time to assume the form of a concave reflector, focusing at his ear.

Sounds may be received and reflected by means of metallic parabolic reflectors, so that many times the volume of sound that naturally strikes the ear will be concentrated, rendering sounds audible that might otherwise be too distant or too faint to be heard. Such reflectors of necessity have a fixed form, and are available under certain conditions only. The accompanying engraving (Fig. 162) represents a sound reflector that may be focused as readily and directed as easily as a telescope. It is, in fact, a portable and adjustable whispering gallery, having many useful applications.

The instrument is very simple, consisting essentially of an airtight drum, one head of which is rigid, the other elastic. This drum, or more properly reflector, is mounted on pivots in a swiveled support, and is provided with a flexible tube having a mouthpiece and stop cock at its free end. Two wires are stretched across the face of the reflector at right angles to each other, and support at their intersection a small plane mirror, the office of which is to determine the position of the reflector in relation to the direction of the sound. A small ear trumpet or funnel, which is shown on the table, is used in connection with the reflector, to increase its effect by gathering portions of the sound that might escape the unaided ear.

The reflector is adjusted by looking through the ear

trumpet toward the small plane mirror, and moving the sound reflector until the source of sound is seen in the mirror. The reflector is then focused by exhausting the air from behind the flexible head until the required degree of

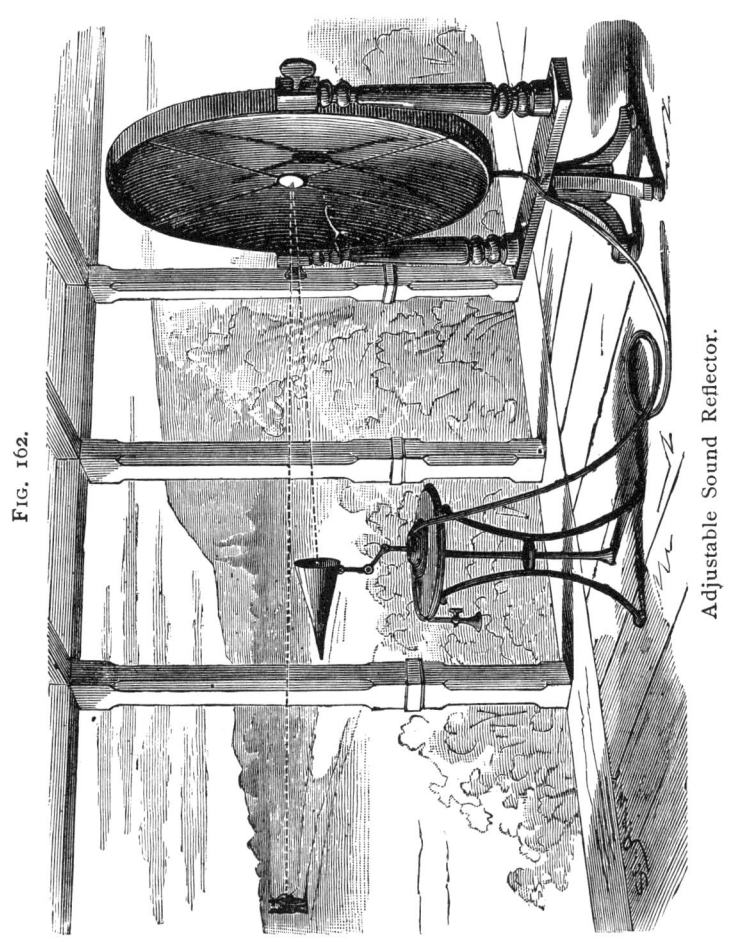

Fig. 162.

Adjustable Sound Reflector.

concavity is reached, which will be when sounds are distinctly heard in the ear trumpet. The air is readily exhausted from the reflector by applying the mouth to the mouthpiece. The details of the construction of the apparatus are shown in the engraving.

Of course, the operation of the instrument may be reversed—that is, sounds made at the focus of the reflector may be projected in parallel lines over long distances, but in practice a speaking trumpet is found to be better for this purpose. The engraving shows but one of the applications of the reflector. It would be a simple matter to provide for a deaf person an instrument on this principle. It could hang on the walls of the parlor unnoticed, as it might take the form of a richly framed picture, and would concentrate a great volume of sound at a single point. The same device

FIG. 163.

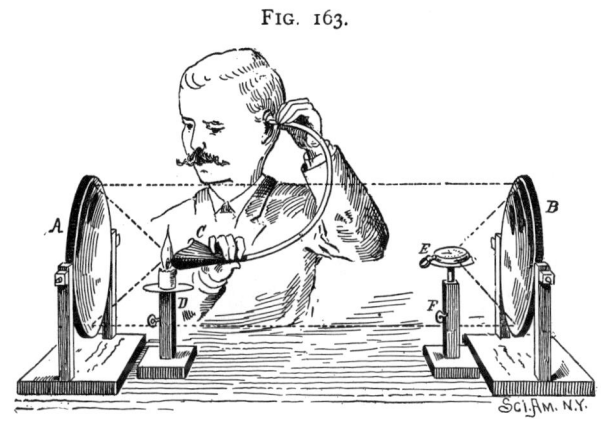

Reflection of Light and Sound.

may also be applied to an auditorium to project the voice of the speaker in any required direction.

To concentrate and project light, heat, and sound by means of concave mirrors is generally supposed to necessitate the use of expensive parabolic mirrors, articles practically out of the reach of amateur experimenters, and not to be found in every institution of learning. To perform most of the experiments possible with concave mirrors, the spun metal reflectors used in large lamps answer exceedingly well. The projection of images and the accurate determination of the foci are the only experiments impossible with such reflectors. The largest size to be found ready made is 10 inches in diameter, with a principal focus of about 8 or 9 inches. The price is $1.50 per pair. To prepare them for use, two common wood screws are secured to them at diametrically opposite points, the heads of the screws being soldered to the edges of the mirrors, so that the screws project radially. Each mirror is provided with a stand formed of a base and two uprights. The wood screws project through the uprights, and are provided with wooden nuts.

To facilitate the experiments to be performed with the concave mirrors, two or three small stands are required. It is desirable that these stands be made adjustable. If nothing is at hand that will answer the purpose, a very good adjustable stand may be made by soldering a disk of tin to the head of a 4 inch wood screw, and inserting the screw in

FIG. 164.

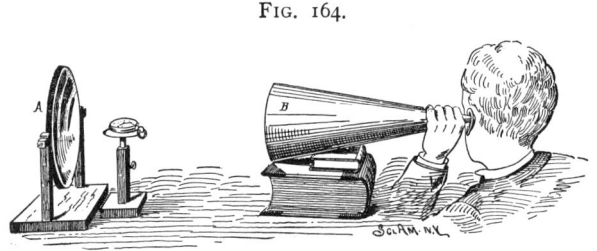

Reflection and Concentration of Sound.

a short column, as shown in the engraving. A paper trumpet, 8 inches in diameter at the larger end and 2 feet in length, is useful, and a rubber tube having a small funnel at one end and an ear piece at the other end is necessary.

To show the concentrating power of one of these common reflectors, place it so that its concave surface faces the sun. Then place a piece of dark-colored cloth in the focus. It is at once ignited.

Place two reflectors, A B, 4 or 5 feet apart, with their concave surfaces facing each other, as shown in Fig. 163. Place a short candle on the stand, D, so as to reflect a parallel beam that will cover the reflector, B, as nearly as possible. Then place a watch, E, in the focus of the reflector, B, upon the stand, F. Now hold the funnel, C, with its mouth facing the reflector, A, and immediately behind the candle, or, better, remove the candle and place the funnel in the position formerly occupied by the candle flame. With the funnel at this point the ticking of the watch will be distinctly heard, but a slight movement of the funnel in either direction will render the ticking inaudible. This experiment shows that the laws governing the reflection of light and sound are the same.

In Fig. 164 the use of the trumpet in connection with a concave reflector is illustrated. The reflector, A, is adjusted to the trumpet, B, by means of the light of a candle placed on the stand in the focus of the reflector. Afterward the candle is replaced by the watch. With this arrangement the watch may be heard twenty or thirty feet away.

TREVELYAN ROCKER.

This apparatus consists of a short piece, A, of lead pipe, about an inch in diameter, and a piece, B, of thick brass tubing, about ¾ inch outside diameter and five or six inches long. The lead pipe is flattened a little to keep it from rolling, and the surface along the side which is to be upper-

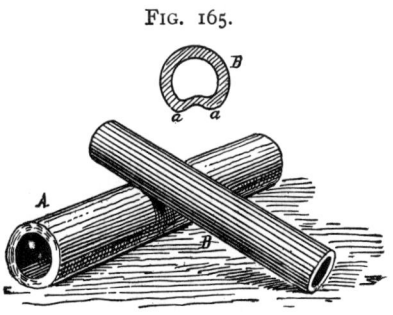

FIG. 165.

Trevelyan Rocker.

most is scraped and smoothed. The brass tubing, B, is filed thin, upon one side, near one end, and the thin part is driven in with the pein of a hammer or a punch so as to leave the longitudinal ridges, *a a*, as shown in the end view in Fig. 165.

When the brass tube is heated and placed across the lead pipe, as shown in Fig. 165, with the ridges, *a a*, in contact with the lead pipe, the brass tube begins to rock, invisibly, of course, but with sufficient energy to give forth a clear musical note. If it does not start of itself, a little jarring will set it going, and it will continue to give forth its sound for some time.

The accepted explanation of this phenomenon is that the contact of the hot brass with the lead causes the lead to suddenly expand and project a microscopic distance upward. These upward projections of the lead alternate between the

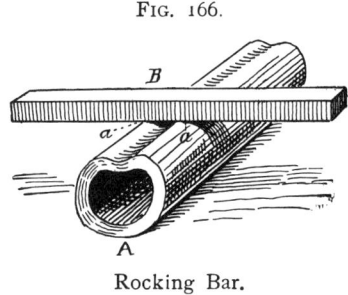

Fig. 166.

Rocking Bar.

two points of contact, and thus cause the tube to rock with great rapidity and regularity.

In Fig. 166 is shown a modification of the experiment, in which the lead is indented to form the two contact surfaces, *a a*, and the heated bar, B, is made to rock at a comparatively slow rate, giving forth a grave note. By careful manipulation, the bar may be made to rock both longitudinally and laterally, thus giving forth a rhythmic combination of the two sounds.

REFRACTION OF SOUND.

In Figs. 167 and 168 is illustrated an adjustable lens for showing the refraction of sound. The frame of the lens consists of three 12 inch rings of large wire, soldered together so as to form a single wide ring with two circumferential grooves. In the central part of the ring, at the bottom, is inserted a standard, and in the top is inserted a short metal tube. Over the edges of the ring are stretched disks of the thinnest elastic rubber, which are secured by a stout thread wound around the edges of the rubber, clamping them in the grooves of the ring.

By inflating the lens through the tube with carbonic acid

Fig. 167.

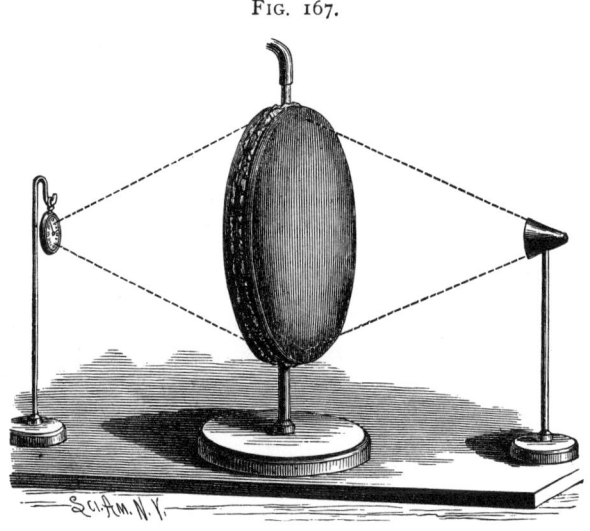

Sound Lens.

gas, it may be focused as desired. A watch placed at the focus upon one side of the lens can be distinctly heard at the focal point on the opposite side of the lens, when it can be heard only faintly or not at all at points only slightly removed from the focus, thus showing that the sound of the ticking of the watch has been refracted by the lens in much the same manner as light is refracted by a glass lens.

Fig. 168.

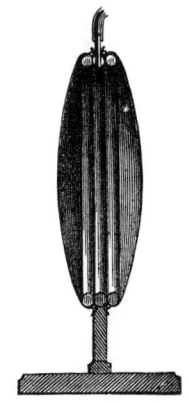

Section of Sound Lens.

SENSITIVE FLAMES.

The sensitive flame, first observed by Dr. Le Conte and afterward developed by Tyndall and Barrett, exhibits some of the curious effects of sound. For its production it is necessary that the gas be under a pressure equal to that of a column of water six or eight inches high. The common method of securing the required pressure is to take the gas from a cylinder of compressed illuminating gas, such as is used for calcium lights. Another method is to take the gas from a weighted gas bag, and still another is to fill a sheet metal tank with gas and displace it with water in the manner illustrated in Fig. 170. The burner is shown at 1, 2, and 3, Fig. 169. It consists of a small tip inserted in the end of a suitable tube. The tip in the present case is made of brass, but those commonly used for this purpose are of steatite. They are superior to the metal ones, but require careful selection.

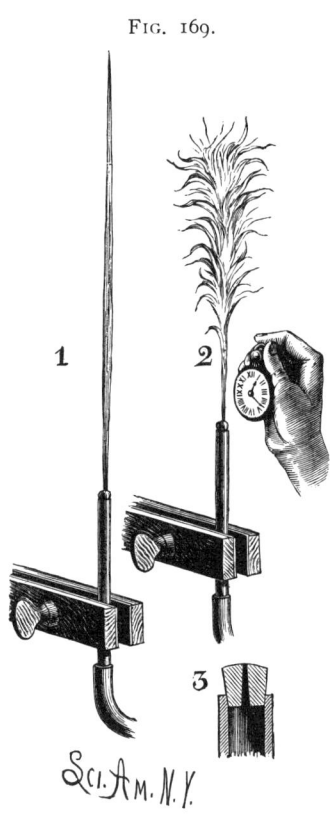

FIG. 169.

Burner for Sensitive Flame.

It has been found that some of the lava pinhole burner tips used in certain kinds of gas stoves answer admirably for this purpose, and cost very little. A tip with a round, smooth hole is to be selected. The bore of the tip is here shown tapering. Its smaller diameter is 0·035 inch The burner is supported in the manner shown at 1 and 2, or in any other convenient manner, and gas under a suitable pressure flows through and is ignited. The flame will be tall and slender, as shown at 1. By regulating the gas pressure carefully, an adjustment will be reached at which the flame will be on the verge of flaring. A very slight increase of pressure beyond this point will cause the flame to shorten and roar. When the flame is at the point of flaring, it is extremely sensitive to certain sounds, particularly those of high pitch. A shrill whistle or a hiss will cause it to flare. The rattle of a bunch of keys will produce the same result. It will respond to every tick of a watch held near it.

Tyndall says that when the gas pressure is increased beyond a certain limit, vibrations are set up in the gas jet by the friction of the gas in the orifice of the burner. These vibrations cause the flame to quiver and shorten. When the flame burns steadily, any sound to which the gas jet will respond will throw it into sympathetic vibration. Experi-

Fig. 170.

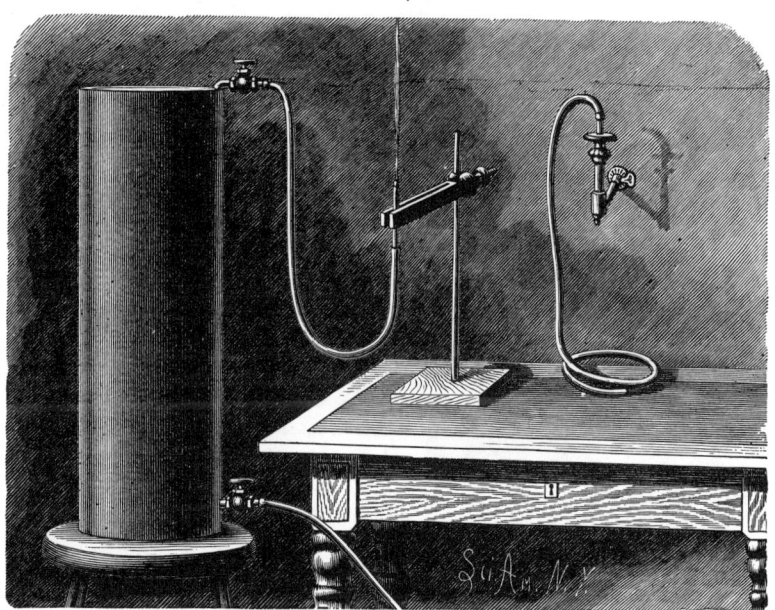

Apparatus for producing Gas Pressure for the Sensitive Flame.

ment has demonstrated that the seat of sensitiveness of the flame is at the base of the flame, at the orifice of the burner.

The method of producing the required gas pressure illustrated in Fig. 170 is available when gas bags or cylinders of compressed gas are not to be had. A tin cylinder of about 15 gallons capacity is provided at the top and bottom with valves. The lower valve is connected with a hydrant, and the cylinder is filled with water, while the upper valve is left open to allow of the escape of air. When the cylinder is filled with water, the supply is shut off and a tube from a gas burner is connected with the upper valve and the gas is turned on. Then the water is allowed to escape from the cylinder, thereby drawing in the gas. When the cylinder is filled with gas, the valves are closed and the lower one is again connected with the hydrant, while the upper one is connected with the pinhole burner. The valves on the cylinder are again opened and water is admitted at the rate required to produce the desired gas pressure. Only two precautions are necessary in this experiment; one is to avoid a mixture of air and gas in the cylinder by driving out all the air, the other is to avoid the straining of the cylinder by water pressure.

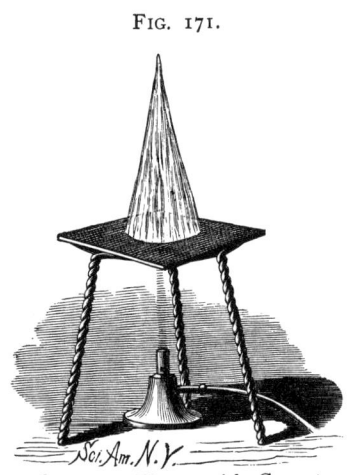

FIG. 171.

Sensitive Flame with Gas at Ordinary Pressure.

Another sensitive flame, which has several advantages over the one described, is shown in Fig. 171. It requires no extra gas pressure, and it is more readily controlled than the tall jet. It was discovered by Mr. Philip Barry, and the discoverer's letter to Mr. Tyndall concerning it is found in Tyndall's work on sound. In the production of this flame a pinhole burner, like that already described, is employed. Two inches above the burner is supported a piece of 32-mesh wire gauze, about 6 inches square. The gas is turned on and lit above the wire gauze. It burns in a conical flame, which is yellow at the top and blue at the base. When the gas pressure is strong, the flame roars continuously. When the gas is turned off, so as to stop the roaring altogether, the flame burns steadily and exhibits no more sensitiveness than an ordinary flame. By turning on the gas slowly and steadily, a critical point will be reached at which any hissing noise will cause it to roar and become non-luminous. Any degree of sensitiveness may be attained by careful adjust-

SOUND. 169

ment of the gas supply. A quiet room is required for this experiment. The rustle of clothes, the ticking of a clock, a whisper, a snap of the finger, the dropping of a pencil, or in

Fig. 172. Fig. 173.

Determining Speed by Resonance.   Siren for Measuring Velocities.

fact almost any noise, will cause it to drop, become non-luminous, and roar. It dances perfect time to a tune whistled *staccato* and not too rapidly.

The flame at its base presents a large surface to the air,

170 EXPERIMENTAL SCIENCE.

so that any disturbance of the air sets the flame in active vibration.

A SIREN FOR MEASURING VELOCITIES.

In this instrument advantage is taken of the well known fact that for every tone a resonator may be provided that will respond to and re-enforce the vibrations producing that tone. The length of a closed resonant tube is one-fourth that of the sound wave to which it responds. The length of an open resonant tube is one-half that of the sound wave to which it responds. It is obvious that a telescopic tube

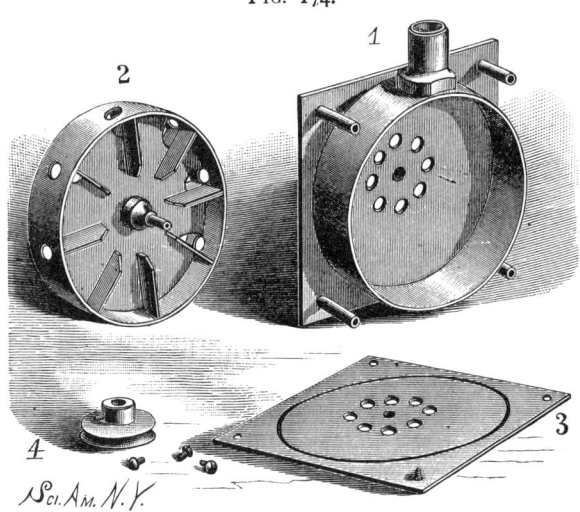

FIG. 174.

Details of the Siren.

may be adjusted to respond to different pitches. Knowing the number of vibrations required per second to produce a certain pitch, it is comparatively an easy matter to determine the rate of any series of regular air vibrations by adjusting the tube to such a length as to cause it to respond to the vibrations.

In Fig. 172 is shown a resonant tube supported over a small fan wheel. The fan has ten blades, so that during one revolution it sends ten puffs of air up the tube. By gradually increasing the velocity of the fan a speed will be reached

SOUND. 171

at which the tube yields a low but distinct musical tone. If, for example, this tone corresponds to middle *c*, it is known that 261 puffs of air are made in the tube, and that since there are ten blades to the fans, the number of revolutions of the fan shaft must be 261÷10=26·1 per second, or 1,566 revolutions per minute.

In Fig. 173 is illustrated a siren constructed on this principle. The parts of this instrument are shown in detail in Fig. 174. It consists of a circular casing containing a rotary fan which draws in air at the center and discharges it

FIG. 175.

Centrifugal Siren.

through an opening in the top of the casing. The blades of the fan are arranged radially upon opposite sides of the disk, and the fan is encircled by a perforated rim, which fits the circular casing and acts as a valve in controlling the escape of air. The perforations of the rim correspond in number and position with the fan blades.

The discharge opening of the casing is provided with a socket for receiving a resonator. The resonator shown in Fig. 173 consists of a pair of tubes made to slide telescopically one within the other, the inner one being graduated to indicate the different lengths required for different pitches, and consequently for different speeds. As the fan revolves, the air drawn in through the holes at the center of the casing is thrown ontward by centrifugal force, thus maintaining a pressure of air at the periphery of the fan. The holes in the rim of the fan allow the air to escape in regular puffs, the frequency of which depends upon the velocity of the fan. These puffs produce sounds varying in pitch and intensity with the speed of the fan, and the resonating tube re-enforces the particular note to which it is tuned, so that when a speed is reached corresponding with the adjustment of the tube, the fact is known by the superior strength of that particular note. Any change of speed may be detected by the lessening of the intensity of the sound and the change of pitch.

The siren is shown in Fig. 175 in connection with mechanism for driving it by hand. It is provided with a revolution counter and with a trumpet-shaped resonator. It is designed to be used in the same manner as the siren of Cagniard Latour, and, like that instrument, it yields sounds under water.

CHAPTER IX.

EXPERIMENTS WITH THE SCIENTIFIC TOP.

Several experiments possessing more or less interest are illustrated in Plate III. This chapter is introduced at this

PLATE III.

Experiments with the Scientific Top.

point on account of the relation of its subject matter to the preceding and succeeding chapters.

The ability of the heavy top to run for a long time and

maintain an equable motion renders it particularly serviceable in experiments requiring uniformity of action.

Two experiments in sound are illustrated: 1, Plate III, showing the adaptation of a simple siren to the top, and 2, Plate III, Savart's wheel. The siren consists of a disk of pasteboard, having four eccentric rows of 3-8 inch holes, there being 12 holes in the inner row, 15 in the next, 18 in the next, and 24 in the outer row. The disk is varnished with shellac to render it waterproof. It is mounted on a chuck fitted to the tapering hole of the top spindle. When the disk is rapidly rotated by the top, and a jet of air is blown upon either row of holes through a flexible tube provided with a small glass or metallic nozzle, a musical sound will be produced by the air pulsations caused by the interruptions of the air jet by the perforated disk. The sounds produced by the different rows of holes are those of the perfect major chord. By holding a card so that its corner will touch the perforated disk at any row of holes, it will be found that the taps of the card will produce the same tones as the puffs of air from the tube. Savart's wheel is simply a toothed disk fitted to the chuck and adapted to be rotated by the top. When the disk is turned very slowly, with the edge of a card held against the teeth, a series of little taps are heard, which do not at all resemble a musical sound; but when the wheel is revolved rapidly by the top, the contact of the card with its periphery produces a sound that may fairly be called musical, the sound being composed of the rapidly repeated taps.

At 3, Plate III, is shown a disk similar to that used for the siren, but having double the number of holes in each circular row. The holes are 1-8 inch in diameter. The disk is blackened to render the effects more conspicuous, and the hole in the center of the disk is eyeleted to prevent wear. A metal disk, secured to a tapering spindle fitted into the top spindle, carries a crank pin 3-16 inch from the axis of rotation. The eyelet of the disk is placed loosely on this crank pin, and when the crank is revolved by the top the disk is gyrated; every part of its surface being made to travel in a circular path 3-8 inch in diameter, when sufficient friction is

EXPERIMENTS WITH THE SCIENTIFIC TOP. 175

applied to it to prevent it from rotating with the top. In this case each perforation of the disk forms a circle, and the circles formed by the entire series of holes interlace, appearing like so many chain links interlocked. By allowing the disk to revolve at different speeds very complicated figures are produced, sometimes like lacework, sometimes like twisted chainwork. Occasionally one part of the figure will appear to turn in one direction while another part turns in the opposite direction. Some of these figures are shown at 4 and 5, Plate III. A similar experiment, developed in a different way, is shown at 7. The black cardboard disk is provided with a central eyelet, which receives the crank pin, as in the case of the perforated disk. On each of two diametrical lines crossing each other at right angles are formed pairs of holes, in which are cemented silvered glass beads or bright spherical steel buttons. The latter were used on the disk illustrated. They are symmetrically arranged, so that the inner four may follow each other in the same path, and the outer four may follow each other in a path of their own.

By treating this disk after the manner of the perforated disk above described, many brilliant and surprising effects may be produced.

By holding one edge of the disk lightly between the thumb and finger, so that it will not revolve, but will be made to gyrate by the little crank, each button will describe a 3-8 inch circle, or a small oval, or an ellipse, as shown at 7. By allowing the disk to slip slowly between the thumb and finger, a series of double scrolls will be produced, as shown at 8.

On varying the speed of rotation by the application of more or less friction to the disk, a great variety of intricate and beautiful figures are produced. Examples are shown at 9, 10, and 11, Plate III. The effect shown at 11 is secured by allowing the edge of the gyrating disk to strike the finger once during each gyration. The luminous curve in this case appears to have a slow retrograde motion.

In Fig. 176 is shown a cardboard disk mounted loosely on the top spindle and provided with two series of black radial bars, the inner series having 13 bars, the outer series having 12 bars. To the chuck inserted in the spindle is secured a black disk having four radial slits.

When the top is revolved and the lower disk is retarded, some very curious illusions will be produced. At times one part of the lower disk will appear to remain stationary, while the other part will appear to revolve. Again, the two series of radial bars will appear to rotate in opposite directions. Viewed in another way they appear curved.

By replacing the slitted disk with the perforated disk, and arranging the perforated disk so that it may be retarded

FIG. 176.

Radial Disks.

by the friction of the finger, some curious effects will be seen. The different rows of holes will appear to advance and recede in a very erratic way. Fig. 177, 12 to 15 inclusive, illustrate the well known and very interesting toy known as the chameleon top. This top is shown in this connection, as the beautiful experiments which have been adapted to it may be transferred with great advantage to the heavier top; 12 shows the top itself, with the black sector lifted out of its normal position to show the colored segments on the face of the top.

When the top is spun with the black sector resting on its face, a great variety of changes of hue may be produced

EXPERIMENTS WITH THE SCIENTIFIC TOP. 177

by retarding the sector, by touching the metallic radially ribbed disk attached to its center. This operation causes it to shift its position on the top, and expose the different colored segments in succession. Persistence of vision causes the segments to appear as circular bands of color, which constantly change.

When the colored paper ellipses shown at 13 are thrown

FIG. 177.

The Chameleon Top.

upon the top and touched by the finger, the colors are curiously blended.

The tricolored disk shown at 14 is to be supported loosely on one of the wires shown at 15. This disk, when revolved, yields some very pretty effects. The wires shown at 15, when inserted in the hollow top spindle and revolved, produce the figures shown in the upper portion of the engraving, appearing like phantom vases, bowls, etc.

When this experiment is adapted to the large top, the wires are replaced by thin nickel-plated tubes, inserted in wooden pins fitted to the spindle of the top. The tubes are provided at their upper ends with small spherical knobs.

In addition to the experiments described, there are of course many others of equal interest which may be performed by means of a heavy top.

The engraving represents an attachment to the " scienti-

FIG. 178.

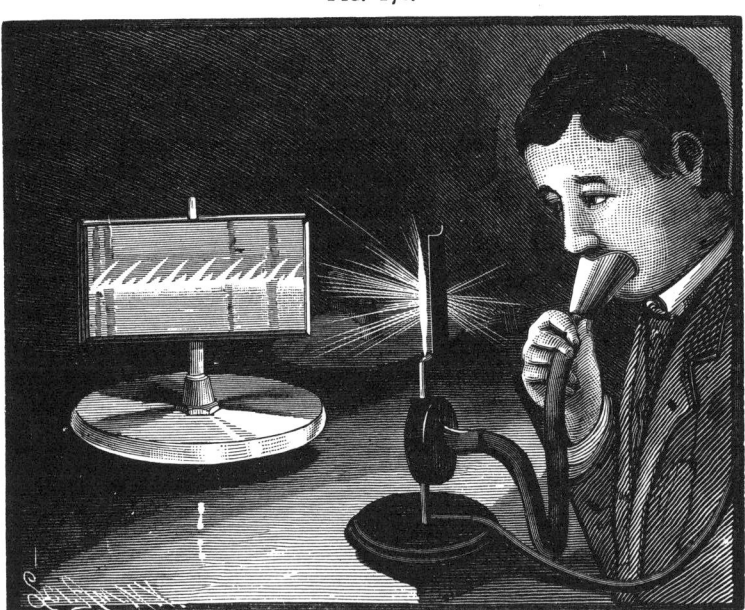

Top with Revolving Mirrors—Koenig's Manometric Flames.

fic top," by means of which the beautiful and instructive experiments of Koenig may be readily repeated. The part of the apparatus carried by the top consists of two pieces of ordinary silvered glass (looking glass), 2½ by 5 inches, secured to opposite sides of a light wooden frame of the same size, and 3-4 inch thick, by means of strips of stout black paper attached to the frame and to the edges of the glasses. The upper and lower edges of the wooden frame are bored at the center to receive the rod inserted in the bore of the top spindle. The frame fits the rod loosely, and is revolved by frictional contact with the rod and the upper end of the top spindle. This arrangement allows the mirror to revolve at a comparatively low rate of speed, the resistance of the air causing the mirror frame to slip on the rod.

It is necessary thus to provide for the slow rotation of the mirrors, as the flame points would be blended into a continuous band of light by the persistence of vision were the mirrors allowed to revolve as rapidly as the top.

The device for producing the variable flame is shown in perspective in Fig. 178 and in section in Fig. 179. It consists of a cell formed of two parts, one inserted in the other, and provided with an air chamber, covered by a diaphragm of very thin soft rubber, a gas pipe entering the lower side of the cell at one end of the diaphragm, and a fine gas burner inserted in the cell upon the same side of the diaphragm. A mouthpiece communicates with the air chamber of the cell through a flexible tube, and the gas pipe leading to the cell is connected with the house supply. The gas burner is provided with a narrow shade, which shields the eye of the observer from the direct light of the flame.

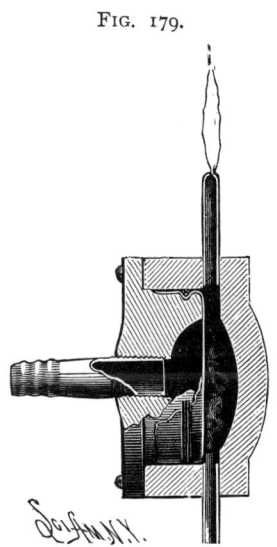

FIG. 179.

Section of Diaphragm Cell.

The top having been set in motion, the mirror is applied and sounds are uttered in the mouthpiece. By viewing the reflection of the flame in the revolving mirror, it will appear as if formed of a regular series of pointed jets, the persistence of the successive images formed on the retina causing them to appear as if produced simultaneously.

The vibrations of the diaphragm due to the sound waves impinging upon it cause the gas to be pushed out of the burner in little puffs, which are not very noticeable when the flame is observed directly, but which are clearly brought out when examined by the revolving mirror.

By employing a double mouthpiece, two sets of flame points of different lengths alternating with each other may be shown. Each vowel sound yields a characteristic series of flame points. A whistle will yield very fine points, while a very low bass note will produce scarcely more than a single point for each half revolution of the mirror.

CHAPTER X.

HEAT.

Heat is the manifestation of an extremely rapid vibratory motion of the molecules of a body. An increase in the velocity and amplitude of the vibrations increases the temperature of the body. A heated mass can impart vibratory motion to the ether which fills space and permeates all bodies, and these wave motions of the ether are able to reproduce in bodies motions similar to those by which they were caused.*

The more obvious effects of heat are expansion, fusion, and vaporization. All bodies increase in volume when heated; gases being the most expansible, liquids next, and solids the least. Heat may partially or wholly balance molecular attraction. Hence it is that, when heated, solids first expand, then (if no chemical action occurs) soften and become liquid, and finally vaporize.† Liquids are changed into vapors, and gases are rarefied.

EXPANSION.

Expansion takes place in all directions. To render this phenomenon apparent, an elongated and attenuated body, such, for example, as a fine wire, is chosen and its linear expansion only is noted. Fig. 180 shows an instrument for exhibiting the linear expansion of a long thin wire, 1 and 2 being respectively front and side views. The instrument is provided with two series of hard rubber pulleys mounted on studs projecting from a board. A fine brass wire (No. 32) attached to the board at one end passes around the successive pulleys of the upper and lower series in alternation, the last end being connected with one end of a spiral spring, which is strong enough to keep the wire taut without

---

* "Heat a Mode of Motion," by John Tyndall, is an interesting popular treatise on this subject.

† Most organic bodies oxidize before the temperature of liquefaction is reached.

stretching it. The other end of the spring is attached to a stud projecting from the board. The pulleys are of different diameters, so that each series forms a cone. By this construction the wire of one convolution is prevented from covering the wire of the next.

The last pulley of the upper series is provided with a boss, to which is attached a counterbalanced index. A curved scale is supported behind the index by posts projecting from the board.

The series of pulleys are 12 inches apart, and there are

FIG. 180.

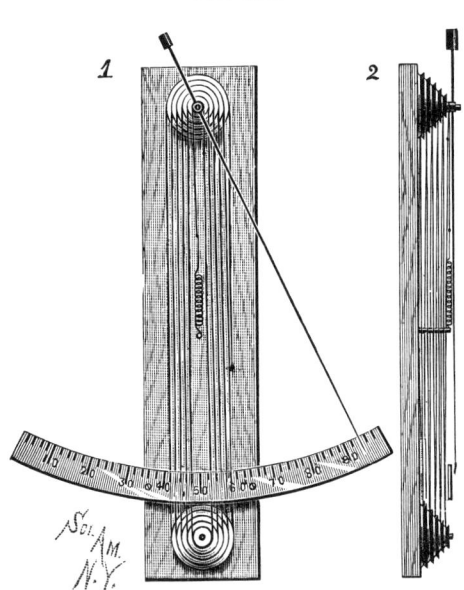

Metallic Thermometer.

ten convolutions of wire, so that a small change of temperature produces sufficient expansion of the wire to cause a perceptible movement of the index. To increase the sensitiveness of the instrument, the wire is blackened by means of smoke or dead black varnish. An electric current passing through the wire heats it sufficiently to cause a deflection of the index; the amount of deflection depending, of course, upon the strength of the current.

SIMPLE THERMOSTAT.

Fig. 181 shows a simple thermostat which is capable of many useful applications. It is represented with an index and scale, but these are not essential for most purposes.

The instrument depends for its operation on the difference between the expansion of brass and steel. The linear expansion of brass is nearly double that of steel, so that when a curved bar of brass is confined at the ends by a straight bar of steel, the brass bar will elongate more than

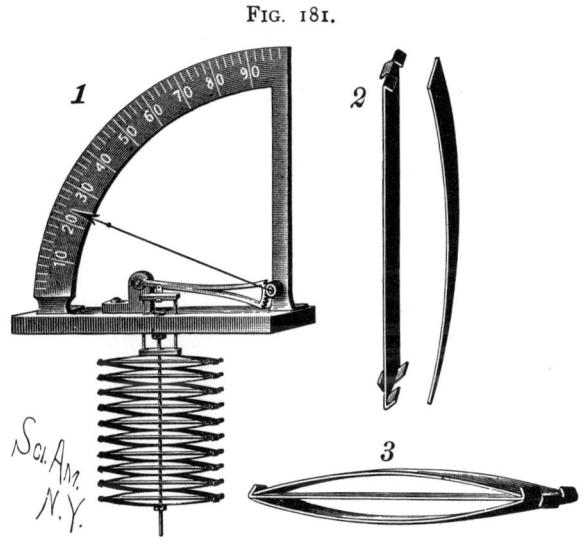

FIG. 181.

Thermostat.

the steel bar when both are heated, and will in consequence become more convex.

At 2 are shown two bars, the straight one being of steel, the curved one of brass. The steel bar is slit for a short distance in two places at each end, and the ears thus formed are bent in opposite directions to form abutments for the ends of the curved brass bars, two brass bars being held by a single steel bar, thus forming a compound bar, as shown at 3. Each compound bar is drilled through at the center. Ten or more such compound bars are strung together loosely upon a rod, which is secured to a fixed support. A stirrup formed of two rods and two cross pieces rests upon the upper compound bar and passes upward through the support. Above the support it is connected by a link with a sector lever which engages a pinion on the pivot of the index. The use to which the thermostat is to be applied will determine its size and construction. It may be used in connection with kilns and ovens and for operating dampers, valves, and electric switches.

AIR THERMOMETER.

The air thermometer, consisting of an air bulb, A, and capillary tube, B, plunged in a colored liquid, shows changes in the volume of air due to expansion and contraction under changes of temperature by the rising or falling of the column of the colored liquid in the capillary tube. It is a sensitive thermometer, but of little practical value, on account of the variability of the volume of air by changes of pressure.

FIG. 182.

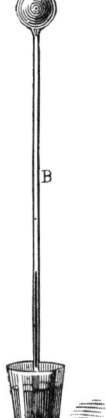

Air Thermometer.

PULSE GLASS.

The pulse glass (Fig. 183) is due to Franklin. It consists of two glass bulbs, formed on opposite ends of a tube bent twice at right angles, the system being partly filled with water, the air having been expelled by boiling the water before sealing the tube. When the bulb which contains the water is held in the hand, and the tube is placed in horizontal position, the rapid evaporation of the water by the warmth of the hand creates a pressure which causes the transfer of the water to the cooler bulb. The quick evaporation of the thin film of water adhering to the sides of the otherwise empty bulb increases the pressure, and causes a rapid ebullition of the water in the other bulb,

FIG. 183.

Pulse Glass.

and at the same time carries off the heat to such an extent as to produce a very decided sensation of cold.*

When the bulb is held at an inclination of about 40°, the water pulsates from one bulb to the other. The interior of the cool bulb becomes quickly dry, and evaporation in it therefore ceases. The water from the other bulb at once flows back into the lower one, to be again expelled by renewed expansion and evaporation.

FIG. 184.

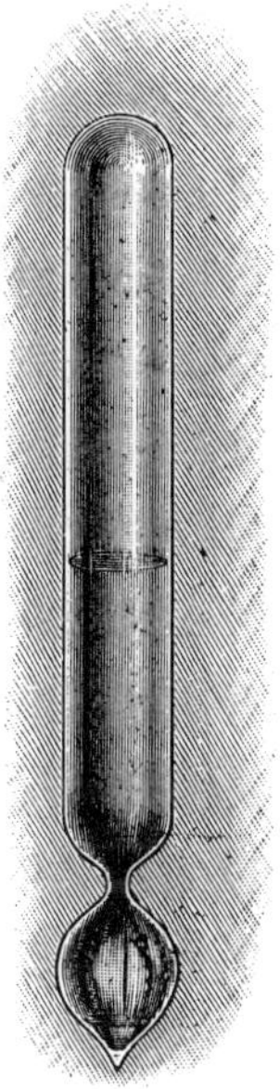

The instrument operates continuously and very regularly when placed in a horizontal position upon a table, with one of the bulbs in the vicinity of a lamp, that is, within eight or ten inches of the flame, the other bulb being placed as far as possible away from the flame and shaded.

The straight form of pulse glass, shown in Fig. 184, exhibits the vaporization of water *in vacuo* to better advantage than the bent form.

When the bulb is held in the hand, the rapid evaporation, by the warmth of the hand, of the water flowing through the narrow neck of the tube and down the inner surface of the bulb creates a pressure of vapor, which finds exit through the neck of the tube, and bubbling up through the main body of the water, is condensed either in the water or above it. Sometimes the tube, when designed for use as a toy, contains the figure of an imp, which the ebullition of the water agitates violently.

THERMOSCOPIC BALANCE.

The action of the thermoscopic balance, shown in Fig. 185, is due to the facility with which liquids evaporate in a vacuum. A small amount of heat is sufficient to vaporize the liquid to the extent required to secure the desired action. The instrument is provided with a glass tube bent twice at right angles, and having a bulb blown on each end. The

---

* This phenomenon is one of latent heat, a subject omitted here, but treated at length in text-books on physics.

tube and the bulbs, like the pulse glass, are partly filled with water, and a vacuum is secured by boiling the water in the bulbs before sealing them. The center of the tube is furnished with V-pivots, which rest in bearings in the top of the forked column. The column also supports a metal screen, which is bright one side and black on the other. Two pins project from the screen to limit the movements of the glass tube and bulbs.

When the instrument is in use, the screen is placed toward the source of heat, and when radiant heat strikes the bulb which is unshielded by the screen, the water in that bulb is vaporized, and sufficient pressure is produced to drive the water upward into the bulb behind the screen. When a little more than half of the water has been in this manner forced from the lower to the higher bulb, the upper bulb preponderates. The tube and bulbs are supported on their pivot so as to secure unstable equilibrium, so that, when the upper bulb begins to descend, it completes its excursion at once, and exposes the full bulb to the radiant heat, at the same time carrying its empty bulb behind the screen, where it cools. The transfer of the water from the full bulb to the empty one now occurs as before. This operation is repeated so long as the bulbs are exposed to the action of radiant heat. The oscillations may be quickened by smoking the sides of the bulbs remote from the screen, and still greater rapidity of action may be secured by concentrating the heat on the bulbs by means of condensers or reflectors.

The principle of the thermoscopic balance has been utilized in the construction of an electric meter. To render it available for this purpose, a coil is inserted in each bulb above the water line and electric connections are provided, by which the current is sent through the coils in alternation as the bulbs tilt. The current thus commuted heats first one coil and then the other, causing the transfer of the water from one bulb to the other in the manner already described. Registering mechanism is provided which records the number of oscillations of the tube. The rapidity of the operation of the instrument is proportional to the strength of the current.

Fig. 185.

Thermoscopic Balance.

CRYOPHORUS.

Wollaston's cryophorus is similar in form and principle to the pulse glass, the only difference being that the tube connecting the two bulbs is made much larger, to avoid choking by ice—a thing sure to occur when the tube is of small diameter—the water vapor which is drawn toward the empty bulb (in a manner presently to be described) being condensed and frozen on the walls of the tube to such an extent as to entirely close it.

The cryophorus in process of construction is partly filled with water, which is boiled in the bulbs before sealing,

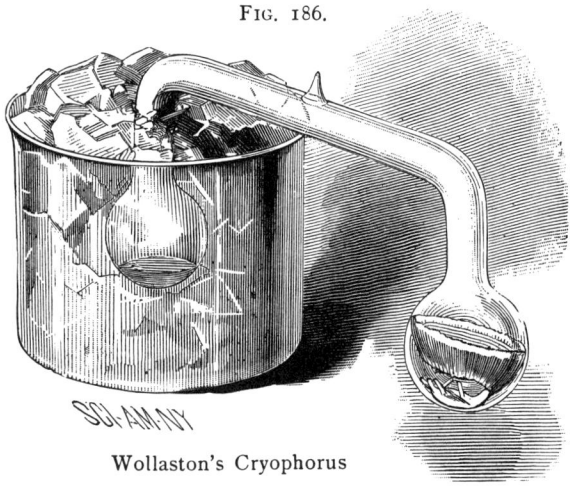

FIG. 186.

Wollaston's Cryophorus

to drive out the air. When the empty bulb of the apparatus is placed in a freezing mixture of ice and salt, for example, the evaporation of the water in the filled bulb, due to the cooling and condensation of vapor in the empty bulb, is rapid enough to carry off the heat to such an extent as to cause the water to freeze. Instead of employing the freezing mixture, a spray of ether or bisulphide of carbon may be projected upon the empty bulb with the same results.

This is a very interesting experiment, illustrating the principle of freezing by rapid evaporation. It also exhibits the change of state of water from gaseous through liquid to solid condition.

HEAT. 189

RADIOMETER.

The radiometer is a heat engine of remarkable delicacy as well as great simplicity. It illustrates a class of phenomena discovered by Crookes, which are difficult to explain in a brief and popular way.*

The instrument consists of a very slight spider of aluminum supporting on the end of each of its four arms a very thin mica plate blackened on one side and silvered on the other side.

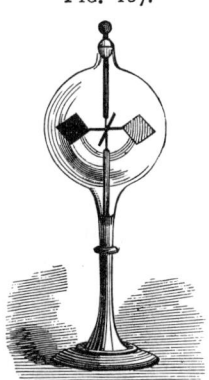

FIG. 187.

The aluminum spider is provided with a jewel, which rests upon a delicate needle point supported at the center of the glass globe.

The spider is retained on its pivot by a small tube extending downward from the top of the globe. When placed in sunlight or near a gas or lamp flame, the vanes revolve rapidly.

An alum cell interposed between the radiometer and the source of light and heat allows the light to pass, but intercepts the heat rays. Under these conditions the vane will not rotate. An iodine cell, which is opaque to light, when arranged in the same way allows the heat rays to go through, and these cause the rotation of the vane.

Radiometer.

TYNDALL'S EXPERIMENT ON RADIANT HEAT.

It often happens that students who desire to test for themselves the experiments of distinguished investigators are prevented from such instructive pleasures by the notion that, for delicate experiments, fine and expensive apparatus is required. Such apparatus is undoubtedly desirable and pleasant to work with, but where it is not to be had, a little courage and ingenuity may provide cheap substitutes which will perfectly answer the student's purpose. The crude apparatus herewith figured illustrates this fact.

* " The Principles of Physics," by Alfred Daniel, contains a clear explanation of the radiometer.

The interesting experiment of Tyndall on radiant heat was suggested to him by Prof. Bell's photophonic experiment, in which musical sounds are obtained by the action of an intermittent beam of light upon a solid body. Referring to this, Prof. Tyndall says:

"From the first I entertained the opinion that these singular sounds were caused by rapid changes of temperature, producing corresponding changes of shape and volume in the bodies impinged upon by the beam. But if this be the case, and if gases and vapors really absorb radiant heat, they ought to produce sounds more intense than those obtained from solids. I pictured every stroke of the beam responded to by a sudden expansion of the absorbent gas, and concluded that when the pulses thus excited followed each other with sufficient rapidity, a musical note must be the result. It seemed plain, moreover, that by this new method many of my previous results might be brought to an independent test. Highly diathermanous bodies, I reasoned, would produce faint sounds, while highly athermanous bodies would produce loud sounds—the strength of the sound being, in a sense, a measure of the absorption. The first experiment, made with a view of testing this idea, was executed in the presence of Mr. Graham Bell, and the result was in exact accordance with what I had foreseen."

The writer has successfully repeated Prof. Tyndall's experiment with the simple apparatus shown in the illustration (Fig. 188). Apparatus already at hand was utilized. A small sized bulbous glass flask, $1\frac{3}{4}$ inches in diameter, was mounted in a test tube holder, and placed behind a rotating pasteboard disk, 12 inches in diameter, having twelve apertures $1\frac{1}{2}$ inches wide and $1\frac{1}{4}$ inches long. Several flasks of the same capacity were provided and filled with the different gases and vapors, and stoppered, to be used at convenience. Near the disk was placed a common gas flame, and into the mouth of the flask was inserted one end of a long rubber tube, the other end being provided with a tapering ear tube, placed in the ear of the listener, whose position was sufficiently remote from the apparatus to avoid any possible disturbance from the revolving disk or the operator. The

Fig. 188.—Apparatus exhibiting the Action of Radiant Heat on Gaseous Matter.

disk being rotated so as to rapidly intercept the thermal and luminous rays of the gas flame and render the rays rapidly intermittent, the effect on the gases and vapors contained by the different bulbs was noted. Dry air produced no sound; moistened, it yielded a distinctly audible tone, corresponding in pitch with the rapidity of the interruptions of the thermal rays.*

Among gases tried, nitrous oxide and illuminating gas yielded the loudest sounds. Among vapors, water and sulphuric ether were most susceptible to the intermittent rays. A candle flame produced distinctly audible sounds in the more sensitive gases, and a hot poker replacing the gas flame yielded the same results.

By using an ordinary concave spun metal mirror, the heat of the flame was satisfactorily projected from a considerable distance. Considering the crudeness of the apparatus and the delicacy of the action which produces the sounds, it appears remarkable that any satisfactory results were obtained, and the experiment shows that any one interested in the finer branches of scientific investigation may often, with the exercise of a little care, enjoy, without material expense, those deeply interesting experiments.

REFLECTION AND CONCENTRATION OF HEAT.

In this experiment the concave mirrors described in a previous chapter are employed in reflecting and concentrating heat.

Instead of placing the watch in the focus of the reflector, B, as in the sound experiment, an air thermometer, E, is supported upon two stands, F F, as shown in Fig. 189, with its bulb in the focus of the reflector. The bulb is smoked over a candle, and when it is nearly cold a drop of water or mercury is introduced into the capillary tube to serve as an index. The candle is removed until the drop in the tube ceases to move. It is then replaced. In a very short time the drop will be pushed outward by the expan-

* The tone to be expected from the gas or vapor when acted on by radiant heat may be determined by blowing through a tube against the apertured portion of the rotating disk.

sion of the air in the bulb. The candle is again removed, and when the drop has returned to the point of starting and ceased moving, a lump, C, of ice is placed on the stand, D,

Fig. 189.

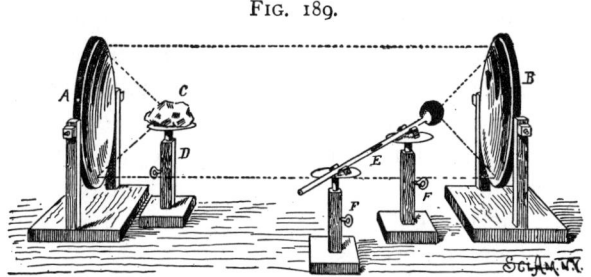

Reflection of Heat.

in the focus of the reflector, A. Immediately the air contracts in the thermometer and draws the drop in. Each of the two bodies is radiating, and receiving heat radiated from the other. But the ice radiates less than the bulb; hence the bulb gives out more than it receives, and the fall of temperature is shown by motion of the index.

Fig. 190.

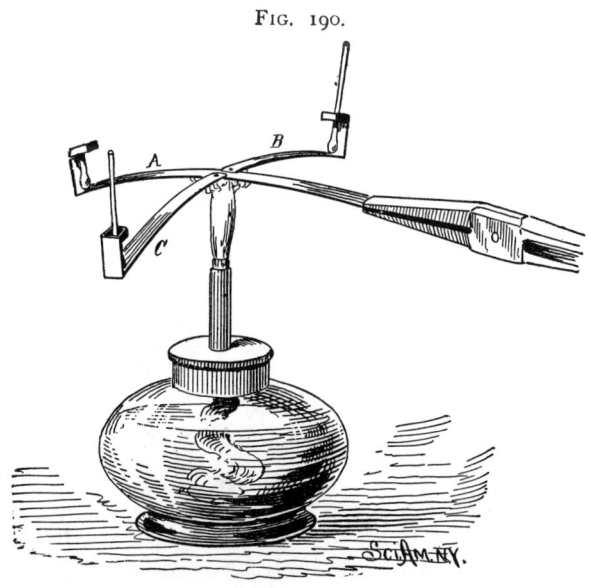

Conduction of Heat.

THE CONDUCTIVITY OF METALS.

The conductivity of metals for heat is admirably shown by the simple device illustrated in Fig. 190. To a strip, A, of iron are attached strips, B C, of brass and copper. The ends of all the strips are bent upward and inward, and the ends of the strips are split and curved to form loops for loosely holding matches, the sulphur ends of which rest upon the strips by their own gravity. The junction of the strips is heated as shown. The match on the copper strip ignites first, that on the brass next, and that upon the iron last, showing that, of the three metals, copper is the best conductor of heat and iron the poorest.

HEAT DUE TO FRICTION.

Every engineer having machinery in charge knows something of this subject. Badly proportioned or poorly lubricated journals often become intensely heated by undue friction. Occasionally a red hot journal is seen. Wherever there is friction there is heat. Often kinetic energy is transformed through friction into heat, which is dissipated by radiation into space, thus causing a loss of energy in a commercial sense, while in a physical sense it still exists, but in another form.

HEAT DUE TO PRESSURE AND COMPRESSION.

Hammering a nail rod until it is red hot and forging a nail without a fire is one of the feats of the blacksmith.

FIG. 191.

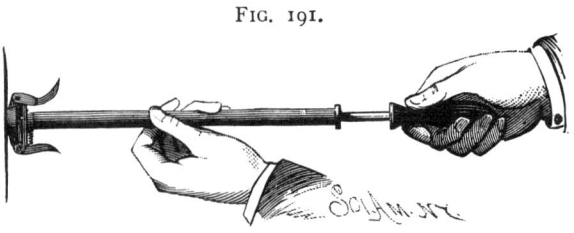

Pneumatic Syringe.

The compression of the iron by the blows of the hammer increases its temperature to such a degree as to render this possible. The impact of a bullet on a hard surface gener-

HEAT. 195

ates sufficient heat to melt the lead of which the bullet is formed. Numerous instances might be given of the generation of heat by the impact of solid bodies.

Gases are also heated by compression. By placing some dry tinder or cotton moistened with ether in the pneumatic syringe (pop gun), Fig. 191, and quickly forcing in the piston, so as to strongly compress the air contained in the barrel of the syringe, the temperature of the air will be raised sufficiently to ignite the tinder or cotton.

FIG. 192.

Candle Bomb.

FORCE OF STEAM.

The candle bomb, shown in Fig. 192, exhibits the explosive power of steam. It consists of a small bulb of glass filled with water and sealed. When the bomb is held in a candle flame by means of a wire loop, the water is converted into steam and an explosion occurs.*

The least expensive machine for applying to mechanical work the force exhibited by the candle bomb is the fifty-cent steam engine, shown in Fig. 193. It is a small and simple machine, but it is far more perfect than the steam engines of our forefathers. It will readily make 800 to 1,000 revolutions per minute. It is a wonderfully inexpensive example of the world's greatest motive power. Its construction is so well known that an extended description seems superfluous.

FIG. 193.

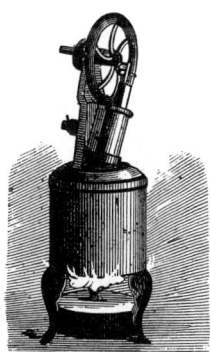

Fifty-cent Engine.

The standard which supports the crank shaft also forms the support of the trunnion of the oscillating cylinder. The piston is connected directly with the crank pin projecting from the fly wheel. The face of the cylinder which contacts with the standard forms the valve for admitting steam to the cylinder and releasing it after use. A passage in the standard conveys steam from the boiler to

---

* A guard of some kind should be placed around the bomb to prevent injury to the experimenter.

the steam ports. A spiral spring on the trunnion draws the cylinder against the standard. The cylinder thus arranged is made to serve as a safety valve. A small alcohol lamp is used as a source of heat.

ASCENSIONAL POWER OF HEATED AIR.

The ascensional power of heated air is exhibited by the draught of every chimney. It is shown by the fire balloon and by the upward tendency of every flame. It is the prime factor in the propelling power of one of the most ancient of motors—the windmill; wind being only air rushing forward to take the place of air which is rising because it is rarefied by heat.

Fig. 194.

Hot Air Motor.

The power derived directly from an ascending column of heated air has never been utilized except as a motor for ventilators, for running mechanical toys, and to some extent for operating small mechanical signs.

The toy motor shown in the annexed engraving is too familiar to require description. It is generally placed over a lamp chimney or at the side of a stovepipe, where the rapidly ascending heated air may impinge on the inclined vanes. The air, acting on the vanes according to the well known law of the inclined plane, produces a lateral movement of each vane, and the vanes being restrained at the center of the wheel while free at their outer ends are compelled to move circularly.

HYGROMETRY.

The toy hygroscope serves to show approximately the hygrometric state of the atmosphere. One of the several forms in which it is made is shown in the annexed engraving. A perforated metal tube, projecting from the back of the figure, contains a short piece of catgut cord, which is fastened in the rear end of the tube by closing the sides of the tube down upon it. The opposite end of the cord projects beyond the front of the figure, and is attached to the arm of the boy. In the hand of the arm thus supported is carried an umbrella. When the air is dry, the catgut cord retains its twist, and the arm holds the umbrella out of the position of use; but when the air becomes moist, the cord swells slightly, and untwists, and in so doing raises the boy's arm and brings the umbrella over his own head and over the head of his companion.

FIG. 195.

Hygroscope.

Another form of the same device consists of a house having two doors and containing two figures—a man with an umbrella and a woman in fair-weather dress; the figures being supported on opposite ends of a bar suspended centrally by a catgut cord. When the cord is untwisted by the action of moisture, the man with the umbrella sallies out; when the cord becomes dry, the man returns indoors and the woman appears.

FIG. 196.

Sensitive Leaf.

These simple, pleasing, and instructive toys illustrate the action of moisture on certain porous bodies, and are of interest, if not of actual use, to the meteorological observer. The action of the sensitive leaf shown in the engraving is also due to expansion by absorption of moisture. The leaf consists of a piece of thin gelatinized paper or gold beater's skin, or even of gelatine, printed in some fantastic design, that of the mermaid being the favorite. When the leaf is laid upon the palm of the hand, the moisture of the hand is absorbed by one side of the leaf, and more in some places than in others, owing to imperfect contact with the hand. The moistened portions rapidly swell, thus warping the leaf, which twists and writhes in every possible direction, as if it were possessed of life. The leaf, being extremely thin, quickly becomes dry, so that the various contortions succeed each other rapidly.

CHEMICAL THERMOSCOPE, HYGROSCOPIC AND LUMINOUS ROSES.

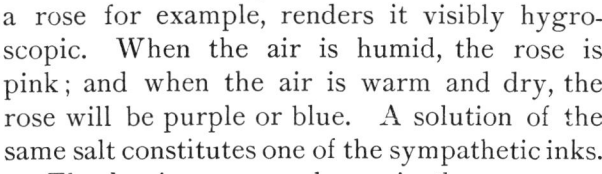

FIG. 197

Chemical Thermoscope.

The chemical thermoscope is made by sealing in a tube a solution of chloride of cobalt in dilute alcohol. When the tube is subjected to a temperature of 40° to 50° Fah., the solution becomes pink, and as its temperature is raised to 90° or 100°, it passes through various shades of purple, and finally becomes blue.

The same salt applied to an artificial flower, a rose for example, renders it visibly hygroscopic. When the air is humid, the rose is pink; and when the air is warm and dry, the rose will be purple or blue. A solution of the same salt constitutes one of the sympathetic inks.

The luminous rose shown in the same vase with the hygroscopic rose is a beautiful example of the wonderful property of storing light possessed by some bodies. The light-storing property is given the rose by a coating of luminous paint, the basis of which is sulphide of calcium. This rose, if exposed to a strong light during the day, will be luminous throughout the night.

The exact nature of the change which takes place in the phosphorescent substance while exposed to the light is unknown. It is supposed to be due to

FIG. 198.

Hygroscopic and Luminous Roses.

some modifying action of the light, rather than chemical action. It has been ascertained that the phosphorescence takes place *in vacuo* as well as in air. Luminous paint has many practical applications. It is used on buoys, guideposts, gates, etc., to render them visible at night. It is applied to match safes with obvious advantage.

CHAPTER XI.

LIGHT.

Various hypotheses have been made regarding the nature and origin of light. The most important of these are the emission or corpuscular theory and the undulatory theory.

The emission or corpuscular theory of light was supported by Newton. It supposes light to consist of exceedingly small particles, projected with enormous velocity from a luminous body. Although this theory seems to have support in many of the phenomena of light, the velocity of light alone, as at present recognized, would seem to render

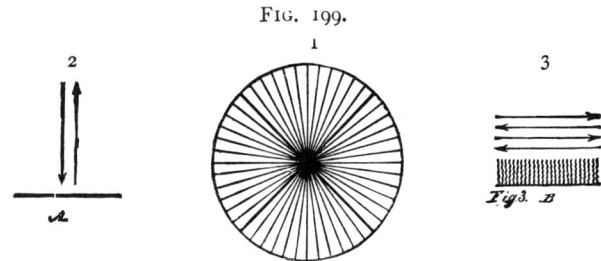

FIG. 199.

Comparison of Sound and Light Waves.

it untenable, however infinitesimal the projected particles might be. Tyndall has said that a body having the weight of one grain, moving with the velocity of light, would possess the momentum of a cannon ball weighing one hundred and fifty pounds and moving with a velocity of 1,000 feet a second; but the most delicate tests known to science have failed to show that light possesses any mechanical force.

The emission theory of light was opposed first by Hooke, Huygens, and Euler, who believed that the propagation of light was due to wave motion. All other eminent scientists supported Newton for one hundred years, but the undulatory theory was finally established beyond a question, by Young and Fresnel.

Sound is propagated by the alternate compression and rarefaction of air, the movements of the waves being parallel with the line of propagation. But not so with light. The vibrations of light are at right angles with its line of progression. These transverse vibrations, in ordinary white light, are in every conceivable direction across the path of the light beam. Their course is represented by Diagram 1, Fig. 199.

We can readily see how the longitudinal vibrations of air would affect the ear drum; 2 shows this action diagrammatically, the horizontal line, A, representing the tympanum, and the two arrows the forward and backward motion of the air wave.

Comparatively recent microscopical research has shown that the retina is studded with fine rods, as shown at B, which are susceptible of being influenced by the lateral movements of the particles in the wave front of a light beam.

The fact that light is wave motion necessitates the assumption of the existence of a medium far more subtile than ordinary matter, which pervades all matter and all space, and is in the interior of all bodies of whatever nature. It is thin, elastic, and capable of transmitting vibrations with enormous velocity. This hypothetical medium is called *ether*. Every luminous body is in a state of vibration, and communicates vibrations to the surrounding ether.

Although light is propagated in straight lines, its direction may be changed by reflection, by any body that will not wholly absorb it. The reflection of light from a mirror is a well known example of this. The direction of light may also be changed by refraction, by causing it to pass from one medium into another having a different density. By holding a strip of plate glass obliquely before a pencil or similar object, the bending of the light beam is shown by the apparent lateral displacement of the object.

Lewis Wright, in his excellent work on light, gives Huygens' explanation of refraction as follows:

"Any beam of light has a wave front across it, and it is obvious that in meeting any refracting surface obliquely, one part of this wave front will meet it before another. Conceive, then, that while the ether permeates the open structure of all matter, it is still hindered in its motions by it, as wind is hindered, but not stopped, by the trees. Then trace a ray, A B (Fig. 200), to the refracting surface, C D, marking off the assumed length of its waves by the transverse lines. The front will be retarded at E before it is retarded at F, and we may assume the retardation is such that the wave in the denser medium is only propagated to G, while in the rarer medium it reaches H. It is plain that the beam must swing round; but when the side, F, also reaches the denser medium, the whole will be retarded alike and the beam

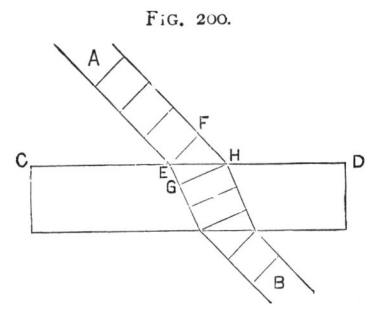

Fig. 200.

Refraction.

will proceed as before, only slower and in a different direction. The theory exactly fits all the phenomena."

As the beam emerges from the denser medium, the reverse of what has been described occurs, and, provided the refracting medium is of uniform thickness and density, the beam proceeds in a path parallel with its former course.

In lenses and prisms the emergent beam takes an oblique path, and in the case of lenses, either convergent or divergent, according to the kind of lens and the position of the lens relative to the object.

PRISMS.

Any refracting body having plane faces inclined to each other is known as a prism. A light beam passing through such a body is permanently deflected. For example, a candle viewed through a prism placed as shown in Fig. 201 will appear to the observer in an elevated position. The light in this case is twice refracted, once on entering the glass, and again on leaving it.

The toy known as the polyprism consists of a plano-convex glass having a number of plane facets on its convex side.

FIG. 201.

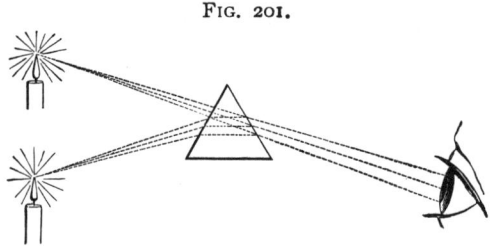

Course of Light through a Prism.

The facets being at slightly different angles with the plane face of the glass, the rays are refracted differently at each facet, thus producing as many images as there are facets. One man seen through this instrument appears like an assemblage. A coin viewed through it is multiplied as

FIG. 202.

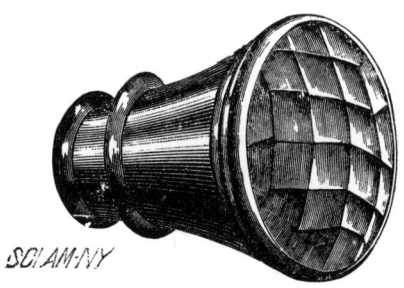

Polyprism.

many times as there are facets, and a grate fire appears like the conflagration of a city.

This toy illustrates in a crude way the principle of the convex lens. The several divisions of the prism are able to so refract a beam of light as to render it convergent, that is to say, each division of the prism will bend as much of the beam as it receives, so that all of the light passing through the prism will be concentrated upon one spot, which will correspond in size with one of the facets. This spot marks the principal focus, a point at which the rays cross, and beyond which they diverge.

LENSES.

A lens may be regarded as an infinite number of prisms of gradually increasing angles arranged around an axis.

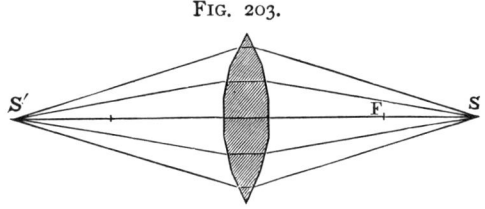

FIG. 203.

Hypothetical Lens.

This idea is illustrated by Fig. 203, in which is shown a hypothetical lens formed of prisms of different angles.

Rays of light proceeding from the point, S, to the lens are refracted differently, those meeting the outer portion of the lens being more deflected than those passing through the inner portions, while the rays coinciding with the axis

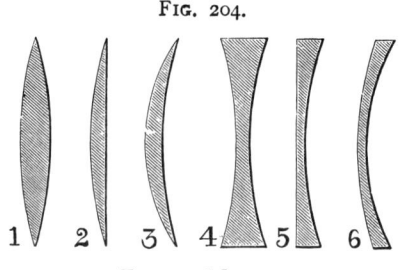

FIG. 204.

Forms of Lenses.

are not refracted. The emergent rays converge to the point, S'. Where there is an infinite number of inclined surfaces, the lens will have spherically convex surfaces.

Of converging or magnifying lenses there are four forms, three of which are shown at 1 2, 3, in Fig. 204; 1 being a double convex lens, 2 a plano-convex, and 3 a convex meniscus. The fourth form, which is a double convex with curved sides of different radii, is known as a crossed lens.

Of diverging or diminishing lenses there are three forms, which are also represented in Fig. 204; 4 being a double concave, 5 a plano-concave, and 6 a concave meniscus.

Parallel rays on entering a double convex lens are re

FIG. 205.

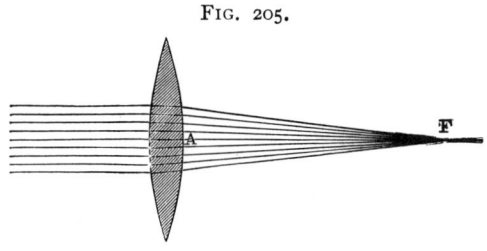

Principal Focus of a Convex Lens.

fracted, and on leaving the lens they are again refracted so that they all converge at the point F, which is the principal focus. The focal length of the lens is the distance from the lens to the focal point.

When light proceeds from a point and is rendered convergent by a lens, as shown in Fig. 203, the point to which the rays converge and the point from which the light emanates.

FIG. 206.

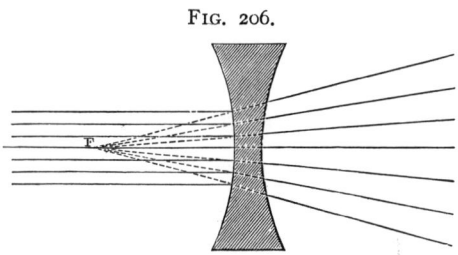

Principal Focus of a Concave Lens.

mark the *conjugate foci* of the lens. Light proceeding from the point, S', will converge to the point, S, and in like manner light proceeding from S will converge to the point, S'.

A concave lens renders a parallel beam divergent, an action which is the reverse of that of the convex lens. If the divergent rays, after passing through a concave lens, are produced backward, as indicated by the dotted lines in

Fig. 206, they will meet in the point, F, which is called the principal focus.

Rays of light which converge toward the point, S′, Fig. 207, before refraction, will, after refraction, converge to the

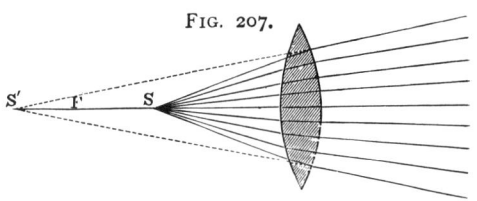

Converging Rays, Convex Lens.

point, S, between the principal focus, F, and the lens, and light emanating from the point, S, will diverge after passing through the lens.

Converging rays passing through a concave lens will

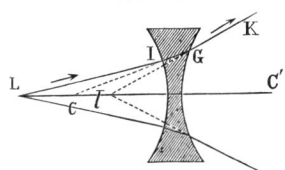

Diverging Rays, Concave Lens.

become less convergent or parallel according to the distance of the point toward which they converge.

Rays proceeding from the point, L (Fig. 208), to and through the concave lens are rendered more divergent. If,

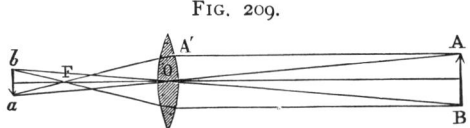

Real and Diminished Image.

in this case, the divergent rays, after passing through the lens, are produced backward, as indicated by dotted lines, they will converge toward the point, l, between the principal focus, C, and the lens.

An object, A B (Fig. 209), placed in front of a convex lens at a distance greater than its principal focal length will have a real image, *a b*, on the other side of the lens This image is inverted and may be either larger or smaller than the object. By holding a double convex lens between the object and a white wall or screen, the image may be seen.

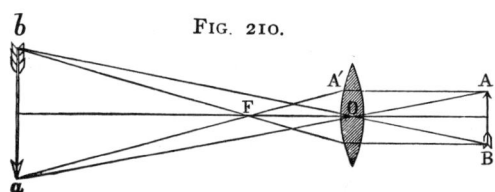

FIG. 210.

Real and Magnified Image.

By changing the relative distances of the object, the lens, and the screen, the size of the image may be varied. In Fig. 209 the object is distant more than twice the focal length of the lens. The photographer's camera exemplifies this principle.

In Fig. 210 is illustrated a case in which the lens is nearer the object, A B. A magnified real image is produced. In this case the distance of the object is greater than the single focal length of the lens, but less than twice its focal length. The projecting lantern exemplifies this principle.

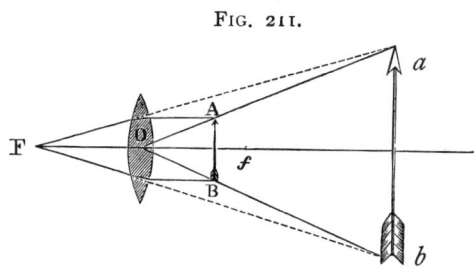

FIG. 211.

Virtual Image, Convex Lens.

When an object, A B (Fig. 211), is placed between the lens, O, and its principal focus, *f*, a virtual image, *a b*, is formed which is erect and magnified, and which appears at a greater distance than the object. This figure illustrates the manner in which objects are viewed by an ordinary magnifying hand glass.

One of the simplest of toys illustrating the action of convex lenses is the water bulb magnifier.

It is a small hollow sphere of glass filled with water and provided with a pointed wire arm for supporting the object to be examined. It is a Coddington lens lacking the central diapraghm. It answers very well as a microscope of low power, and illustrates refraction as exhibited by glass lenses. It receives the rays from the object placed within its focus, and refracts them, rendering them convergent upon the opposite side of the bulb; but all of the rays do not converge exactly at one point, so that the image, except at the center of the field, is distorted and indistinct. This effect is spherical aberration.

FIG. 212.

Water Bulb Magnifier.

MIRRORS.

The convex cylinder mirror shows an ordinary object very much contracted in one direction.

The pictures accompanying these mirrors are distorted to such an extent as to render the object unrecognizable until viewed in the mirror, which corrects the image.

By tracing the incident ray from any point in the picture to a corresponding point in the image in the mirror, then tracing the reflected ray from the same point in the mirror to the eye, it will be found that in this, as in all other mirrors, the simple law of reflection applies; that is, that the angle of incidence and the angle of reflection are equal.

The concave cylindrical mirror (Fig. 214) is the reverse of the mirror just described. It produces a laterally expanded image of a narrow picture, and while the convex cylindrical mirror disperses the light from a distant source, the concave mirror renders it convergent; but, as in the case of the water bulb, the reflected rays do not focus at a single point, but cross each other, forming caustic curves. These curves may be exhibited by placing an ordinary cylindrical concave mirror edgewise on a white surface, and arranging a small light, such as a candle or lamp, a short

Fig. 213.

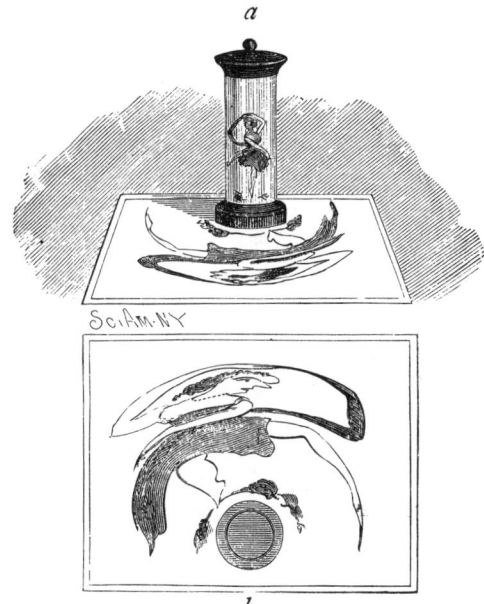

a, Convex Cylindrical Mirror.   *b*, Distorted Picture to be viewed in Mirror.

Fig. 214.

Concave Cylindrical Mirror, Caustics.

distance from the mirror, as shown in the engraving. The same phenomenon may be witnessed by observing a glass partly filled with milk, arranged in proper relation to the light. The inner surface of the glass serves as a mirror, and the surface of the milk serves the same purpose as the white paper. A cylindric napkin ring will show the curves under similar conditions. In fact, any bright concave cylindrical surface will do the same thing.

A convex spherical mirror distorts to a remarkable

FIG. 215.

Spherical Mirror.

degree. A silvered glass globe held in the hand yields an image something like that shown in the engraving.

The size of the image depends upon the distance of the mirror, and is always less than that of the object. The farther the object is, the smaller is its image. This explains the distortion of the image, which appears to be behind the mirror.

The spherical concave mirror produces effects which are the reverse of those just described if the object be nearer than the principal focus. In this case, as in the other, the virtual image appears behind the mirror, and is a magnified

LIGHT. 211

one. The image which appears in front of the concave mirror may be either larger or smaller than the object itself, depending upon the position of the object relative to the mirror and the observer.

It is inverted, and is formed in the air. A candle placed between the center of curvature of the mirror and the principal focus forms an inverted image in air, which is larger than itself.

PHANTOM BOUQUET.

The phantom bouquet, an interesting and very beautiful optical illusion, is produced by placing a bunch of flowers

FIG. 216.

Concave Mirror, Phantom Bouquet.

(either natural or artificial) in an inverted position, behind a shield of some sort, and projecting its image into the air by means of a concave mirror. A magnifying hand glass answers the purpose, if of the right focal length, and a few books may serve as a shield. Two black-covered books are placed upon one end and arranged at an angle with each other, and a third book is laid horizontally on the ends of the standing books. The bouquet is hung top downward in the angle of the books, and a vase is placed on the upper book, over the hanging bouquet.

The concave mirror is arranged so that the prolongation of its axis will bisect the angle formed by lines drawn from the top of the vase and the upper part of the suspended bouquet, and it is removed from the bouquet and vase a distance about equal to its radius of curvature.

A little experiment will determine the correct position for the mirror. When the proper adjustment is reached, a wonderfully real image of the bouquet appears in the air over the vase. It is necessary that the spectator shall be in line with the vase and mirror. With a good mirror and careful adjustment, the illusion is very complete. The bouquet being inverted, its image is erect. A very effective way of illuminating the bouquet, which is due to Prof. W. Le Conte Stevens, of Brooklyn, is shown in the engraving. It consists in placing two candles near the bouquet and behind the shield, one candle upon either side of the bouquet. In addition to this, he places the entire apparatus on a pivoted board, so that it may be swung in a horizontal plane, allowing the phantom to be viewed by a number of spectators.

This simple experiment illustrates the principle of Herschel's reflecting telescope. In that instrument the image of the celestial object is projected in air by reflection and magnified by the lenses of the eyepiece.

MULTIPLE REFLECTION.

The kaleidoscope is one of the most beautiful and inexpensive of optical toys. It can be purchased in the ordinary form for five or ten cents. It is sometimes elaborately mounted on a stand and provided with specially prepared objects. It consists of a tube containing two long mirrors commonly formed of strips of ordinary glass, arranged at an angle of 60°, with a plain glass at the end of the mirrors, then a thin space and an outer ground glass, the space being partly filled with bits of broken glass, twisted glass, wire cloth, etc. The mirrors may be arranged at any angle which is an aliquot part of 360°. When the mirrors, $a\ b$, are inclined at an angle of 60°, as in the present case, the object, c, together with the five reflected images, will form a hexagonal figure of great beauty, which may be changed an infinite number of times by turning the instrument so as to cause the bits of glass, etc., to fall into new positions.

The images adjoining the object are formed by the first reflections of the object. The images in the second sectors are formed by second reflections, and two coincident images

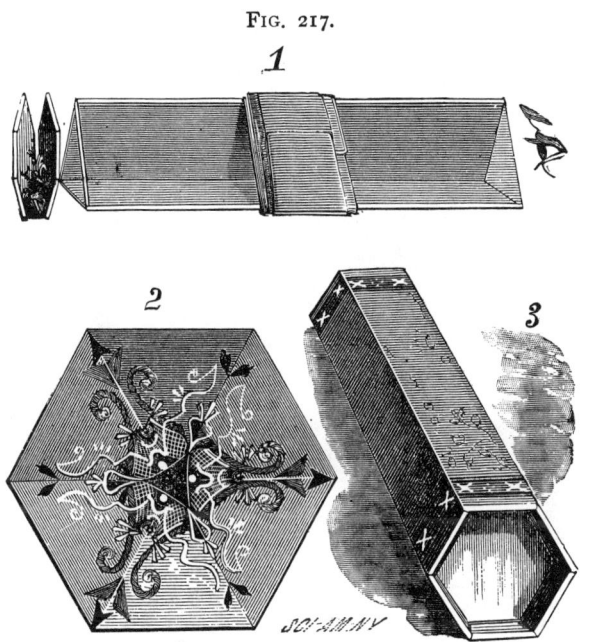

Fig. 217.

1, Parts of Kaleidoscope.   2, The Figure.   3, Kaleidoscope.

in the sector diametrically opposite the object are formed by third reflections.

In most kaleidoscopes a third mirror is added, which multiplies the effects, and in the best instruments an eye lens of low power is provided.

ANALYSIS AND SYNTHESIS OF LIGHT.

An ordinary glass prism, such as may be purchased for fifty cents, is sufficient for the resolution of a beam of white sunlight into its constituent colors. By projecting the dispersed beam obliquely upon a smooth, white surface, the spectrum may be elongated so as to present a gorgeous appearance. It is not difficult to understand that whatever is exhibited in the spectrum must have existed in the light before it reached the prism, but the recombining of the colors of the spectrum so as to produce white light is of course conclusive.

The colors of the spectrum have been combined in several ways, all of which are well known. Newton's disk does it in an imperfect way by causing the blending, by persistence of vision, of surface colors presented by a rotating

FIG. 218.

Simple Rocking Prism.

disk. Light from different portions of the spectrum has been reflected upon a single surface by a series of plane mirrors, thus uniting the colored rays forming white light. The colored rays emerging from the prism have been concentrated by a lens upon a small surface, the beam resulting from the combination being white. Besides these methods, the spectrum has been recombined by whirling or rocking a prism; the movement of the spectrum being so rapid as to be beyond the power of the eye to follow, the retina receiving the impression merely as a band of white light, the colors being united by the superposing of the rapidly succeeding impressions, which are retained for an appreciable length of time.

The engravings show a device to be used in place of the ordinary rocking prism. It is perfectly simple and involves no mechanism. It consists of an inexpensive prism, having attached to the knob on either end a rubber band. In the present case the bands are attached by making in each a short slit and inserting the knobs of the prisms in the slits. The rubber bands are to be held by inserting two of the fingers in each and drawing them taut. The prism is held in a beam of sunlight, as shown in Fig. 218, and with one finger the prism is given an oscillating motion. The band of light thus elongated will have prismatic colors at opposite ends, but the entire central portion will be white. To show that the colors of the spectrum pass over every portion of the path of the light, as indicated by the band, the prism may be rocked very slowly.

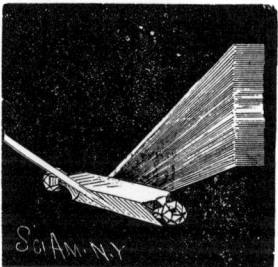

Fig. 219.

The Spectrum.

An ordinary prism may be made to exhibit several Fraunhofer's lines by arranging it in front of a narrow slit, through which a beam of sunlight is admitted to a darkened room. One side of the prism in this experiment must be adjusted at a very small angle with the incident beam. The spectrum will contain a number of fine dark lines, known as Fraunhofer's lines.

These lines tell of the constitution of the sun. The principle illustrated by this experiment is the one upon which the spectroscope is based.*

SIMPLE METHOD OF PRODUCING THE SPECTRUM.

Color is a sensation due to the excitation of the retina by light waves having a certain rate of vibration. Those

* For further information on this subject the reader is referred to "Studies in Spectrum Analysis," by J. Norman Lockyer.

having the highest rate capable of affecting the eye are perceived as violet, while those of the lowest rate are perceived as red. According to Ogden Rood's "Modern Chromatics," the rate of the former is 757 billions of waves per second,

Fig. 220.—Simple Apparatus for producing the Spectrum.

that of the latter is 395 billions of waves per second, and between these extremes are ranged waves of every possible rate, representing as many colors. When light waves of all periods are mingled, there is no color—the light is white.

Newton discovered a way of resolving white light into its constituent colors. He made exhaustive experiments with prisms, first producing the gorgeous array of colors known as the spectrum, then recombining the colored rays by means of another prism producing white light. He found that the colors of the spectrum were simple, $i.\ e.$, they could not be further decomposed, and he also demonstrated that the red rays were the least and the violet rays the most refrangible.

The solar spectrum is always a delight to the eyes of every person having normal eyesight, and it is a simple matter to produce it by means of a prism. When a prism is not available, it may be produced in the manner illustrated by Figs. 220 and 221. This method is inexpensive, and yields a large spectrum. The materials required are a piece of a plane mirror, five or six inches square, a dish of water, and a sheet of white paper or a white wall. The mirror is immersed in the water and arranged at an angle of about 60°; this angle, however, may be varied to suit the direction of the light. The incident beam received on the mirror is refracted on entering the water and dispersed.

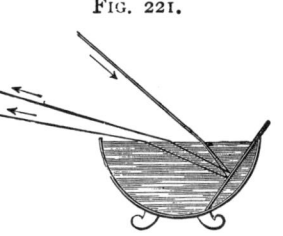

FIG. 221.

Diagram of Spectrum Apparatus.

It is further dispersed upon emerging from the water. By causing the reflected beam to strike obliquely upon the white paper or wall, the spectrum thus produced may be made to cover a large surface.

Should the sun be too high or too low, the proper direction may be given to the incident beam by means of a second mirror held in the hand. The diagram, Fig. 221, shows the direction of the rays.

Some very interesting absorption experiments may be made in connection with this simple apparatus. For example, colored glass, or sheets of colored gelatine, may be placed in the reflected beam. If red be placed in the path of the beam, red light, with perhaps some yellow, will pass through, while the other colors will be absorbed, and will not, therefore, appear on the wall. With the other colors the same phenomenon is observed. Each colored glass or gelatine is transparent to its own color, but opaque to other colors. It will be observed that few bodies have simple colors.

In a similar manner a piece of red paper or ribbon placed in the red portion of the spectrum will reflect that color, but if placed in some other part of the spectrum it will appear dark, the other colors being absorbed or quenched by the colored surface. It is seen by these experiments that when light passes through a colored glass or film, it does not retain all its colors. It is simply a matter of straining out every color except that to which the glass or film is transparent. In reality only a small part of all the light striking the colored glass passes through it.

In the above experiment it is essential to avoid all jarring of the water, as ripples upon its surface defeat the experiment. If it is possible to so place the dish as to avoid jarring, the ripples may be prevented by suspending a transparent plane glass horizontally, so that its under side will just make contact with the surface of the water.

NEW CHROMATROPE.

A novel toy which illustrates some of the phenomena of color is illustrated by Fig. 222. Upon the spindle, A, is secured a star, B, formed of two triangular pieces of pasteboard arranged so that their points alternate. One triangle is red, the other bluish green—complementary colors, which produce white when they are blended by the rotation of the star. In the angles of one of the stars are secured wire nails, which serve as pivots for the three disks, C, as shown at 1 and 4. Each disk is divided into three equal parts, which are colored respectively red, green, and violet. The disks overlap at the center of the star, B.

Around the spindle, A, is wound a cord which passes through the loop formed in the star frame in which the spindle is journaled, and is provided at its end with a button, D. By pulling the cord, the star, B, is whirled first in one direction and then in the other. As the series of disks, C, turn, the colors are blended in different ways, according to

LIGHT. 219

the relative arrangement of the different sections. All the phenomena of the blending of surface colors are illustrated by this simple toy. At times the center will be a fine purple, while the outer part is green. At other times some portions of the color disk presented by the rotating disks are white, showing that a proper mixture of the three primary colors yields white light.

At the instant of the change of rotation from one direc-

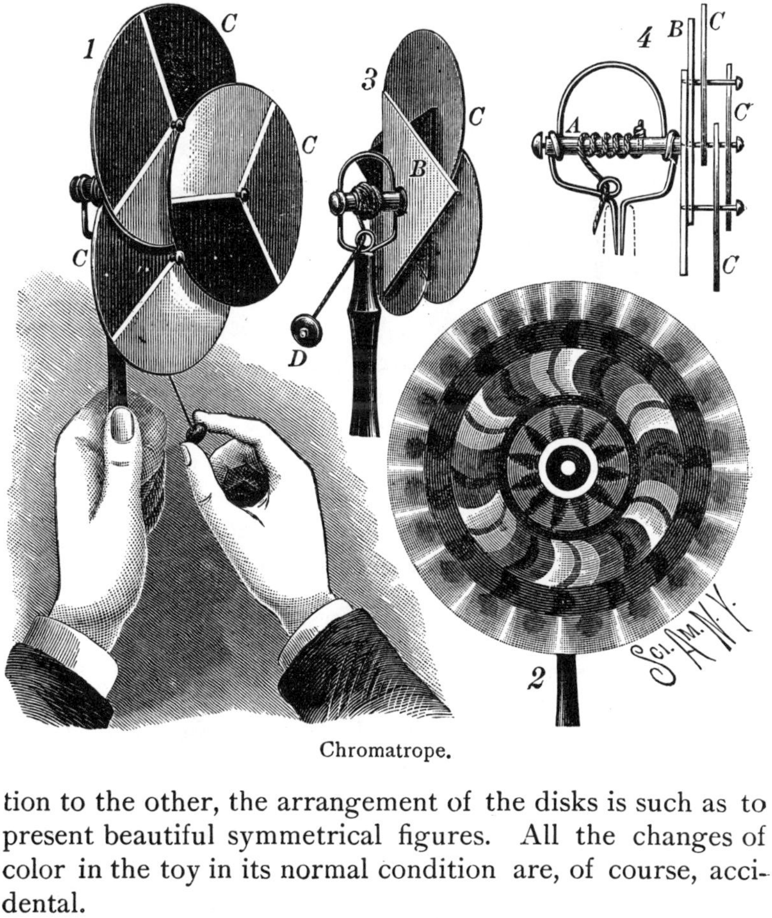

Chromatrope.

tion to the other, the arrangement of the disks is such as to present beautiful symmetrical figures. All the changes of color in the toy in its normal condition are, of course, accidental.

When it is desired to try the blending of any of the colors, when arranged in a particular way, the disks may be prevented from turning on their pivots by stretching over each disk a small rubber band.

The maker of this simple toy has succeeded in securing colors which produce remarkably good effects.

PERSISTENCE OF VISION.

The zoetrope, or wheel of life, is a common, but interesting, optical toy. It depends for its curious effects upon the persistence of vision. It consists of a cylindrical paper box mounted on a pivot, and having near its upper edge a series of narrow slits, which are parallel with its axis. Against the inner surface of the wall of the box is placed a paper slip, carrying a number of images of the same object arranged in as many different positions, each image differing slightly from the adjoining images, the successive positions of the several images being such as to complete one entire motion or series of motions.

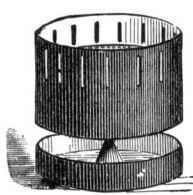

Fig. 223.

Zoetrope.

When these pictures are viewed through the slits, as the box is turned, the eye glimpses the figures in succession, and retains the image of each during the time of eclipse by the paper between the slits and until the next figure appears. The images thus blend into each other, and give the figure the appearance of life and action.

Some very interesting studies for the zoetrope have been produced by the aid of instantaneous photography.

IRRADIATION.

Brilliantly illuminated white surfaces and self-luminous bodies, when emitting white light, appear to the eye much larger than they really are. In nature examples of this phenomenon are presented by the sun, moon, and stars. The sun, viewed with the naked eye, appears very much larger than when the light is modified by a smoked glass. The crescent of the moon appears to project beyond the moon's periphery; and the stars, which are mere points of light even when viewed through the largest telescope, appear to the eye to have a disk of some size.

This phenomenon—known as irradiation—is due to the stimulation or sympathetic action of the nerves of the retina adjoining those which actually receive the image.

The ends of pieces of iron heated to incandescence by the blacksmith for welding seem to be unduly enlarged—an appearance due to irradiation.

Without doubt the most striking illustrations of irradiation are to be found in electric illumination. The electric arc, which is no larger than a pea, appears to the eye as large as a walnut; and the filament of an incandescent lamp, which is scarcely as large as a horsehair, appears as large as

Fig. 224.

An Example of Irradiation.

a small lead pencil. In viewing an ordinary incandescent lamp, it is difficult to believe that the delicate filament is not in some way immensely enlarged by the electric current or by the heat, but the experiment illustrated by the engraving shows that the size of the filament is unchanged, and proves that the effect is produced in the eye.

The experiment consists merely in holding a smoked or darkly colored glass between the eye and the lamp. The glass cuts off a large percentage of the light, and enables the eye to see the filament as it really is.

The effects of irradiation are different in different persons, and they are not always the same in the same person.

INTENSITY OF LIGHT.

It is estimated that 5,500 wax candles would be required to illuminate a surface twelve inches distant as strongly as it would be illuminated by the sun, while the light of a single candle at a distance of 126 inches would equal that of the full moon. The relative intensities of the light of the sun and moon are as 600,000 to 1.

Light from different sources can be compared and measured by the photometer, several forms of which have been devised. The usual way of determining the intensity of light from any source is to compare it with a standard of illumination, a "sperm candle weighing $\frac{1}{6}$ pound, and burning 120 grains an hour," being commonly used for this purpose. Thus it is that a gas flame or an electric lamp is rated at a certain candle power.

Owing to the divergence of luminous rays, the intensity of light decreases rapidly as the illuminated surface is removed from the source of light. This may be readily shown by holding a screen, say 12 inches square, half way between a lamp and the wall. The shadow of the screen on the wall will be 24 inches square. If the light falling on the screen be allowed to proceed to the wall, it will cover the area which was before in the shadow of the screen. This area being four times as large as that of the screen, it is seen that the light which was received on the screen must, when distributed upon a surface four times as great, be reduced in intensity to one-fourth of that falling on the screen. It is thus shown that the intensity of light is inversely as the square of the distance; that is, when the distance of the illuminated surface from the source of light is doubled, it receives one-fourth the amount of light; at three times the distance, one-ninth, and so on.

The law of inverse squares may be demonstrated by the extemporized photometer, shown in Fig. 225. In front of a white cardboard screen is supported an opaque rod. The sources of light to be compared are arranged so as to cast separate shadows of the rod on the screen. If the sources of light when equally distant from the screen form shadows of the same depth, their illuminating power is the same.

When, however, the intensities of the two lights differ, the shadows will differ, and it will be necessary to remove the stronger light to a greater distance to secure shadows of equal depth.

In the experiment illustrated, the single candle being distant one yard from the screen, it is found that the group of four candles must be placed two yards from the screen

FIG. 225.

Photometer.

to secure shadows of the same intensity. Nine candles would require removal to a distance of three feet, and so on. All the candles of the group must be in the same line in the direction of the rod. The eye is able to detect a difference of one-sixtieth in the values of the shadows, provided the lights be of the same color.

OPTICAL ILLUSIONS.

It is sometimes difficult, even for the practiced eye, to accurately estimate distances and dimensions, and to correctly appreciate forms. Very much depends upon the relation of the object viewed to surrounding objects. Two straight parallel lines of equal length would be appreciated by the eye in accordance with the facts, but when a light line is drawn perpendicular to a heavy one of the same length, as in Fig. 226, the eye at once accords the greater length to the lighter line.

In the case of two like parallel lines joined at the ends in one case with outwardly convergent lines and in the other with outwardly divergent lines (Fig. 227), the apparent difference in the length of the lines is considerable.

It often happens in engineering drawing that a sectional

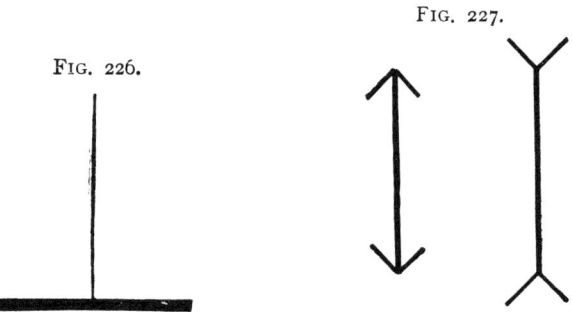

view will present some curious distortions, which give the drawing the appearance of being incorrect, but which in reality are only illusions. Fig. 228 is an example taken from such a drawing.

In Figs. 229 and 230 are shown examples of line combinations in which series of oppositely disposed oblique lines are joined to parallel lines. In Fig. 229 the latter appear to bend outwardly and in Fig. 230 they seem to bend inwardly;

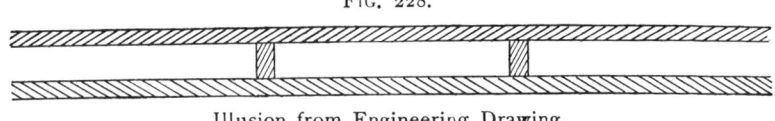

Illusion from Engineering Drawing.

but by looking at the diagrams lengthwise, or through partly closed eyes, the parallel lines appear as they really are.

A more marked example of the effects of oblique lines on a series of parallel lines is shown in Fig. 231.

In Fig. 232 the single oblique line extending above the

FIGS. 229 AND 230.

Apparent Deviation by Oblique Lines.

FIG. 231.

Parallel Lines appearing Alternately Convergent and Divergent.

FIG. 232.

Apparent Displacement of a Single Oblique Line.

black bar appears to be a prolongation of the lower oblique line below the bar. That such is not the case may be shown by placing a card against the line above the bar or sighting it endwise. It will thus be shown that it is a prolongation of the upper of the two lines below the bar.

The curious optical illusions shown in Figs. 233 and 234 were published some time since in a French scientific journal.*

Fig. 233 represents two pieces of paper or cardboard cut into the shape of arcs of a circle. Which is the larger of the two? To this the answer will certainly be: " It is No. 2." But if No. 1 be placed under No. 2, the answer will be just the reverse. The fact is that both are exactly of the same size, as may be seen by measuring them, or by laying

FIG. 233.

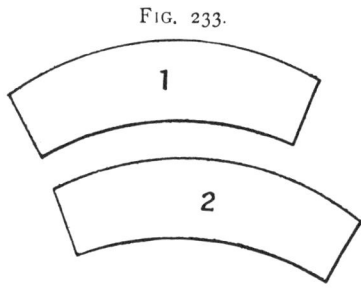

Curious Optical Illusion.

one upon top of the other. When the two figures are placed so close together that their edges touch, the illusion is still greater.

Which is the tallest of the three persons figured in the annexed engraving? If we trust our eyes, we shall certainly say it is No. 3. But if we take a pair of compasses and measure, we shall find that we have been deceived by an optical illusion. It is No. 1 that is the tallest, and it exceeds No. 3 by about 0.08 inch.

The explanation of the phenomenon is very simple. Placed in the middle of the well calculated vanishing lines the three silhouettes are not in perspective. Our eye is accustomed to see objects diminish in proportion to their

* *La Nature.*

distance, and, seeming to see No. 3 rise, concludes therefrom that it is really taller than the figures in the foreground.

Fig. 234.

An Optical Illusion.

The origin of the engraving is no less curious than the engraving itself. It serves as an advertisement for an English soap manufacturer, who prints his name in vanishing perspective between each of the decreasing lines, and places the cut thus formed in a large number of English and American newspapers.

Here is a row of letter S's and one of figure eights, taken at random.* At a casual inspection the reader might say the letters were symmetrically made—that is, the top and bottom lobes of the figures and letters the same size—though upon a close inspection he would either say that it was

S S S S S S S
8 8 8 8 8 8 8

doubtful whether any difference existed or he would notice the true relation that exists, the top lobe being the smaller.

FIG. 235.   FIG. 236.

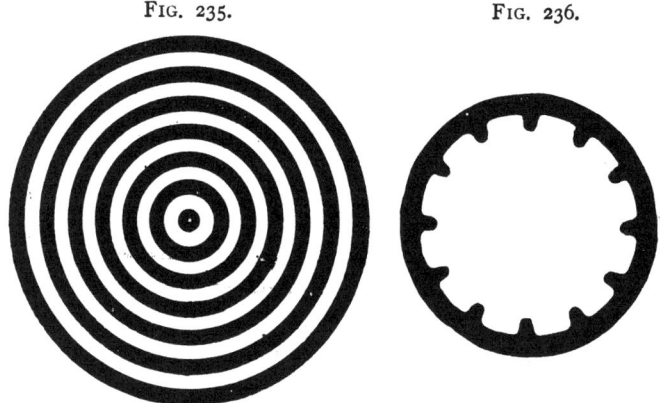

Professor Thompson's Optical Illusion.

Let him, however, turn this page upside down, and the most cursory glance possible will show him their shapes, and the dissimilarity between the upper and lower halves will strike him with astonishment if he never tried the experiment before.

One of the most interesting of optical illusions is that devised by Prof. Silvanus P. Thompson. This is illustrated by Figs. 235, 236, and 237. The first of these figures is composed of a series of concentric rings about a twentieth of an inch wide and the same distance apart. If the

* Mr. G. Watmough Webster, in *British Journal of Photography*.

illustration is moved by hand in a small circle without rotating it, *i. e.*, if it is given the same motion that is required to rinse out a pail, the circle will revolve around its center in the same direction that the drawing moves.

A black circle (Fig. 236) having a number of equidistant internal teeth is provided for the second experiment, the drawing being moved in the manner above described, but in a contrary direction.

In Fig. 237 is shown a combination of the toothed and concentric circles.

By means of photographic transparencies Mr. Thomp-

FIG. 237.

son has shown these figures on a screen on a large scale, and by moving the plates as before described, the figures on the screen were made to rotate.*

When viewed in a microscope under certain conditions, the minute markings of some of the diatoms appear as hexagons, while under other conditions, and with a first-class objective, they appear spherical.

M. Nachet, the French microscopist, has published a

* A. O., on p. 133, vol. 41, *Scientific American*, furnishes an explanation of the phenomena of these circles.

curious optical illusion which, he thinks, accounts for the markings on the diatoms appearing as hexagons.

The circular spots (Fig. 238) are arranged as nearly as

FIG. 238.

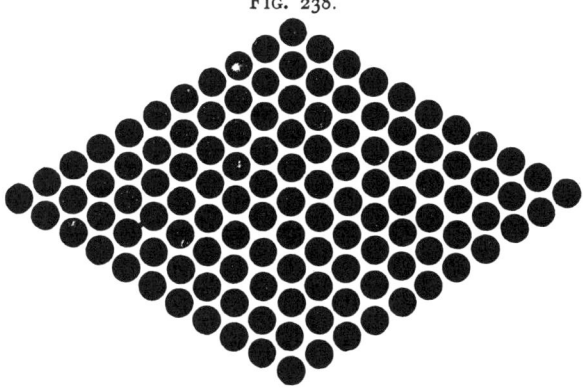

possible like the markings on the diatom called *Pleurosigma angulatum.* If the figure is viewed through the eyelashes with the eyes partly closed, the circles will appear as hexagons.

FIG. 239.

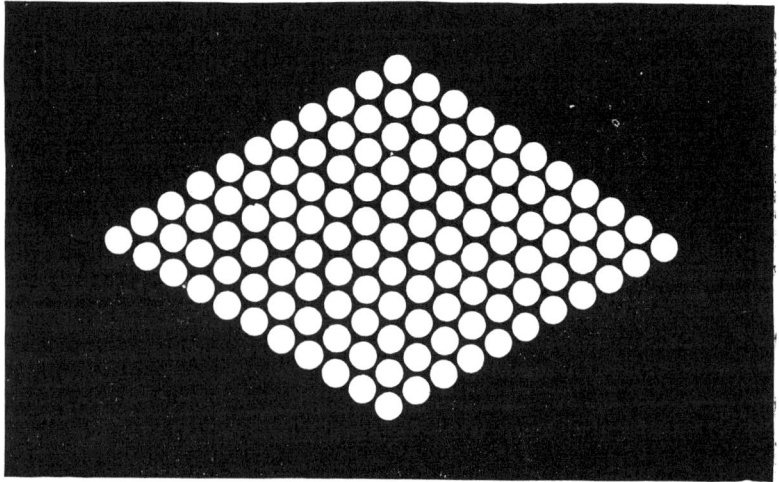

In Fig. 239 is shown a negative reproduction of Fig. 238, in which the spots are white on a black ground. When these figures are compared, the white spots, on account of irradiation, appear much larger than the black ones, although they are of exactly the same size.

Fig. 240 illustrates an interesting illusion observed by Mr. J. Rapieff, the well known electrician. The apparatus consists of semicircular and circular wire loops, provided with axles, by which they may be twirled between the thumbs and fingers. The lower row of figures shows some of the

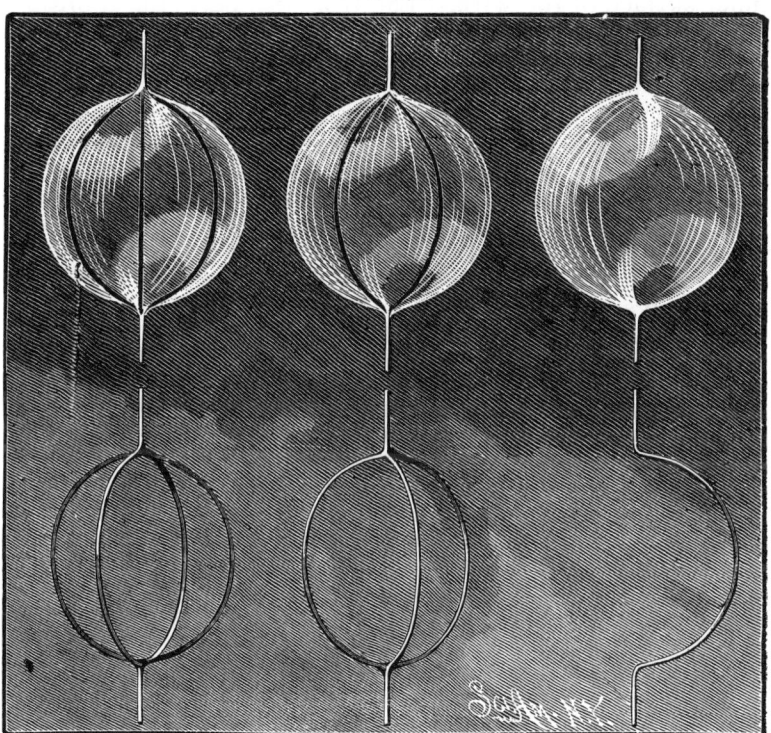

FIG. 240.

Rapieff's Optical Illusion.

loops used in the experiment, while the upper figures represent the effects produced. The wire has a polished surface. When the single semicircular loop is twirled, the only effect is to produce a gauzy glimmer of spherical form, as shown in the upper right hand figure. When three of the loops are joined together, each extending from the other at an angle of 120°, the figure produced is similar to that already described, but with two perfectly distinct curved black lines extending from one axle to the other, as shown in the upper central figure. When four loops are joined at right angles to each other, three jet black lines are shown, as indicated in the upper left hand figure. A circular loop shows a single black line.

This curious effect is produced by holding the apparatus so that the light is reflected as much as possible from the inner surface of the wire. The result is due to the eclipsing of the bright surface by the shaded portion of the upper loop as it passes between the eye and the lower loop. The whole of the loop is not eclipsed at the same instant, but persistence of vision causes the entire eclipse to be seen at once.

Success in this experiment depends upon holding the loops in the right position relative to the light, as well as the provision of the proper background. The loops should be held over a dark ground, with the axles parallel with the plane of vision.

CHAPTER XII.

POLARIZED LIGHT.

Glass, like all uncrystallized bodies, is said to be single refracting, because it diverts the ray in one direction only. By placing a rhomb of Iceland spar over a small black spot formed on a piece of white paper, two images of the spot appear, showing that the beam of light has been split up into two rays, one of which is called the ordinary ray, the other the extraordinary ray. As the rhomb is turned, the extraordinary ray moves around the ordi-

FIG. 241.

Iceland Spar.

nary one, and the image of the spot produced by the extraordinary ray appears nearer to the observer than the spot itself. This property of splitting the ray transmitted through the crystal, which was first noticed and commented on by Erasmus Bartholinias, in 1669, is known as double refraction. It is possessed by many crystalline bodies in a greater or less degree. Both rays emerging from the spar have acquired peculiar properties.

Newton, after investigating the properties acquired by light in its passage through the spar, concluded that the particles had acquired characteristics analogous to those of magnetized bodies, that is, they had become two-sided, and were, in fact, polarized.

Light, in the state of two-sidedness as observed by Newton, is still known as polarized light. By inserting the double refracting crystal known as tourmaline between the eye and the rhomb of spar, and turning it, the ordinary and extraordinary rays will be extinguished and will reappear in alternation. All vibrations, except those executed parallel with the axis of the tourmaline, are quenched. A Nicol prism (to be described later on) will do the same thing. When the Nicol is turned, the black spots seen by the two rays become alternately visible and invisible. One-quarter of a revolution of the prism is sufficient to extinguish one ray, and bring the other out; and a further turning of the prism through another quarter of a revolution

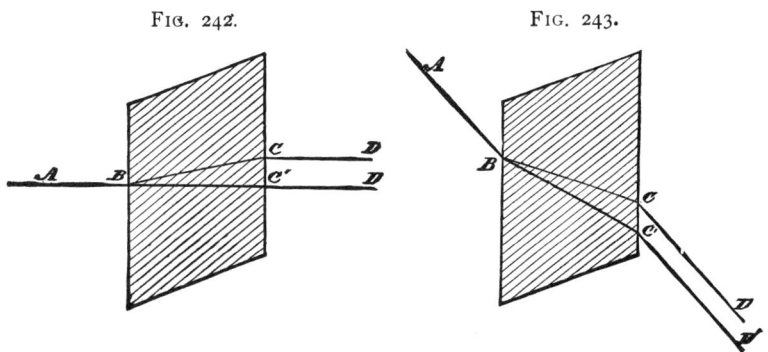

Course of Light through Iceland Spar.

reproduces the extinguished spot and effaces the visible one. This experiment shows that the vibrations of the two rays are in planes at right angles to each other. A beam of light in which all of the transverse vibrations are parallel with a single plane is plane-polarized. Both of the beams emerging from the spar are therefore plane-polarized, but in different planes.

The course of the light through the rhomb of Iceland spar when the incident ray is perpendicular to one of the faces of the crystal is shown in Fig. 242. The ordinary ray, A, passes straight through the crystal on the line, A C′, while the extraordinary ray is bent away from the ordinary ray, on the line, B C.

POLARIZED LIGHT. 235

When the incident ray enters the side of the rhomb at an angle (as shown in Fig. 243), the ordinary ray follows the law of refraction, and the extraordinary ray is bent away from the ordinary ray, as in the other case.

Fig. 244.

The most perfect instrument for polarizing light and analyzing it after its polarization is the Nicol prism, made from a rhomb of Iceland spar, and named after its inventor. In this prism, the ordinary ray is disposed of, and the extraordinary ray alone is used.

The prism which is shown in Fig. 244 consists of a rhomb of Iceland spar, divided through its axis on the line, D D, with its ends cut off at right angles to this line. The two halves of the prism are cemented together by Canada balsam, whose index is between that of the two indices of the spar, so that the ordinary ray, B C', meets the film of balsam at an angle which is sufficiently oblique to secure the reflection of this ray to one side, where it is lost, while the extraordinary ray, B C, passes through the balsam, and

Nicol Prism.

Fig. 245.

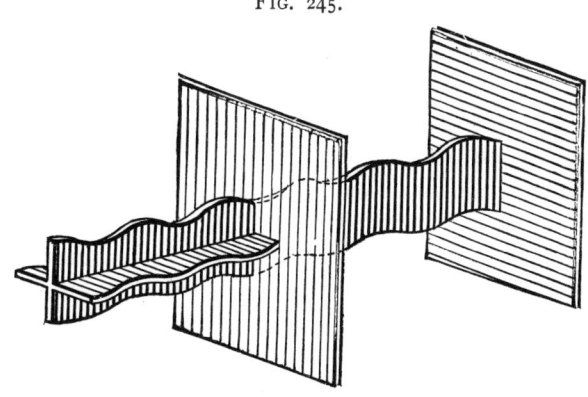

Action of Tourmaline Crystals.

onward through the other half of the prism perfectly polarized.

To observe the effects of polarization, an analyzer is required. Anything that will act as a polarizer will also serve as an analyzer, and since the Nicol prism is unsurpassed as a polarizer, it will answer equally well for an analyzer.

Perhaps the action of polarized light cannot be better illustrated than by a representation of a hypothetical beam of light and two tourmaline plates (Fig. 245). Here is shown the beam of light with vibrations traversing the path of the beam in two directions. On reaching the first tourmaline plate, those vibrations which are parallel with the axis of the tourmaline crystal (represented by the parallel lines) are readily transmitted, but all the vibrations in any other direction are extinguished. The beam now polarized passes on to the second tourmaline plate, and the axis of the crystal being arranged at right angles with the plane of vibration, it is extinguished; but if the axis of the

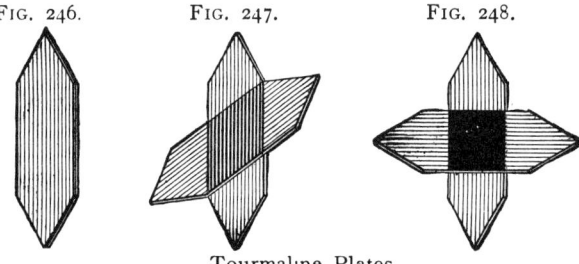

FIG. 246.   FIG. 247.   FIG. 248.

Tourmaline Plates.

second tourmaline is parallel with the plane of vibration, the light will pass through.

If the axes of the tourmalines are arranged at an angle of 45° with each other, the light is only partly extinguished.

These effects of the two tourmaline plates are illustrated by the annexed diagrams, Fig. 246 showing the crystals with their axes arranged parallel with each other, Fig. 247 showing them arranged at an angle of 45°, and Fig. 248 shows them crossed or arranged at right angles with each other, exhibiting a complete extinction of the ray at the intersection of the crystals.

If, now, when the polarizer and analyzer cross, a double refracting crystal be inserted between them, the light passing the polarizer will be made to vibrate in a different plane, and will therefore prevent the complete extinction of the beam by the analyzer.

Besides those means of polarizing light already described, there are others which should be examined. Light is polarized by reflection at the proper angle from almost every object; glass, water, wood, the floating dust of the air, all under certain conditions will polarize light.

That the light beam becomes polarized may be readily ascertained by receiving it through a double-refracting body and an analyzer.

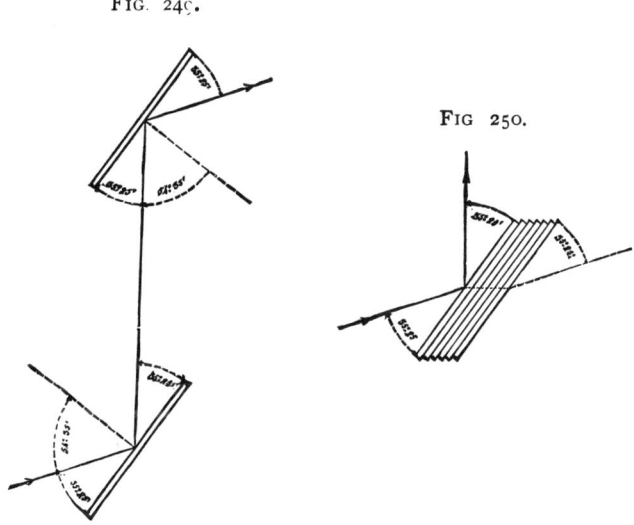

Polarization by Reflection and Refraction.

Two plates of unsilvered glass, receiving and reflecting light, as indicated in Fig. 249, act respectively as polarizer and analyzer.

For every substance there is an angle at which the polarization is at a maximum. For common window glass the angle the ray must make with the normal is $54° 35'$. This is called the polarizing angle. It depends upon the index of refraction of the glass, and is such that the reflected and transmitted rays are at right angles to each other.

Balfour Stewart explains polarization by reflection as follows:

"It is imagined that in the reflected ray the vibrations are all in a direction perpendicular to the plane of reflection, so that the portion of the incident ray consisting of vibrations in the plane of reflection has not been reflected at all. If, therefore, we allow an ordinary ray of light (Fig. 249) first to be reflected from a plate of glass, at the polarizing angle, and if the reflected ray be again made to impinge upon another surface of glass at the same angle, the latter will then be the analyzer, and if its plane be parallel to the polarizer, as in the figure, the light will be again reflected in the direction indicated by the arrow. If the analyzer be turned round the first reflected ray as an axis, until its plane is at right angles to the polarizer, it will be found that the light is no longer reflected. For the reflected ray consists entirely of vibrations perpen-

FIG. 251.

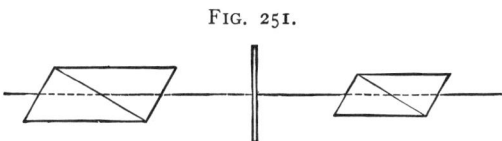

Arrangement of Polarizer, Analyzer, and Object to be Examined.

dicular to the first plane of incidence. But vibrations perpendicular to the first plane of incidence will be in the second plane of incidence, which is at right angles to the first, and therefore they will not be reflected from the second surface."

A series of thin plates (Fig. 250), at the proper angle, polarizes light in a marked degree. These plates will also act in a similar manner when the light is transmitted through them, a part of the light in each of these cases being reflected and a part transmitted, both the reflected and transmitted beams being polarized, but in planes at right angles to each other. A single black glass plate is a good polarizer, but a bundle of glass plates backed with black is perhaps better. The arrangement of the polarizing and analyzing prisms with reference to the object to be examined is shown in Fig. 251.

The beam of polarized light may be apparently depolarized by a body which will produce no color, but will simply render the field bright when the polarizer and analyzer are crossed, as shown by the insertion of a rather thick piece of mica between the polarizer and analyzer.

By placing thinner pieces of mica in the same position, various colors are produced. When the polarized beam encounters the thin mica, it is resolved into two others at right angles to each other, the waves of one being retarded with reference to the other; but as long as these rays vibrate at right angles to each other, they cannot interfere. The analyzer reduces these vibrations to the same plane, and renders visible the effects of interference due to the retardation of the waves of one part of the beam. The thick plate of mica gives no color, because the different colors were superposed and blended together, forming white light.

In a slice of Iceland spar cut at right angles to the axis of the crystal, the ray is not divided as it is when the light passes in any other direction through the crystal, and if the slice be placed in a parallel beam of polarized light, no marked effect is produced; but when the beam is rendered convergent, by a lens interposed between the polarizer and the crystal, beautiful interference phenomena are developed.

When the polarizer and analyzer are crossed, a system of colored rings intersected by a black cross appears.

The arms of the cross are parallel with the planes of the polarizer and analyzer. On these lines no light can pass, but between them the colors of the rings increase in intensity toward the middle of the quadrants inclosed by the arms where the interference is most marked. Turning the polarizer or analyzer causes complementary colors to change places, and brings out a white cross instead of the dark one.

SIMPLE EXPERIMENTS IN POLARIZED LIGHT.

It is ever a source of pleasure to the student of science to be able to explore an unfamiliar realm by means of commonplace and readily accessible things, which, if not already possessed, may be had almost for the asking.

There is scarcely a branch of scientific research more prolific in the development of expensive apparatus than that of light, yet there is nothing in the domain of physics capable of being better illustrated by apparatus of the most simple and inexpensive character. The subject of polarized light, as intricate and difficult as it may at first appear, may be illustrated by apparatus costing less than a dime, in a manner that can but excite the wonder and admiration of one inexperienced in this direction.

A small piece of window glass and a black-covered book constitute the apparatus for beginning the study of this interesting subject, and with a glass bottle stopper, a glass paper weight, or a piece of mica, the effects of polarized light may at once be shown.

The book is placed horizontally near a source of light,

FIG. 252.

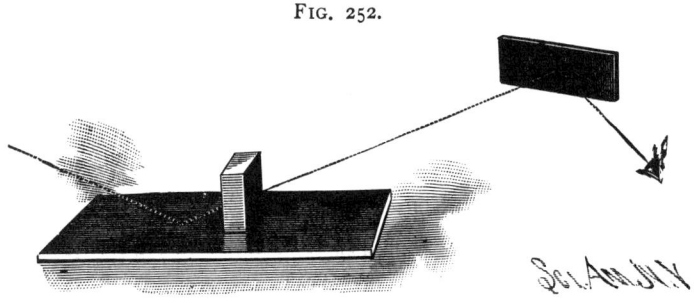

Polarization by Reflection from Blackened Glass.

such as a window or a lamp, so that a broad beam of light will fall obliquely on it, and upon the book is placed the object to be examined, which may be either of those named.

Now, by viewing the reflected image of the object in the piece of window glass, with the glass arranged at the proper angle, it is probable that colors will be seen in the object. If no colors appear, it is due to one of three causes: either the object is incapable of depolarizing the light polarized by reflection from the book cover, or it is too thick or too thin to produce interference phenomena, or the eye of the observer and the glass employed for the analyzer are not in a correct position relative to the object and the polarizer (the book cover).

The glass, if thoroughly annealed, will produce no effect on the polarized beam, but most thick pieces of glass, such as paper weights, ink stands, heavy glass bottle stoppers, and the like, are either unannealed or only partly annealed, and are thus under permanent strain, which is readily indi-

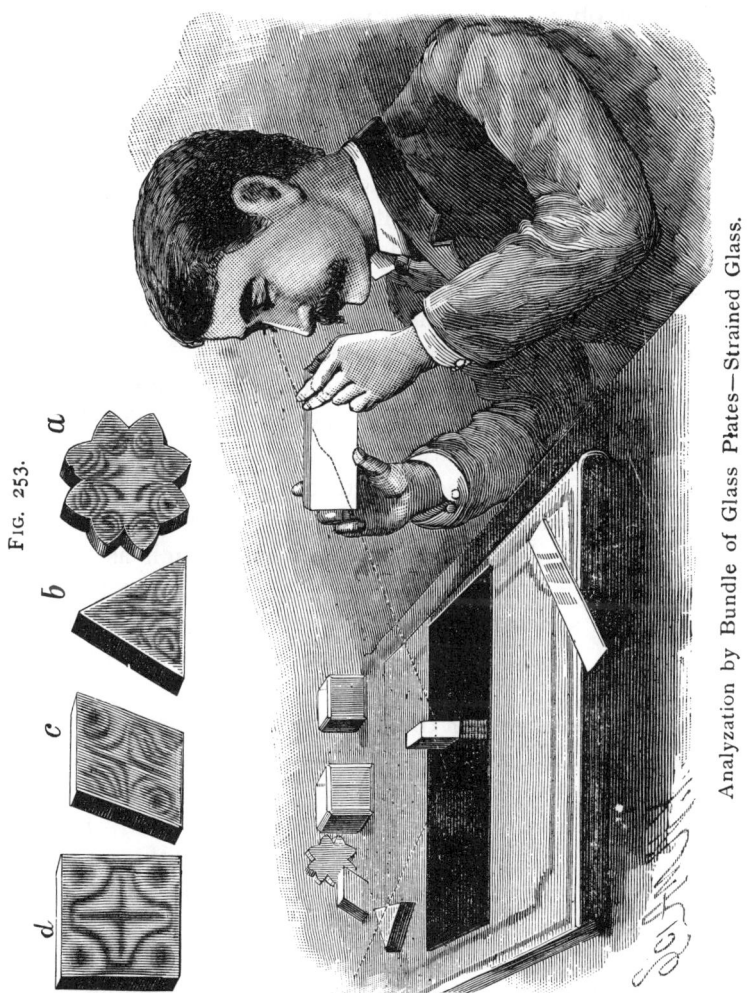

Fig. 253. Analyzation by Bundle of Glass Plates—Strained Glass.

cated by their action on polarized light. A plate of mica of suitable thickness exhibits bright colors when examined by polarized light, particularly when the plate is either bowed or inclined.

To render the polariscope thus described more efficient, a plate of glass may be placed on the book, when the superior reflecting surface will at once make itself manifest in the increased brightness of the colors and improved definition of the object. A still greater improvement may be made by blacking one side of each glass with asphaltum varnish or any other convenient black varnish or paint, using in the experiments the unblackened surfaces, as shown in Fig. 252.

The angle which the incident light beam should make with the polarizer or horizontal blackened plate is 35° 25', and the polarized beam should strike the analyzing plate at the same angle to secure the maximum effects; but it is unnecessary to measure the angles, as they may be easily determined by the appearance of the object.

With the two plates of blackened glass much may be learned with regard to the properties of polarized light. Plates of mica of various thicknesses and forms, inclined at various angles, bowed and turned in their own planes, pieces of quartz, bodies of glass such as those already mentioned, and odd-shaped pieces of unannealed glass, such as may be picked up at glass works, are easily secured objects. Brazilian pebble spectacle lenses often show gorgeous colors when turned at different angles in the beam of polarized light.

The best position for the polarizing plate is near a window, with the broad light of the clear sky shining upon it.

By turning the analyzing plate on the axis of the light beam, some curious effects may be observed. When the plates are at right angles with each other, the polarized beam will be nearly quenched,* and when they are parallel with each other, the reflection of the sky will be quite bright.

The employment of a blackened glass reflector for an analyzer is attended with some difficulty, on account of the necessity of changing the position of the eye for each new

---

* With black glass reflectors employed as polarizer and analyzer, the extinction of the light is not quite complete, even when they are arranged accurately at the polarizing angle.

position of the analyzer. A bundle of six or eight plates of ordinary glass is more convenient, but not quite so efficient.

FIG. 254.

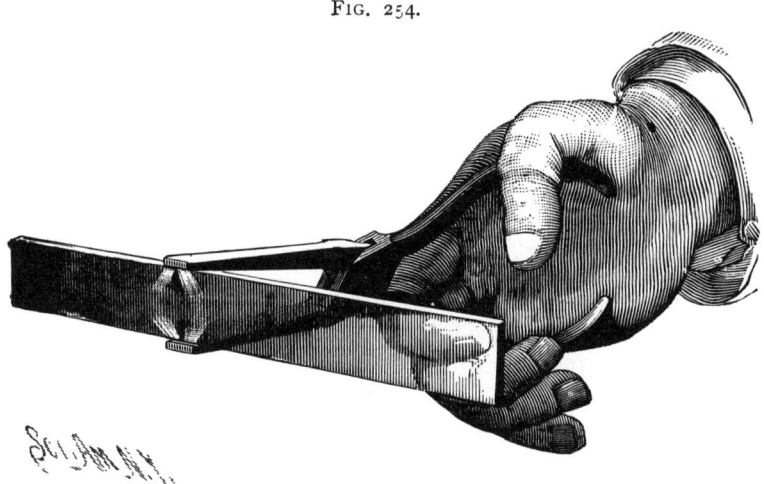

Glass Strained by Pressure.

These plates will be used as shown in Fig. 253, the light passing through them to the eye instead of being reflected.

FIG. 255.

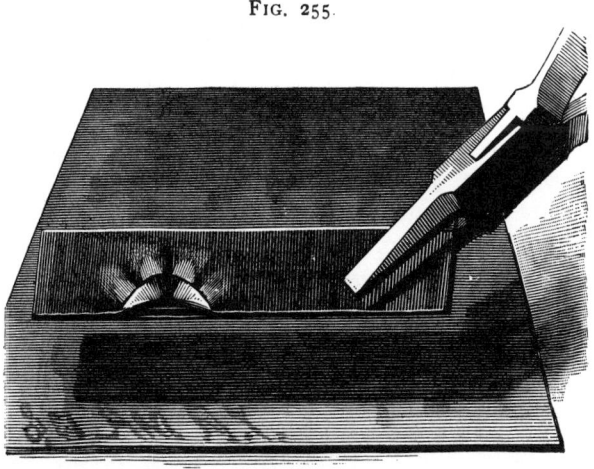

Glass Strained by Heat.

The plates may be turned at any angle without changing the position of the eye.

The most perfect analyzer, however, is the Nicol prism. A very small one will answer perfectly for this class of experiments, and is not expensive. But to return to our experiments; when the analyzer and polarizer are crossed and the field is dark, if a few pieces of mica of various thicknesses and shapes are held between the analyzer and the black glass plate, and bowed and inclined at different angles, a great variety of tints will be observed, and if held in one position while the analyzer is turned, another effect will be noticed.

Among the objects which may be examined in this way are the paper weights, stoppers, and other thick, partly annealed pieces of glass, a piece of glass held edgewise in a hand vise or pair of pliers, and put under compression, as shown in Fig. 254. A piece of glass held edgewise for a moment in a small gas or candle flame, and then placed in the polarized beam, shows the strain by a light figure, like that represented in Fig. 255, or it may assume other forms, according to circumstances. As the glass cools, the figure fades away.

Small glass squares and triangular and diamond-shaped plates, about three-quarter inch across, suspended by a fine wire in the flame of a Bunsen burner or alcohol lamp until their corners begin to fuse, and then cooled in air, become permanently strained, and exhibit symmetrical figures formed of dark and light spaces, but show little color on account of their thinness. By superposing several such plates, color effects may be seen.

The beautiful *verre trempe*, or strained glass blocks, a few examples of which are represented at *a, b, c, d,* in Fig. 253, are similar in character to what has just been described. They vary in thickness from one-fourth inch to one-half inch, and even thicker. They are expensive objects, but exceedingly beautiful and interesting.

In Fig. 256 is shown a method of polarizing and analyzing with a single bundle of plates. It is, in principle, a Norremberg doubler. The light strikes the under surface of the bundle of plates at the polarizing angle, and is reflected downward in a polarized state, passing through the object

POLARIZED LIGHT. 245

which rests upon the horizontal silvered mirror. It is then reflected back through the object, and passes through the bundle of plates to the eye of the observer; the plates, as before stated, serving to analyze the polarized beam.

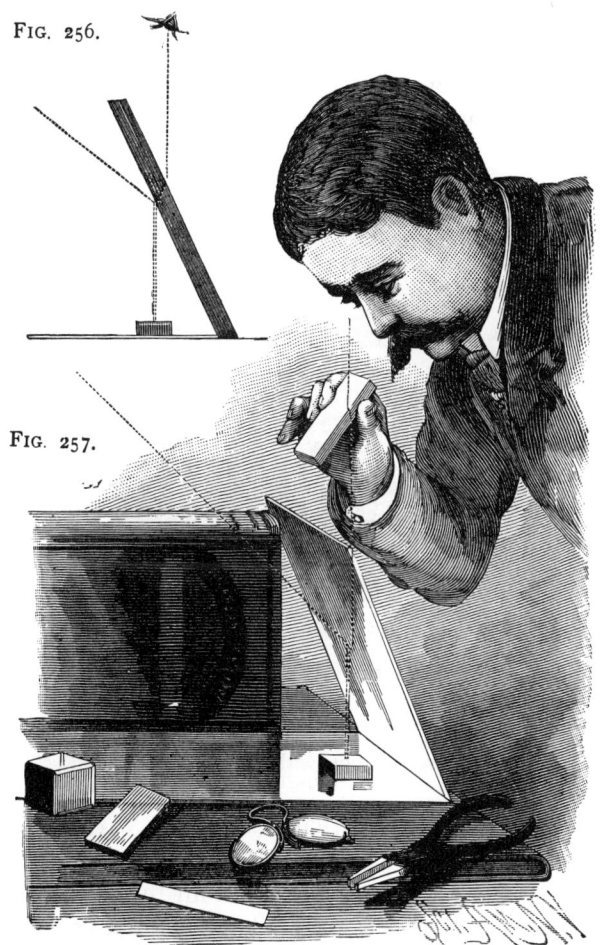

Fig. 256.

Fig. 257.

Simple Form of Norremberg Doubler.

A Norremberg doubler, which answers a good purpose, may be made by leaning a clear plate of glass upon the edge of a book, over a piece of ordinary looking glass, and employing a bundle of glass plates as an analyzer, as shown in

Fig. 257. Here the polarization is effected by the single plate of glass, and the analyzation by the bundle of plates held in the fingers. Equipped with this instrument, the student of polarized light may proceed a long way with his investigations.

In this instrument the objects to be examined are laid upon the horizontal mirror, and the inclined plate is arranged with reference to the light so that it will reflect the broad light of the sky downward. The position of the

FIG. 258.

Double Polarization with Single Glass Plate.

single plate and bundle of plates may be varied to secure the best effects.

In Fig. 258 is shown an arrangement by which the object and the blackened glass both act simultaneously as polarizer and analyzer. By placing a specimen of strained glass edgewise on the blackened glass, as shown in the engraving, the light, striking the strained glass at about the polarizing angle, is reflected from the back surface of the glass and partly polarized. The beam thus polarized is reflected downward obliquely, and at the same time depolarized by the strained body of the glass; it is reflected upward to the eye and analyzed by the blackened glass mirror, thus producing an image which is apparently below the surface of the mirror. The image seen in the strained glass itself is produced by the reverse of what has just been described. The light is polarized and reflected by the black glass mirror, and passes through to the back surface of the strained glass, which reflects it back through the body of the glass; the glass then acts as both object and analyzer.

When the polarizer, analyzer, and object are each movable, different effects will be produced by rotating any of them. As a means of exhibiting complementary colors, nothing can excel the polariscope, since the colors produced in the successive changes resulting from turning the analyzer or polarizer are necessarily complementary to each other.

MICA OBJECTS FOR THE POLARISCOPE.

A few simple objects easily prepared from mica are here shown. The material is of course procurable everywhere, and it requires little more than a glance at the engravings to enable any one to prepare the objects. Doubtless many

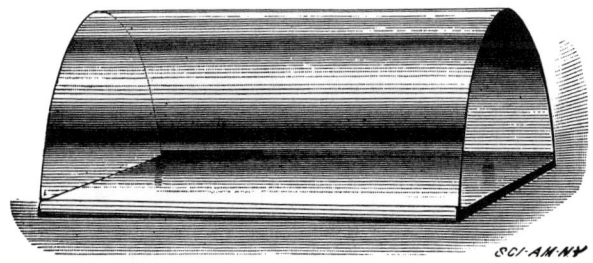

Mica Semi-Cylinder.

other forms than those illustrated will suggest themselves to the student.

The simplest form is shown in Fig. 259. It consists of a thin plate of mica bowed into approximately semi-cylindrical form, and secured by its edges to a plate of glass by means of narrow strips of gummed paper. The size is immaterial; the glass plate may be 1½ inches wide by 3 inches long. This object exhibits fine bands of prismatic color when viewed in the polariscope. Two such semi-cylinders, when crossed, exhibit the intricate figure shown in Fig. 260, with all the splendid colors of the spectrum.

The object shown in Fig. 261 is formed of a disk of mica having a sector cut out and the radial edges overlapped, forming a low cone. The overlapping edges are best fast-

FIG 260.

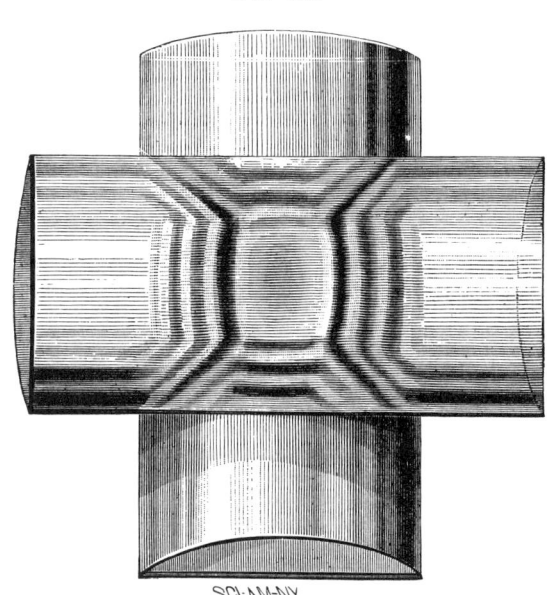

Mica Semi-Cylinders Crossed.

ened together by small tin clips inserted in holes in the mica and bent downward on opposite sides. The clips are not noticeable, and are efficient in holding the edges together. Cement will not answer the purpose, as it adheres to the surface only, and it must be remembered that mica splits almost indefinitely.

The cone thus made has the appearance in the polariscope of a huge circular crystal of salicine. The colors of the cone may be heightened by mounting it on a sheet of mica, as shown in the engraving. The cone is first placed in the polariscope, with the polarizer and analyzer crossed, and turned until it appears brightest, when the lower edge is marked. The mica sheet is then placed in the polariscope,

FIG. 261.

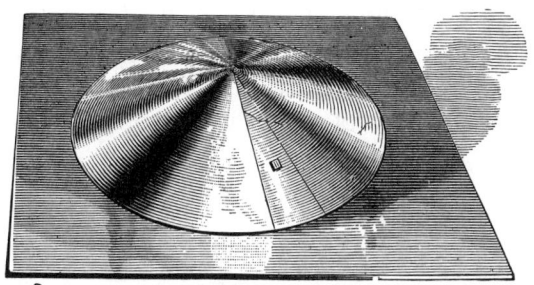

Mica Cone.

and turned and marked in a similar way. The cone is then cemented by its edges to the sheet, the marked edges of both members being arranged in the same direction.

The Maltese cross shown in Fig. 262 is revoluble. The

FIG. 262.

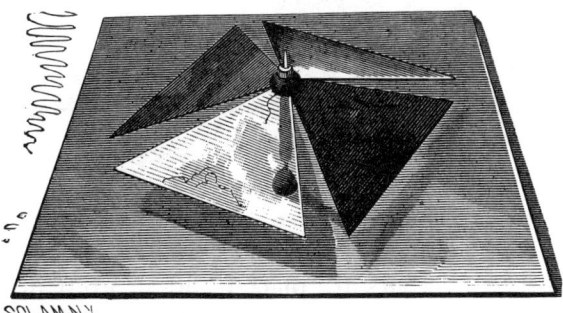

Maltese Cross.

first step toward the preparation of this object is to secure a pin head downward on a square of glass with sealing wax or other cement. A small paper tube which will fit the pin loosely is then made, and a little head of sealing wax is formed around the tube near one end. A piece of mica is selected which exhibits fine colors in the polariscope, and four equilateral triangles are cut from it, either with their corresponding sides cut upon the same base line, or with one side of each cut from one side of a square, or they may be cut and mounted haphazard.

To the apex of the angle designed for attachment to the paper tube a small drop of sealing wax is applied, and with the tube on the pin the first triangle is attached by holding it in the required position by means of a pair of tweezers, and then fusing the wax on the mica and that on the tube

FIG. 263.

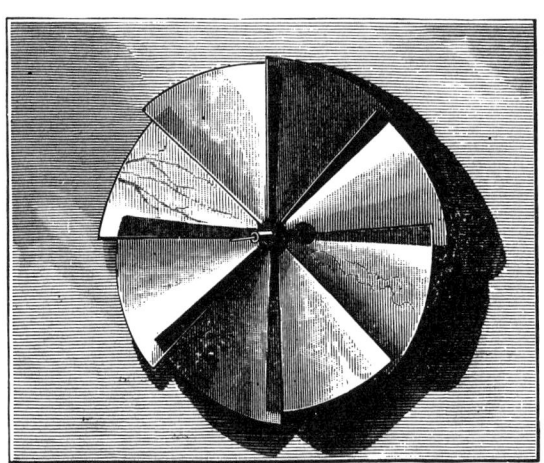

Mica Wheel.

simultaneously by means of a small heated wire, such as a knitting kneedle.

The other members are placed and secured in a similar way, care being taken to arrange the triangles symmetrically, and at a slight angle with the plane of rotation of the object, as shown in the engraving.

The wheel shown in Fig. 263 and the star shown in Fig. 264 are prepared in a similar way. The sections of the wheel are cut from a circular piece of mica, and cemented in place on the paper tube after the fashion of a propeller

POLARIZED LIGHT.   251

wheel or wind wheel. Each ray of the star is made of two scalene triangles of mica oppositely arranged with respect to each other, and inclined in opposite directions, the longer and shorter sides of adjacent triangles being fastened at the periphery of the star by a minute drop of sealing wax.

In Fig. 264, beside the star are shown two somewhat similar objects, formed of strips of mica, pivoted together on a small rivet, one object having the pivot in the center

FIG. 264.

Star, Fan, and Crossed Bars of Mica.

of the strips, the other having it at the end, giving the object an appearance similar to that of a folding fan.

Any of these objects may be viewed by means of the black glass polarizer in connection with either of the forms of analyzer already described or in the simple form of Norremberg doubler. These objects are also very satisfactory when projected on the screen.

POLARISCOPES.

One of the simplest and best instruments for a certain class of investigations in polarized light is the Norremberg doubler, named after its inventor, and shown in a very simple form in Fig. 265.

To one edge of a wooden base, 6 in. square and three-fourths of an inch thick, is secured a vertical standard, 1 in. square and about 15 in. high, and to the top of the standard is attached an arm extending over the center of the base, and apertured to receive the short tube containing the analyzing prism or bundle of glass plates. The tube may be made of paper, hard wood, or metal, and it should be fitted with a shoulder, so that it will turn readily in the aperture of the arm. To the standard below the arm is fitted a stage formed of a thin piece of wood centrally apertured and blackened.

The stage is notched to receive the standard, and is attached to a short vertical bar 1 in. wide. A clip of wood extending across the back of the bar, and two small clips secured to the sides of the short vertical bar, bear with sufficient friction on the standard to hold the stage in any desired position.

About 6 in. above the base a grooved wooden strip is pivoted to the standard, by means of a common wood-screw passing loosely through the grooved strip and tightly through the standard. A wooden knob is turned on the end of the screw, and serves as a nut to bind the grooved strip in any desired position. The strip, screw, and knob are shown in detail at 2, Fig. 265.

Into the groove of the strip is wedged or cemented a plate of glass, 4 by 9 in. A fine piece of ordinary window glass will answer, but plate glass is preferable.

Upon the base is laid a square of ordinary looking glass, or, better, a piece of plate mirror.

The tube, shown in detail partly in section at 3, is provided with an inner tube of pasteboard or wood, divided obliquely at an angle of $35°\ 25'$ with the axis of the tube, and upon the oblique end of one-half of the tube are placed twelve or fifteen well cleaned elliptical microscope cover glasses, which are held in place by the other half of the divided tube. This bundle of glass plates, if of good quality and well cleaned, forms a very good analyzer; but instead of this, if it can be afforded, a small Nicol prism should be secured and mounted in a centrally apertured cork, the latter being inserted in the analyzer tube, as shown at 4.

The object to be examined may be laid either on the stage or on the mirror below. If viewed on the stage, the usual effects will be observed; but if laid on the mirror, it is traversed twice by the light, once by the incident beam and once by the reflected beam. This is particularly noticeable in thin films of mica and selenite, and it serves as an excellent means for selecting eighth and quarter wave plates, which are useful in the study of circular and elliptical polarization.*

It is quite difficult to produce a perfectly uniform thin film of selenite, owing to the brittleness of the material. For this reason mica is generally used, as it possesses considerable flexibility and toughness. The common method of cleaving off thin films of mica is to split off a moderately thin plate and then separate the laminæ at one of the corners by bending it between the thumb and fingers. A medium sized sewing needle secured point outward in a slender handle is probably the best instrument for teasing the laminæ apart; but after the separation begins, the thin end of the ivory handle of an ink eraser seems to serve the purpose exceedingly well.

A score or so of plates are split, and examined one by one in the Norremberg doubler, by laying them on the mirror and turning them in their own planes, while the polarizer and analyzer are crossed. Should the plates exhibit any unevenness under the test, they should be at once rejected. Such as exhibit an even tint should be preserved carefully, and examined further to determine which, if any,

---

* The writer intends to deal sparingly with the theoretical part of this subject, especially the portion relating to circular and elliptical polarization, it having been treated extensively in many physical works and in books especially devoted to light and optics. Daniel's "Physics," prominent among works of its class, "Light," by Lewis Wright, and "Polarization of Light," by William Spottiswoode, are excellent books, bearing directly on the subject. The writer knows of no better means of securing a good knowledge of polarized light than by reading these three books.

possess the required qualities. Not every piece of mica will split evenly, therefore it may be necessary to make several trials before success is attained.

Should the film, when placed on the stage, exhibit a dull

FIG. 265.

Simple Norremberg Doubler.

plum color, slightly inclined toward red, when the polarizer and analyzer are parallel, it produces a difference of phase of half a wave length, and is called a half wave film. As a matter of course, if two films of like thickness, superposed and arranged with their axes in the same direction, produce the same color under the same circumstances, they are one-fourth wave films; and if a pair of films exhibit the same color when similarly arranged on the mirror of the doubler, they may be regarded as eighth wave films, as the polarized beam passes twice through the film to produce the same tint. These films should be carefully mounted between glass plates, either dry or in benzole balsam, the latter being preferable.

The practical application of the eighth and quarter wave films will be treated further on. Beautiful and instructive designs made from thin films are described and illustrated in Wright's "Light," to which reference has been made.

The only simple device for exhibiting the rings and brushes of wide-angled crystals is the tourmaline tongs (Fig. 274), of the kind commonly employed by opticians for testing spectacle lenses; but the dark color of ordinary tourmaline renders a polariscope of this kind objectionable.

A system of lenses devised by Norremberg, and improved by Hoffman, is at present employed for observing the phenomena of wide-angled crystals; but it is a matter of some difficulty to secure exactly such lenses as are required for the apparatus as constructed by Hoffman. Very good results, however, may be obtained by the employment of lenses designed for other purposes. Reference is made to the hemispherical condensing lenses used by microscopists, and ordinary meniscus (periscopic) spectacle lenses. Six lenses in all are required. The converging and collecting systems are exactly alike, but they are oppositely arranged with respect to each other. In the present case the two systems are adapted to a Norremberg doubler, Fig. 266, substantially like that described in a former part of this article, the main difference being that the instrument now illustrated is made principally of metal.

The tube of the upper system of lenses is prolonged upward beyond the upper lens, Fig. 267, to receive a Nicol prism, E, or other analyzer, which is mounted in a short inner tube arranged to revolve in the outer tube.

Fig 266.

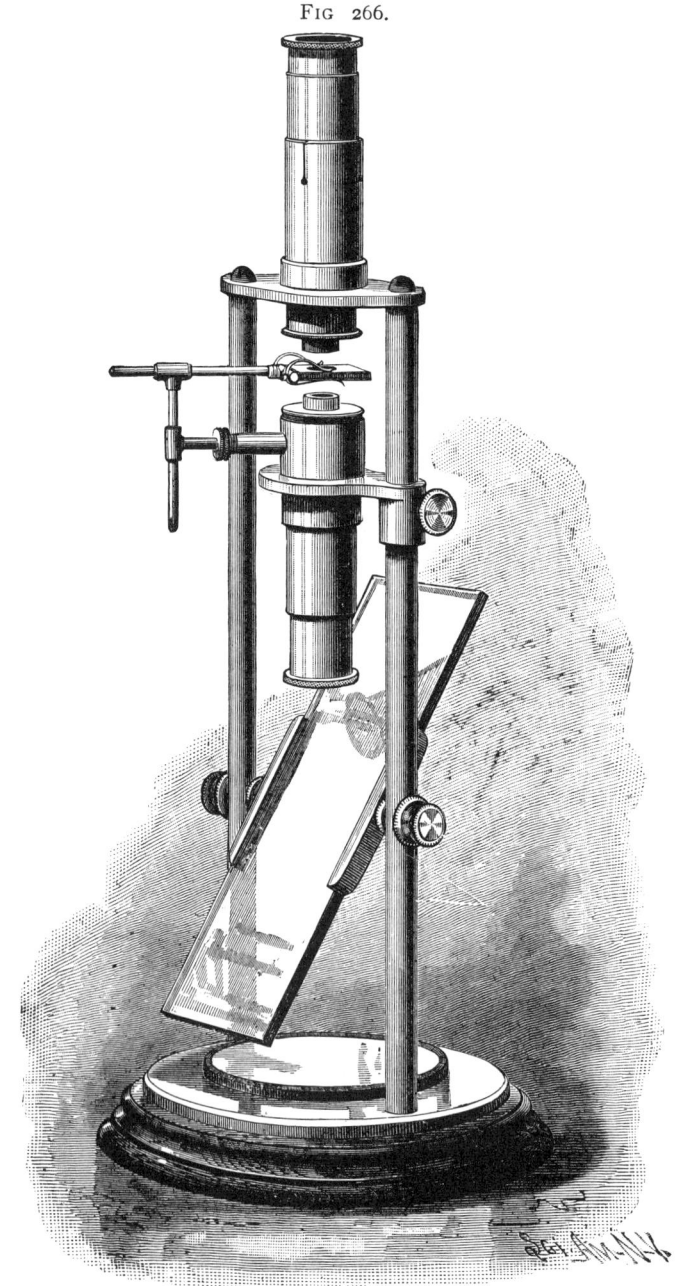

Polariscope for exhibiting Wide-angled Crystals.

POLARIZED LIGHT. 257

The lower system of lenses is contained by a tube fitted to the stage of the doubler. The arrangement of the lenses and analyzer is shown in Fig. 267. The two systems of lenses being alike, a description of one will answer for both. The object, A, to be observed is held between the adjacent ends of the two tubes in the universal holder shown in Fig. 266.

The lens, B, next the object is nearly a hemisphere, about eleven-sixteenths inch in diameter and three-eighths inch focus. The second lens, C, a meniscus (periscopic) spectacle lens of 3 inch focus, is arranged with the concave face one-sixteenth inch from the convex side of the hemisphere. Beyond the 3 inch meniscus, $3\frac{1}{2}$ inches distant, is placed a biconvex spectacle lens, D, of 4 inch focus. The inner surfaces of the tubes are made dead black by the application of a varnish formed of lampblack and alcohol, in which only a trace of shellac has been dissolved.

The tubes may have any suitable diameter, and the proportions of the doubler may be about the same as indicated by Fig. 266, which is one-quarter actual size. The tubes and lenses shown in Fig. 267 are one-half size. The exact proportions, except as to the focal lengths and distances apart of the lenses, are immaterial. The lower system of lenses must produce a very convergent beam of light, while the upper system is

FIG. 267.

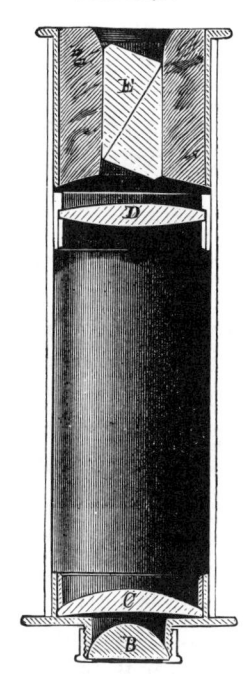
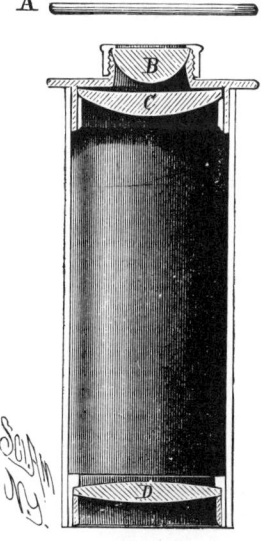

Longitudinal Section of Tubes of Polariscope.

arranged to collect the rays after they pass through the crystal, and bring them within the range of vision.

The angle between the optic axes in some crystals is so small as to permit of seeing them readily. Niter and carbonate of lead are examples of such crystals; but there are other crystals whose angle is so great as to render it exceedingly difficult to exhibit them, and in some crystals the angle is so wide as to render it impossible to see both axes at once. The only method of exhibiting them is by tilting the crystal first in one direction and then in the other, and viewing them separately.

Figs. 268 to 273, inclusive, represent the figures shown by several crystals in the instrument illustrated. The drawings, having been made directly from the objects by the aid of the instrument, are correct in form and proportion, but the beautiful coloring is necessarily absent.

Fig. 268 shows the rings and brushes exhibited by calcite in a convergent beam of polarized light, with the polarizer and analyzer crossed. With the polarizer and analyzer parallel, the dark cross is replaced by a white one.

Niter is shown in Fig. 269 as it appears when the analyzer is crossed. With the analyzer parallel with the polarizing plate, the dark brushes are replaced by light ones. Turning the crystal in its own plane produces different effects.

In Fig. 270 is shown a figure produced by a slice of quartz cut at right angles to the axis of the crystal, and examined in the instrument with the analyzer arranged at an angle of 45° with the polarizer. Crystals of quartz vary in their effects on the polarized beam, some requiring the turning of the analyzer to the right and others to the left to produce like results. For this reason the plates are called right or left handed, according to the direction in which the analyzer is required to be turned.

By superposing a right hand quartz on a left hand quartz, the beautiful spirals discovered by Airy, and named after their discoverer, may be exhibited. These spirals are shown in Fig. 271.

In Fig. 272 is shown the figure produced by the inter-

POLARIZED LIGHT. 259

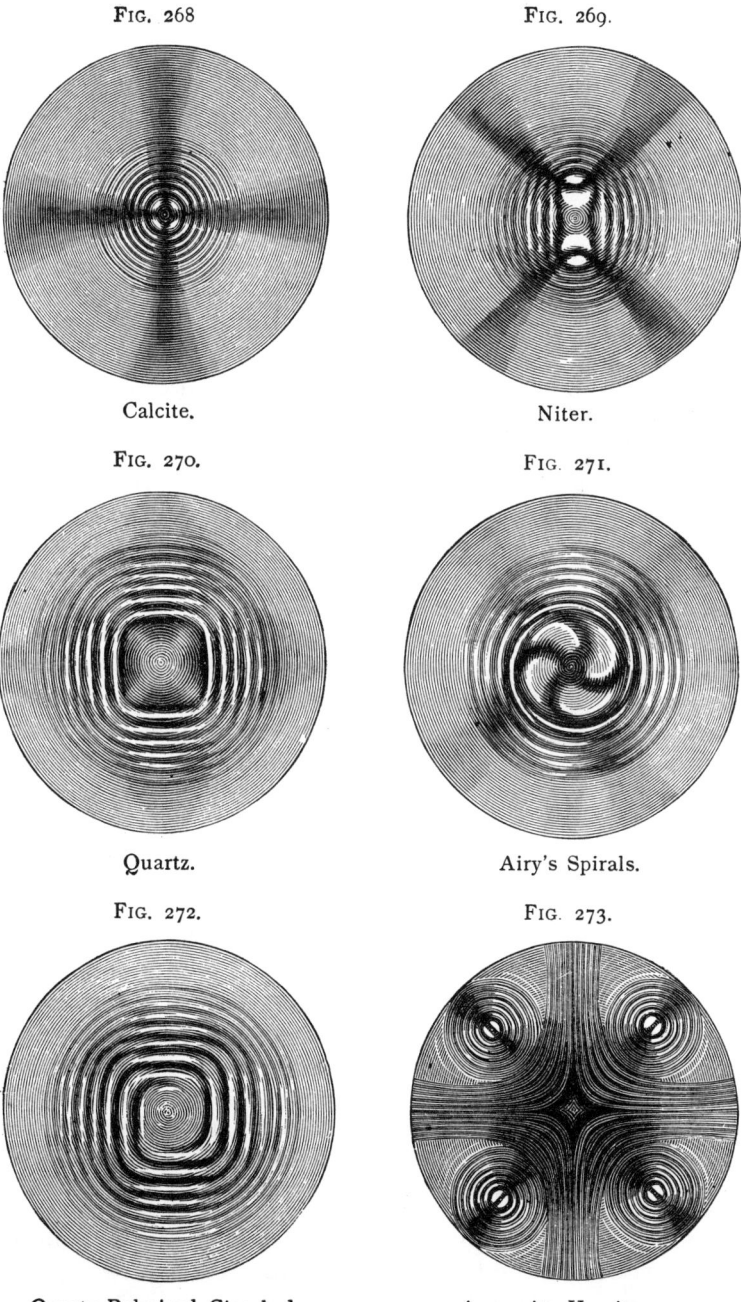

FIG. 268 — Calcite.

FIG. 269. — Niter.

FIG. 270. — Quartz.

FIG. 271. — Airy's Spirals.

FIG. 272. — Quartz Polarized Circularly.

FIG. 273. — Aragonite Hemitrope.

position of a quarter wave mica film between the polarizer and a plate of quartz viewed in the instrument. This altered appearance is due to circular polarization, a phenomenon treated extensively in the literature of the subject, but requiring an explanation too elaborate for the space at command.

Calcite polarized circularly shows singularly broken up and disjointed rings, the brush-like cross being absent, and when analyzed circularly, or viewed through a quarter wave plate, as well as through the analyzer, the rings appear perfect, and there are no transverse markings.

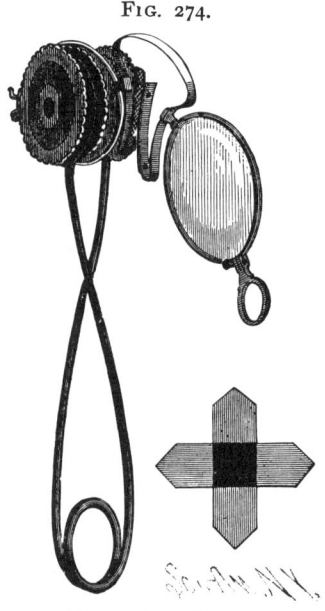

FIG. 274.

Tourmaline Tongs.

Fig. 273 shows the intricate figure produced by aragonite hemitrope, or a pair of crystals arranged at right angles with each other. Somewhat similar figures are produced by crossed plates of mica.

The following is a list of some additional objects which may be viewed in the instrument:

Sulphate of nickel, sugar, aragonite, bichromate of potash, chrysoberyl, chrysolite, topaz, anhydrite. Instead of employing the Norremberg doubler for polarization, the lower tube may be prolonged, and a large Nicol prism inserted and arranged like the analyzer.

In Fig. 274 is shown the tourmaline tongs, the simplest polariscope known. It consists of two plates of tourmaline, cut parallel to the optic axis of the crystal, and mounted in cells arranged to turn in eyes formed at the extremities of the looped wire. When the plates are parallel, light passes through them; but when they are arranged at right angles with each other, the light is completely extinguished. If a plate of quartz crystal, a Brazilian pebble spectacle lens for

POLARIZED LIGHT.

Fig. 275.

Polariscope for Large Objects.

example, be placed between the tourmalines arranged in this way, the light will again pass, showing that it has been depolarized by the rock crystal.

This has been accepted as an infallible test of the genuineness of quartz lenses. In the hands of an expert it is undoubtedly valuable, but glass lenses may be put under strain by heating them and allowing them to cool rather quickly. They will then, to some degree, act on the polarized beam like the true crystal.

This form of polariscope is useful in the examination of crystals generally, but on account of the natural dark color of the tourmaline, the utility of the instrument is limited.

In Fig. 275 is shown a polariscope designed for the examination of large objects, such as glassware, etc. It consists of a bundle of 16 glass plates, about 20 or 24 inches square, arranged with reference to the Nicol prism employed as an analyzer at an angle of 35° 25'. Behind the series of plates is hinged a board covered with black velvet, which may be raised up parallel with the glass plates when it is desired to polarize the beam by reflection.

The analyzer, a Nicol prism, is mounted in a revoluble tube, supported by the small adjustable standard. Articles to be examined are placed on the small table between the polarizer and analyzer.

The light for the polariscope should be taken through either a white paper or cloth screen or a plate of ground glass. Any strain in the article examined will exhibit itself by its depolarizing effect on the polarized beam.

SIMPLE POLARISCOPE FOR MICROSCOPIC OBJECTS.

The examination of microscopic crystals by the aid of the polariscope is an exceedingly interesting part of the study of polarized light. The indescribable play of colors, and the variety of exquisite forms of the smaller crystals, render this branch of the subject very fascinating. But to undertake the examination of this class of objects in the usual way requires a microscope with the addition of a polariscope, which calls for an outlay of at least fifty dollars, besides the cost of the objects, and while it is believed that

POLARIZED LIGHT. 263

such an outlay would be indirectly, if not directly, profitable, it is not necessary to expend a fiftieth of that amount to arrive at very satisfactory results.

The cost of the compact and efficient little instrument shown in Fig. 276 is as follows:

One pocket magnifier, having two lenses $1\frac{1}{2}$ inches and 2

FIG. 276.

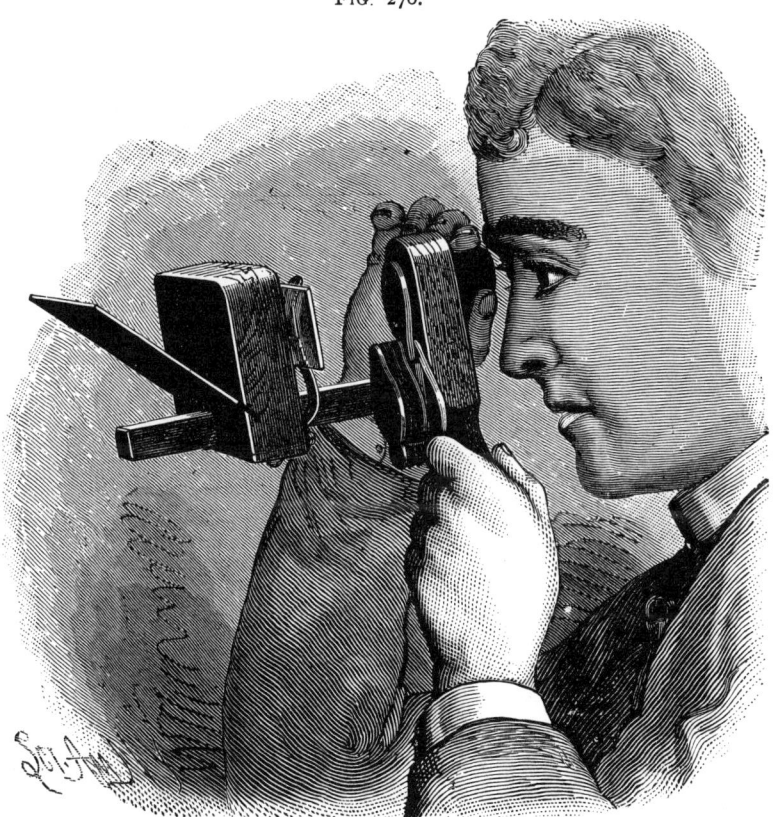

Polariscope for Microscopic Objects.

inches focus respectively, giving when combined a $\frac{3}{4}$ inch focus, 50 cents; eighteen elliptical microscope cover glasses for analyzer, 38 cents. The cost of wood for the principal parts, the pasteboard tubes, the glass for the polarizer, and the metal strips for the slide-holding springs, can hardly be counted, and the labor must be charged to the account of recreation; so that less than one dollar pays for an instrument that will enable its owner to examine almost the entire range of microscopic polariscope objects with a degree of satisfaction little less than that afforded by the use of the best instruments.

The form, proportions, and material of the body of the instrument are entirely matters of individual taste. In the

FIG. 277.

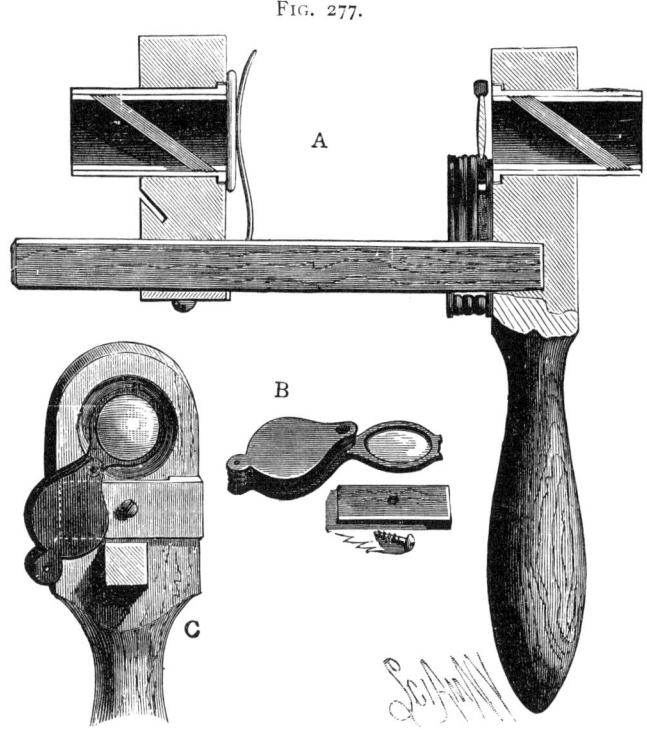

Longitudinal Section of Polariscope and Details. Half Size.
A, Longitudinal Section. B, Magnifier and Clamp. C, Cross Section showing Clamp and Magnifier.

present case, the hand piece and sliding stage are made of ⅛ in. mahogany, the handle being formed on the hand piece by turning. The stage is 2½ in. square, and has in its lower edge a half inch square, transverse groove, which receives the square rod projecting from the hand piece at right angles. The rod is held in the groove by a wooden strip fastened to the lower edge of the stage by two wood

POLARIZED LIGHT. 265

screws, so that it bears with a light friction on the under side of the rod.

The hand piece and stage are both pierced above the rod with holes which are axially in line with each other. The diameter of the holes is governed by the size of the cover glasses. Those in the instrument shown are of the exact size and form of the annexed diagram (Fig. 278).

These cover glasses are procurable from any dealer in supplies for microscopists. Eighteen of them, at least, are required. The paper tube inclosing these glasses is a little more than $\frac{11}{16}$ in. internal diameter; its outside diameter is $\frac{7}{8}$ in. and its length is $1\frac{3}{4}$ in. A narrow paper collar is glued around one end of the tube, and both the hand piece and the stage are counterbored to receive the collar, as shown in the sectional view, A, Fig. 277. To the tube thus described is fitted an internal paper tube, which is about $\frac{1}{8}$ in. shorter than the outer tube. The inner tube is divided diagonally at an angle of 35° 25', which is the complement of the polarizing angle for glass (54° 35'). The oblique surfaces thus formed, when placed in the tube in opposition to each other, support them between the glass plates at the polarizing angle. The simplest way to arrange the angles of the tubes and other parts of the polariscope is by the employment of a triangle of cardboard like that illustrated in Fig. 279. In fact, a copy of the triangle here shown may be used.

FIG. 278.

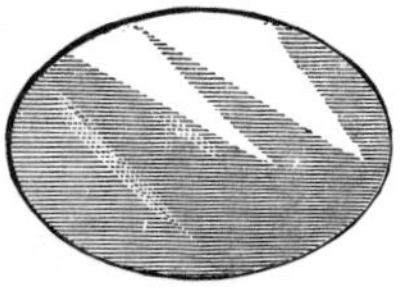

Elliptical Cover Glass

It is sometimes a matter of considerable difficulty to clean the thin cover glasses without the risk of breaking a large percentage of them. An effective device for holding the glasses while they are being cleaned is shown in Fig. 280. It consists of a piece of thin Bristol-board, having an elliptical aperture loosely fitting the edges of the glass to be cleaned, and a plain card glued to the back of the apertured card, and forming the bottom of the shallow recess into which the glasses are dropped for cleaning. The holder may be pressed down upon the table by the fingers of one hand, while the glass is rubbed with a soft linen handkerchief, after being breathed on. Glasses that cannot be easily and thoroughly cleaned in this way are worthless for this purpose.

Before the glass plates are put together, they are dusted with a camel's hair brush to remove any adhering lint and dust. The paper tubes are made dead black inside and outside.

The front of the stage is provided with a pair of thin brass springs, which serve to clamp the object slide with a light pressure to the stage. In the back of the stage, below the central aperture, is formed a groove for receiving the

FIG. 279.

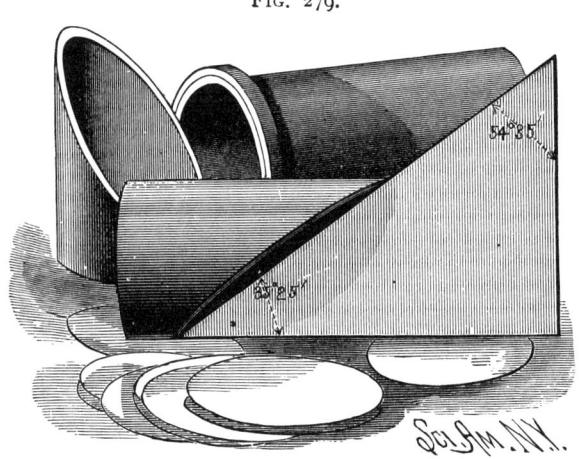

Triangle and Paper Tube. Full Size.

black glass polarizing plate. The groove supports the black glass at an angle of 54° 35′ with the plane of the stage, or at an angle of 35° 25′ with the holes in the stage and hand piece. The polarizing plate may consist of a plate of polished black glass, but it is generally more convenient to employ an ordinary piece of glass blackened on one side. A thin pine wedge cemented to the back of the plate causes it to bind in the groove of the stage.

To the inner face of the hand piece is clamped an ordinary pocket magnifier by means of the wooden clip. At C is shown the arrangement of the magnifier relative to the analyzer. Any convex lens of suitable focus may be pressed into the service. The face of the stage and other parts of the instrument visible through the analyzer are blackened.

The object to be viewed is placed on the stage and focused, when the instrument is held so that the black glass polarizing plate reflects the light through the object and through the analyzer. The analyzer is then turned, and the object observed. To heighten the color effects, a plate of selenite or mica may be placed immediately behind the

FIG. 280.

Holder for Glass.

object, or between the stage and black glass plate. Mica plates of suitable thickness are selected by trial in the instrument, and preserved for future use.

It is sometimes desirable to rotate the polarizer. When the black glass plate is used, this is impracticable, but on removing this plate, and inserting in the stage a polarizer consisting of a tube containing plates like the analyzer, the effects of rotating the polarizer may be observed. To render the rotation of the paper tubes smooth and uniform, their bearings in the hand piece and stage are rubbed over with the point of a soft lead pencil, imparting to them a thin coating of plumbago, which diminishes friction and prevents sticking. The objects which may be examined by the aid of this instrument are very numerous. Many of them are easily prepared, and some need no preparation at all. The chemical salts mentioned below may be prepared for observation by allowing their solutions to evaporate on a slip of glass: Alum, bichromate of potash, bichloride of mercury, boracic acid, carbonate of potash, carbonate of soda, citric acid, chlorate of potash, hyposulphite of soda, iodide of potassium, nitrate of ammonia, nitrate of copper, nitrate of soda, oxalic acid, prussiate of potash (red), prussiate of potash (yellow), sugar, sulphate of copper, sulphate of iron, sulphate of nickel, sulphate of potash, sulphate of soda, sulphate of zinc, tartaric acid.

Slips of glass, 1×3 inches, are convenient for this purpose. A circle about ¾ inch diameter is formed on each slip with a piece of paraffin or wax, and while the slips are supported in a level position, a few drops of a rather strong solution are placed in each circle, and the slips are allowed to remain quietly until the crystals form.

For methods of covering and preserving these crystals, as well as for hints on the preparation of the more difficult crystals, the reader is referred to the chapter on microscopy.

The following vegetable and animal substances may be examined by polarized light:

Cuticles, hairs, scales from leaves, fibers of cotton and flax, starch grains, thin longitudinal sections of wood, oiled; spicules of sponges and gorgonia, cuttlefish bone, hairs, quills, horn, finger nail, and skin. These objects should be thin and translucent or transparent. It is necessary in some cases to increase their transparency by soaking them in oil or some other suitable liquid. Many rock sections and sections of minerals may be studied advantageously by the aid of polarized light, but since the object sare quite difficult to prepare, no list of them is given.

PRACTICAL APPLICATIONS OF THE POLARISCOPE.

The practical applications of the polariscope are few but important. In chemistry, its most prominent use is in the determination of sugars. In medicine, it finds an application in the examination of diabetic urine. In geology and mineralogy, it is of utility in determining the origin and nature of rocks and minerals. In photometry, it forms the basis of several photometers. In photography, the polariscope, or at least a part of it—the Nicol prism—has been employed for reducing the glare of highly illuminated objects. In a similar way, the Nicol prism has been used for extending the field of vision in a fog. It forms an important part of the water telescope. It has also been used to some advantage in viewing paintings unfavorably situated in galleries. In the trades the polariscope has proved useful in detecting strains in glass. By opticians, it has for years been recognized as a test for the genuineness of Brazilian pebble lenses for spectacles. It has also proved of great utility to the microscopist in the examination of minute structures.

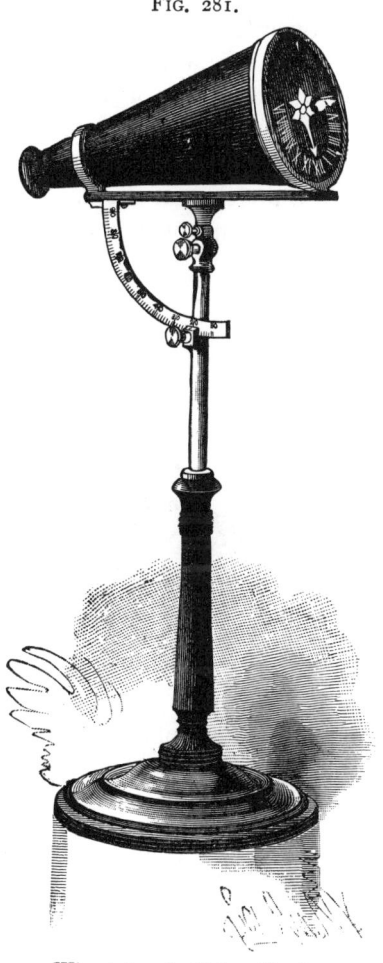

Fig. 281.

Wheatstone's Polar Clock.

The polariscope has recently been applied in France to determining the temperature of incandescent iron and other metals. The color of a glowing mass of metal varies according to its temperature, and a ray of the light when polarized is rotated by a plate of quartz to a degree dependent upon the color. The degree of rotation is measured by the polariscope, and an empirical scale of temperature is thus obtained, which has been found very useful and reliable in metallurgical operations.

One of the most curious uses of polarized light is the indication of the time of day. Sir Charles Wheatstone devised a polar clock in which a Nicol prism in connection with atmospheric polarization is made to indicate the time of day. Several forms of this instrument have been made; one of them is shown in Figs. 281 and 282.* Atmospheric polarization, according to Professor Tyndall, is due to the reflection of light from the fine particles of matter floating in the air. By examining the sky on a clear day by means of a Nicol prism and a plate of selenite or

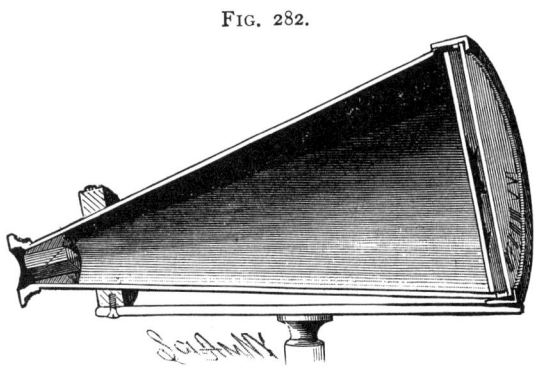

Fig. 282.

Longitudinal Section of Polar Clock.

other crystal, polarization will be detected without difficulty. The brightest effects are noticed at a point 90° from the sun. By directing a Nicol prism to the north pole of the heavens —a position always at right angles to the sun, or approximately so—and turning it round, the colors of the crystal plate, viewed through the prism, will change in a definite order, or, if the position of the Nicol be fixed, the movement of the sun will produce similar changes of color. The polar clock is based upon this principle.

The inventor describes this instrument as follows: "At the extremity of a vertical pillar is fixed, within a brass ring, a glass disk, so inclined that its plane is perpendicular to the

* Other forms are described in Spottiswoode's "Polarization of Light."

polar axis of the earth. On the lower half of this disk is a graduated semicircle, divided into twelve parts (each of which is again subdivided into five or ten parts), and against the divisions the hours of the day are marked, commencing and terminating with VI. Within the fixed brass ring containing the glass dial plate, the broad end of the conical tube is so fitted that it freely moves round its own axis; this broad end is closed by another glass disk, in the center of which is a small star or other figure, formed of thin films of selenite, exhibiting, when examined with polarized light, strongly contrasted colors; and a hand is painted in such a position as to be a prolongation of one of the principal sections of the crystalline films. At the smaller end of the conical tube a Nicol prism is fixed so that either of its diagonals shall be 45° from the principal section of the selenite films.

The instrument being so fixed that the axis of the conical tube shall coincide with the polar axis of the earth, and the eye of the observer being placed to the Nicol prism, it will be remarked that the selenite star will in general be richly colored; but as the tube is turned on its axis the colors will vary in intensity, and in two positions will entirely disappear. In one of these positions, a smaller circular disk in the center of the star will be a certain color (red for instance), while in the other position it will exhibit the complementary color.

This effect is obtained by placing the principal section of the small central disk $22\frac{1}{2}°$ from that of the other films of selenite which form the star. The rule to ascertain the time by this instrument is as follows: The tube must be turned round by the hand of the observer until the colored star entirely disappears, while the disk in the center remains red; the hand will then point accurately to the hour.

"The accuracy with which the solar time may be indicated by this means will depend on the exactness with which the plane of polarization can be determined. One degree of change in the plane corresponds with four minutes of solar time."

SUGGESTIONS IN DECORATIVE ART.

Occasionally, evidences of the use of the microscope in decorative art are seen, and every microscopist knows that

Salicine Crystals.

there are thousands of beautiful forms lost to unaided human vision which are revealed only to the user of the

Sulphate of Cadmium.

microscope.* These minute forms are always exquisite in their construction and finish, often symmetrical and graceful

Fig. 285.

Santonine.

in form, and quite as often finely colored. All this is true of microscopic objects in general, but it is especially true of

Fig. 286.

Lithic Acid.

* See also chapter on microscopy.

polariscopic microscope objects. Some of these are, to a certain extent, artificial. The crystals, for example, are the result of manipulation, but the laws of crystallization are natural, so that, after all, we are indebted to nature even for these objects.

In the present instance, a few striking examples of crystallization have been selected as the basis of some suggestions in decorative art. These crystals, as exhibited by polarized light in the microscope, are shown in the annexed engravings, necessarily divested of their principal charm—

FIG. 287.

Border Dado or Frieze.

that of color. The forms only are shown. The reader can imagine these figures invested with most gorgeous colors combined in a perfectly harmonious way. In respect to color, the polariscope never errs. Whatever colors are presented are correctly related to each other. This feature alone is of great value to the designer and colorist. The circular crystals of salicine, shown in Fig. 283, are always interesting. The play of the radial bands of color as the polarizer or analyzer is revolved gives each disk the appearance of having an actual rotation of its own.

POLARIZED LIGHT. 275

In Fig. 284 are shown the delicate, feathery crystals of sulphate of cadmium, in which the coloring, as exhibited by polarized light, is scarcely more beautiful than the exquisite forms. The shapes of the different crystals vary somewhat, but there is a characteristic feature pervading them all.

In Fig. 285 are shown crystals of santonine in a variety

FIG. 288.

Panel with Ornamentation of Crystals.

of forms—some like spears of grass, others resembling heads of grain, and still others like ferns and various leaves, while the larger crystals or aggregation of crystals has a radial arrangement.

In Fig. 286 are shown crystals of lithic acid, which adjoin each other, and form a solid field, having strongly contrasting bands of light and dark color.

276   EXPERIMENTAL SCIENCE.

Fig. 287 will be recognized as a part of a dado, frieze, or border, formed of lithic acid as a ground, crystals of platino-cyanide of barium as the division of the panels, and crystals of sulphate of cadmium as rosettes upon the centers of the panels.

Fig. 288 shows a panel formed in part of the same crys-

FIG. 289.

A Composite Border.

tals, with a crystal of salicine planted at the intersection of two of the slender platino-cyanide of barium crystals, and small crystals of kinate of quinia forming flowers.

In Fig. 289 is shown a border formed of crystals of santonine, arranged on a ground of neutral tint, with a row of circular crystals of sulphate of copper and magnesia above a row of crystals of kinate of quinia, arranged on a dark ground.

Fig. 290 shows a pattern having a background of stearic

Pattern with Background of Stearic Acid and Crystal Leaves, Stalks, and Flowers.

acid, branches of platino-cyanide of barium, leaves of platino-cyanide of magnesium, and flowers of salicine.

What has been shown in the engravings constitutes only a hint of what may be done in this direction. The number of beautiful crystals and other polariscope objects available for this purpose is very large.

CHAPTER XIII.

MICROSCOPY.

The world of the minute existing beyond the range of the unaided vision is little realized by those who take no interest in microscopy. The beauty and perfection of the smaller works of nature can never be fully known through the medium of literature or art; the objects themselves must be observed by the student personally.

In every pond and stream may be found microscopic forms of life. In every plant and flower, upon leaves and stalks, among the sands and rocks, almost everywhere in all seasons, may be found objects of absorbing interest to the student of microscopy. Animals and insects, food and manufactured articles, yield objects which may be examined microscopically with pleasure and profit. Chemistry and mineralogy afford attractive fields, and the physicist finds the microscope a necessity in his investigations. In fact, one so inclined cannot fail of finding interesting and instructive objects with little difficulty.

Microscopical investigations may be carried on by the aid of an ordinary inexpensive microscope, but this, in the natural course of things, will give place to a more perfect instrument and a complete list of accessories, provided the student becomes interested in the subject. A fine instrument is desirable on account of its wider range of usefulness, its superior optical powers, and the facility with which it may be adapted to different classes of objects. It has the further important advantage of being less fatiguing to the eyes.

The simplest and cheapest of all microscopes is represented in Fig. 291. It consists of a thin piece of glass, having attached to it one or two short paper tubes, which are coated with black sealing wax, and cemented to the glass with the same material.

By aid of the small stick water is placed, drop by drop, in the cells until the lenses acquire the desired convexity. Objects held below the glass will be more or less magnified, according to the diameter and convexity of the drop.

A convenient stand for the water lens is shown in Fig. 292. The detail views are vertical sections of the lenses, showing the screw for adjusting the convexity of the drop.

The stand is made of wood. The sleeve that supports the stage slides freely upon the vertical standard. A wire having a milled head passes through the upper end of the

FIG. 291.

Simple Water Lens Microscope.

standard, and has wound upon it a strong silk thread, one end of which is tied to a pin projecting from the stage-supporting sleeve. An elastic rubber band is attached to the lower end of the sleeve, and to a pin projecting from the standard near the base, to draw the table downward. The stage is raised or lowered by turning the milled head.

Two standards project from the bed piece for receiving the corners of a rectangular piece of silvered glass which forms the reflector.

The water cell consists of a brass tube about $\frac{3}{8}$ inch long and $\frac{1}{8}$ to $\frac{3}{16}$ inch internal diameter, having in one side a screw for displacing the water to render the lens more or less convex. A thin piece of glass is cemented to the lower end of the tube, and the inside of the tube is blackened.

Several bushings may be fitted to the upper end of the tube to reduce the diameter of the drop, and thus increase the magnifying power of the lens.

Water containing animalcules or a solution of a salt for crystallization may be placed on the under surface of the

FIG. 292.

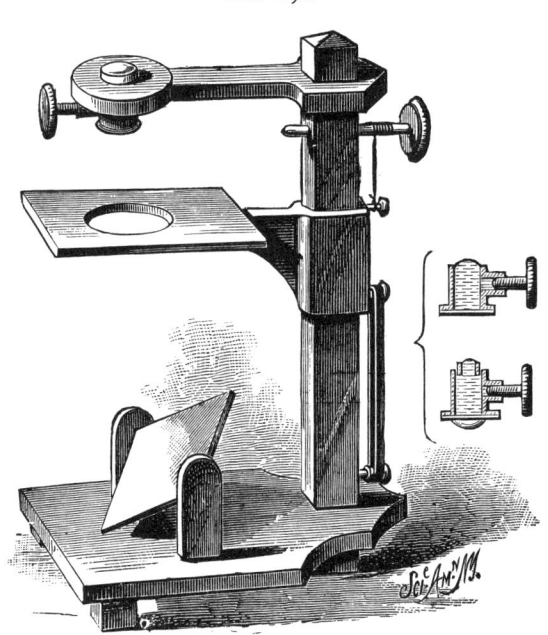

Water Lens Microscope Complete.

glass, when the lens may be focused by turning the adjusting screw. The lens may be adjusted to magnify objects placed on the movable stage by rendering it less convex, thus increasing its focal length.

Air bubbles forming on the upper surface of the glass may be readily displaced by means of a cambric needle.

The water lens microscope or any lens or combination of lenses through which an erect virtual image is seen, magnified, is known as a simple microscope, while a compound microscope is an instrument in which a lens, or system of lenses, known as an objective, forms a real and greatly enlarged image of the object, and in which this image is itself magnified by a second lens or system of lenses, known as the eyepiece or ocular.

Fig. 293.

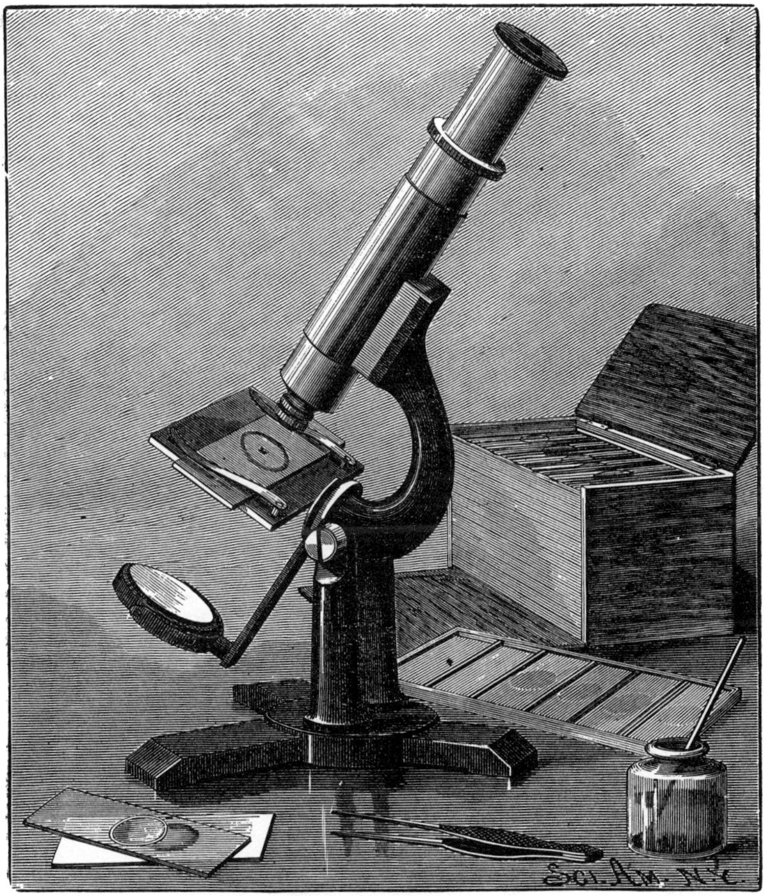

Compound Microscope.

An inexpensive compound microscope is shown in Fig. 293. This instrument, when closed, is 8 inches high, and has a draw tube which permits of extending it to a height of 11 inches. The foot and arm are of japanned iron. The tubes are well finished and lacquered. It has an achromatic objective divisible into two powers. The mirror may be swung over the stage for the illumination of opaque objects.

FIG. 294.

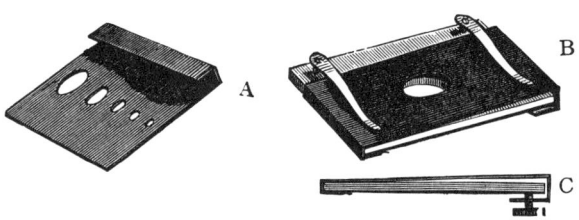

Diaphragm and Fine Adjustment.

To the instrument as received from the manufacturer is applied a home-made diaphragm, as shown at A, in Fig. 294, and a fine adjustment, as shown at B C, in the same fig-

FIG 295.

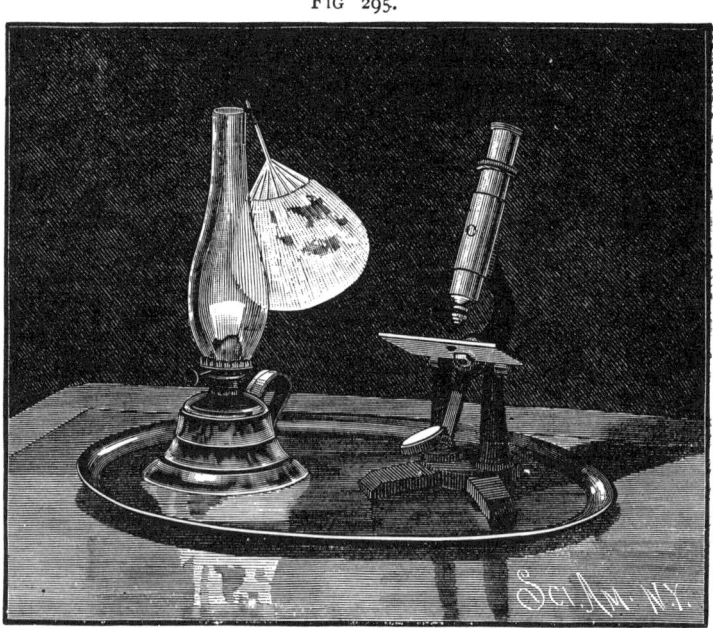

Substitute for Revolving Table.

ure. The diaphragm consists of a piece of perforated thin sheet metal, extending along the under surface of the stage, and neatly bent over the outer edge of the stage, so as to be self-supporting—the perforations of the metal being respectively one-sixteenth, one-eighth, three-sixteenths, one-fourth, and five-sixteenths inch diameter, all arranged on a longitudinal line of the metal plate intersecting the axial line of the microscope tube, so that the centers of the holes of the diaphragm may be made to coincide with the center of the hole in the stage.

The attachment for fine adjustment is made by bending one end of a thin metal plate twice at right angles, so that it will spring on the edge of the stage and clamp the stage tightly. The opposite end of the metal plate is bent in a similar manner, but the space between the body of the plate and the bent-over end is made wider, to permit of a small amount of movement of this end of the plate. In the portion of this end of the plate extending under the stage is inserted a fine screw with a milled head, by means of which the free end of the plate may be made to move either up or down through a small distance. The body of the plate is inserted under the stage clips, and the object slide is inserted between the clips and the movable plate.

The instrument has no rack adjustment, but the main tube slides easily and smoothly in the guide tube, so that little or no difficulty is experienced in focusing. Besides the instrument and accessories, only the following articles will be required to begin in earnest the study of microscopic objects: A small pair of spring forceps, a bottle for objects, a few concaved glass slides, a few thin cover glasses, a glass drop tube, a small kerosene lamp; and if the investigator desires to entertain his friends with the microscope, he will need a Japanese or tin tray, large enough to contain both microscope and lamp, as shown in Fig. 295, so that the relation of both may be preserved while the tray is moved to bring the instrument into position for different observers, by simply sliding the tray on the table.

A little caution as to illumination is necessary, as the beginner is generally unsparing of his eyes, using far too much light. A blue glass screen placed between the mirror and source of light, or between the mirror and the stage, modifies the light so as to greatly relieve the eyes.

284   EXPERIMENTAL SCIENCE.

The lamp should be provided with a shade of some sort to prevent the light from passing directly from the lamp to the eyes. A small Japanese fan suspended from the chim-

FIG. 296.

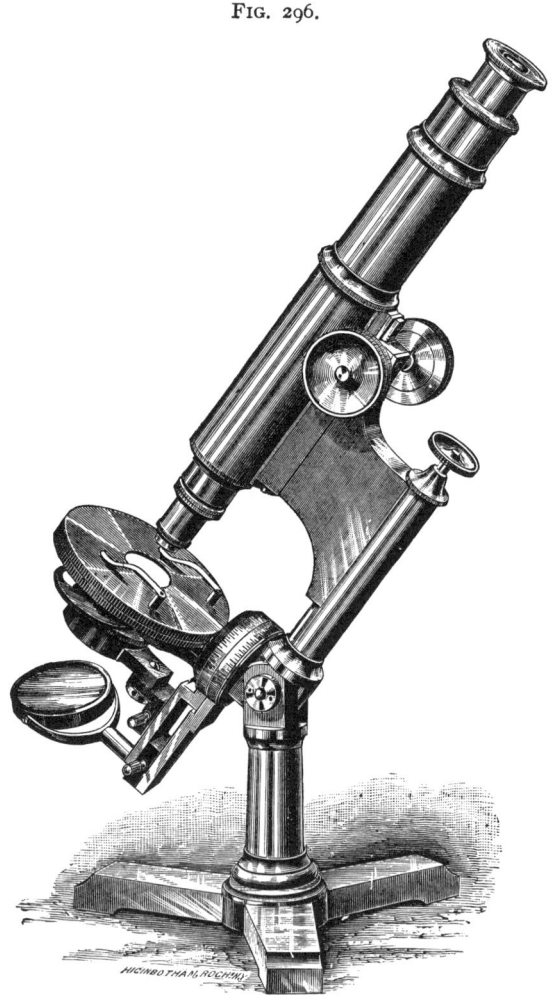

A Modern Microscope.*

ney by a wire, as shown, forms a very desirable shade. Most objects viewed by transmitted light in an instrument of this class require an absolutely central light, that is,

* Bausch & Lomb Optical Co.'s "Universal."

the light must be reflected straight upward through the object and through the tube.

When opaque objects are examined, the mirror is raised above the stage and made to concentrate the light on the object. Different angles of illumination should be tried, as some objects are greatly relieved by their shadows, while

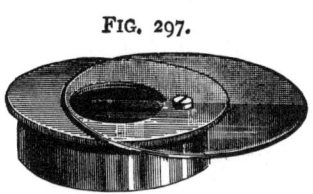

Fig. 297.

Light Modifier.

Fig. 297a.

Iris Diaphragm.

others require illumination as nearly vertical as possible. Experience will soon indicate the right magnification for different objects. This may be varied by taking off or putting on the lower half of the objective, also by drawing out or pushing in the draw tube.

For truly scientific microscopical work a better instru-

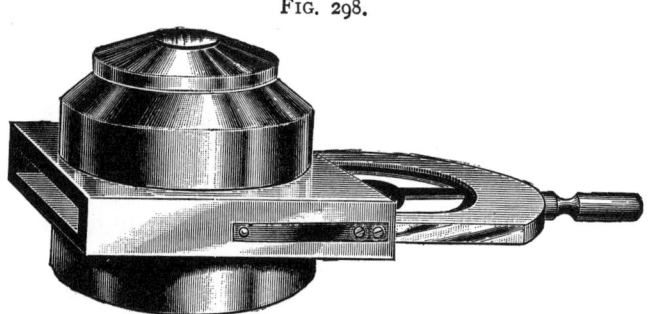

Fig. 298.

Sub-Stage Condenser.

ment than that already described will be needed. The microscope shown in Fig. 296 is perfectly adapted for general use. The main tube has two draw tubes by which any desired tube length may be secured. The coarse adjustment is effected by means of a rack and pinion; and a micrometer screw is used for the fine adjustment. The stage, which is revoluble, is made thin to allow of the greatest obliquity of illumination. The arms which support the sub-stage and the mirror turn upon the same axis, and are capable of being moved independently. The mirror may be swung above the stage for the illumination of opaque objects.

The sub-stage is adapted to receive any of the accessories, such as the light modifier shown in Fig. 297, the condenser represented in Fig. 298, and other desirable and indispensable appliances. A stand of this character is perfectly adapted to objectives of the highest class. All adjustments required to secure any angle of illumination, any position of the object, or any degree of fineness of focalizing, can be made quickly and with precision. The possessor of a microscope of this quality will always feel a degree of satisfaction which the poorer instrument can never give.

A larger, more complete, and at the same time much more expensive microscope is shown in Fig. 304, in connection with light-intermitting apparatus. This microscope has, in addition to the features already described, complete mechanism for centering the stages, a rack and pinion for the sub-stage adjustment, a graduated circle on the stage, a graduated head on the micrometer screw, graduations upon the pillars for the angle of inclination of the tube, and graduations at the base for measuring angles of objectives. A microscope of either of these grades, with a complement of fine objectives, eyepieces, and other necessary accessories, will yield all the results attainable at this stage of microscopy.

The graduated blue glass light modifier above referred to consists of a disk of flashed glass ground and polished so as to give all shades between white and dark blue, both transparent and translucent. This disk is pivoted upon an adapter (Fig. 297), so that it may be turned to receive any desired quality of illumination. It may be used in conjunction with the condenser shown in Fig. 298. This condenser is fitted to the sub-stage, and is provided with several stops and diaphragms, by which the light may be controlled. This condenser has a very wide angle, and is adapted for use in connection with objectives of all grades; but its efficiency is specially noticeable when it is used in connection with objectives of high numerical aperture in the examination of difficult objects and the resolution of tests.

The iris diaphragm shown in Fig. 297a is of great value in ordinary work. As its name indicates, its aperture may be expanded or contracted to adapt it to a particular object. It shuts off much superfluous light, thus saving the eyes; at the same time improving definition of the object.

For further information regarding microscopes and their accessories the reader in referred to the literature of the subject. Of this there is an ample supply.*

GATHERING MICROSCOPIC OBJECTS.

Objects for microscopical examination are gathered by means of a wide-mouthed bottle clamped in tongs attached to a long handle, cane, or even a fishing rod. By this device mud can be removed from the bottom, the stems and leaves of aquatic plants can be scraped so as to remove animalcules, and objects can be readily dipped from pools and shallow places. The under surface of plants and of grasses hanging over into the water may be scraped with the bottle, and more or less of the matter adhering thereto will be secured. Occasionally a long leaf like that of the flag may be lifted from the water and traversed by the bottle with good results. Small twigs and dead leaves floating in the water are often found teeming with life. The thousands of animalcules and forms of minute plant life found in water will afford the most zealous student a life-long supply of objects for examination. A wide-mouthed bottle or jar is provided with a perforated cork, in which is inserted a funnel for receiving the material; and another funnel, inverted and placed within the jar or bottle, with its nozzle extending

---

* "The Microscope and its Revelations," by Carpenter; "How to Work with a Microscope," Beale; "How to See with a Microscope," Smith; and "Practical Microscopy," by George E. Davis, are among the excellent works on the subject.

upward through the stopper, is used for concentrating the material. Over the lower end of this funnel is stretched a piece of thin muslin, and to the upper end is applied a short piece of rubber pipe, which is retained in a curved position by a thread tied around the neck of the bottle. The material gathered is poured into the funnel, the water escapes through the strainer, and the objects are retained in the bottle.* The hooked knife shown in the engraving is of great

FIG. 299.

Implements for gathering Microscopic Objects.

utility in cutting and fishing out parts of aquatic plants and submerged branches and roots, which are often teeming with microscopic life.

It would be futile to attempt anything more than the mere mention of a few of the interesting objects that may be seen to advantage in a small microscope. In Fig. 300 the engraver has beautifully shown some of the common objects which are easily secured, readily examined, and always interesting.

* This device is due to Mr. Stephen Helm.

MICROSCOPY.

Fig. 300.

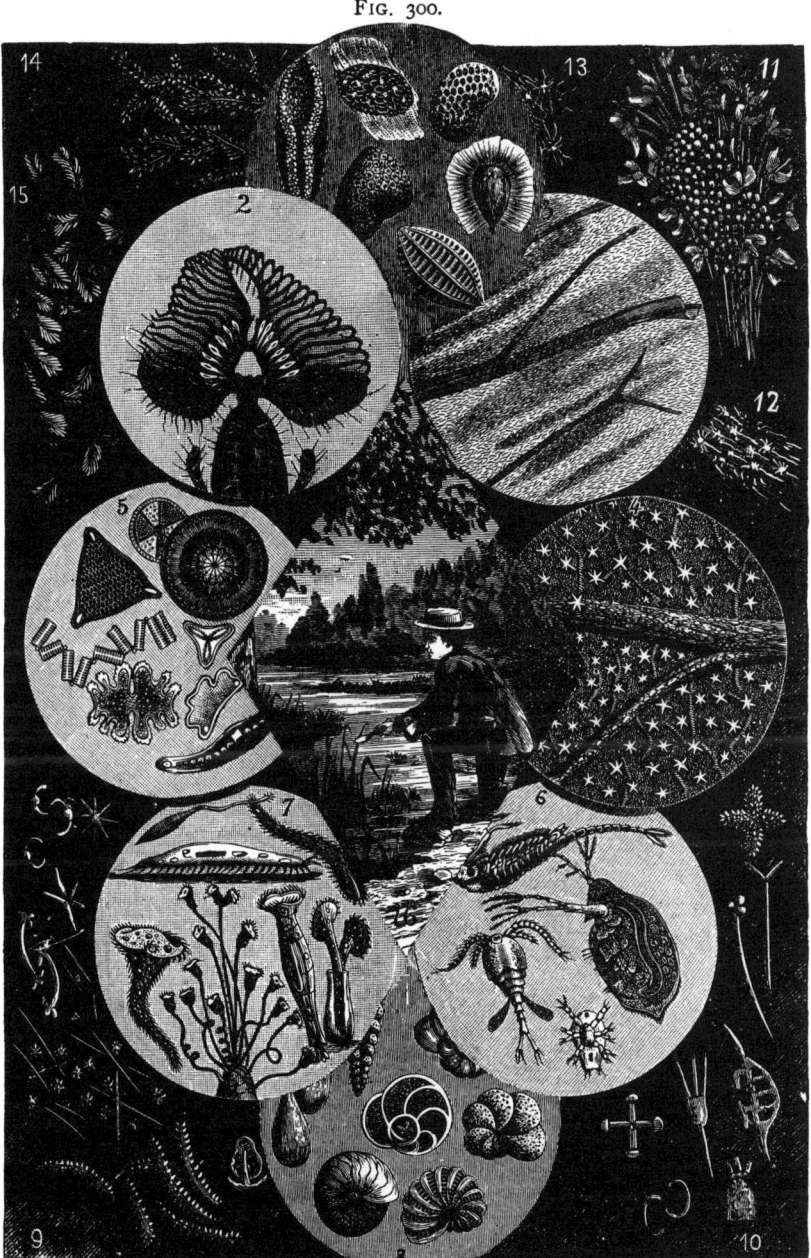

1. Seeds. 2. Tongue of Fly. 3. Bee's Wing. 4. Deutzia Leaf. 5. Diatoms and Desmids.
6. Entomostraca. 7. Infusoria Rotatoria. 8. Foraminifera. 9. Spicules. 10. Spicules
and Plates. 11. Pollen of Marsh Mallow. 12. Plant Hairs. 13. Shepardia Canadensis.
14. Crystals of Silver. 15. Fern Gold 16. Gathering Objects.

Various Microscopic Objects.

At 1 in this engraving are shown various seeds; the lace-covered one at the top being the seed of the *Nemesia compacta*. The seed in the center is that of heather. That on the right of the lace-covered one is the seed of the poppy. The fringed one below it is that of the climber. At the bottom of the disk the seed of sorrel is shown at the left, and portulacca at the right. The remaining seed at the left is that of eucharidium.

No. 2 represents the proboscis of the blowfly as it appears in the field of the microscope, except that the intricate structure of the pseudo-trachea is not shown in the cut as it appears in the microscope.

No. 3 shows the doubling hooks of a bee's wing, which enable the insect to connect the wings of each pair so that they may be used as a single wing.

No. 4 shows the silicious stellate hairs on the back of a deutzia leaf. The upper half of 5 shows several forms of diatoms, and the lower half is filled with desmids.

In 6 branchipus is shown at the top, cyclops at the left, a young cyclops at the bottom, and daphnia or the water flea at the right. These are common in almost every pond.

In disk 7 are shown on the left the stentor, so named on account of its trumpet-like form; in the center the beautiful and sensitive vorticella, and upon the right of the vorticella common rotifer, and upon the extreme right the sheathed trumpet animalcule. All of these have cilia around their margins, which by their peculiar vibratory motion give the bell-shaped mouths the appearance of rotation. In the common rotifer, and in the animals shown in disk 6, the internal organs may be readily seen in operation.

In the upper part of disk 7 are shown a few of the hundreds of forms of life found in water in which animal or vegetable matter has been infused.

In disk 8 are represented a number of the exquisite little shells of foraminifera. At 9 are shown various spicules of sponges, sea urchins, etc. At 10 are shown sponge spicules and the anchor of *Synapia inherens;* 11 shows the pollen of marshmallow, and 12 and 13 are examples of plant hairs;

14 shows arborescent crystals of silver, and 15 the fern-like crystals of gold.*

TRANSFER OF OBJECTS TO SLIDE.

The objects are transferred from the bottle to the concavity of the slide for examination in the manner shown in

FIG 301.

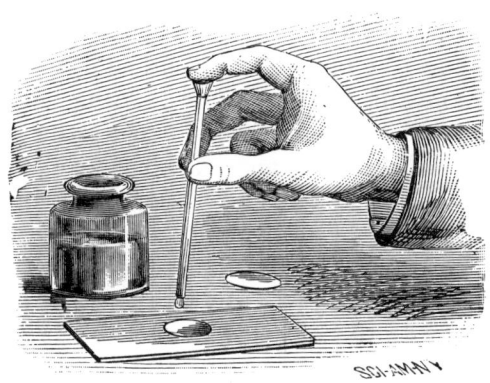

Transferring Objects to the Slide.

Fig. 301. The drop tube, which has a funnel-shaped top, is stopped by the finger at the upper end, while its lower end is inserted in the water in the bottle above the matter to be removed. The finger is then removed and some of the

FIG. 302

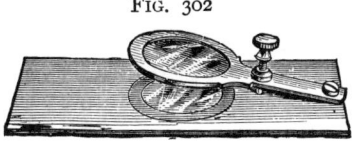

Compressor.

water, together with the objects carried by it, rushes upward into the tube. While the lower end is still in the water, the finger is again placed on the tube and this is withdrawn from the bottle and held over the cavity of the

* The following books are recommended to the beginner in microscopy: Wood's "Common Objects for the Microscope;" "One Thousand Objects for the Microscope," by M. C. Cooke; "Evenings at the Microscope," by Gosse; and "Practical Microscopy," by George E Davis.

slide, as shown in the engraving, when a drop or so of the water is forced out by pressing down the end of the finger on the top of the tube; the soft end of the finger acting as a sort of diaphragm in forcing out the required amount of water. Care must be taken to avoid getting solid matter upon the slide around the edge of the cavity, as it will prevent the cover glass from seating itself properly. The cover glass is placed over the cavity and pressed down lightly to squeeze out the surplus water, when the slide may be inserted under the clips of the stage and examined.

A more convenient device for holding animalcules is represented in Fig. 302. It is known as the compressor, and serves to lightly hold any object placed between the glass in the oblong plate and the glass in the adjustable arm. In any position it retains a drop of water.

To confine living objects to the field of vision, it is common to place between the glasses of the compressor a few fibers of cotton or a piece of fine lace.

MICROSCOPIC EXAMINATION OF CILIATED ORGANISMS BY INTERMITTENT LIGHT.

Every observing person has noticed that moving objects appear stationary when viewed by a flash of light; examples of this are seen during every thunder storm occurring in the night. The wheels of a carriage, a moving animal, or any moving thing, seen by the light of the lightning, appears perfectly stationary, the duration of the light being so brief as to admit of only an inappreciable movement of the body while illumination lasts.

If by any means a regular succession of light flashes be produced, the moving body will be seen in as many different positions as there are flashes of light. If a body rotating rapidly on a fixed axis be viewed by light flashes occurring once during each revolution of the body, only one image will be observed, and this will result from a succession of impressions upon the retina, which by the persistence of vision become blended into one continuous image. In this case no movement of the body will be apparent, but if the flashes of light succeed each other ever so little slower than the rotary period of the revolving body, the body will appear to move slowly forward, while in reality it is moving rapidly; and should the light flashes succeed each other more rapidly than the revolutions of the revolving body, the body will appear to move slowly backward, or in a direction opposite to that in which it is really turning.

These curious effects are also produced when the number of the light flashes is a multiple of the number of revolutions, or *vice versa*.

The combined effect of interrupted illumination and persistence of vision may be practically utilized for examining objects under motion which could not otherwise be satis-

FIG. 303.

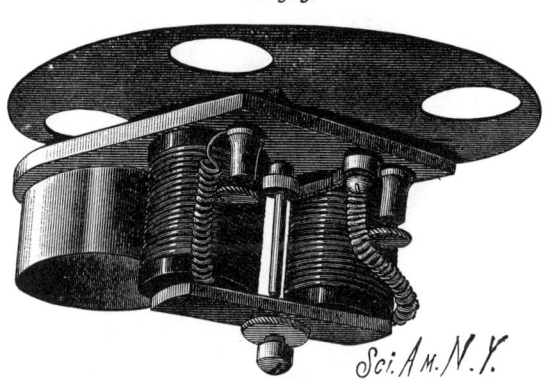

Light Interrupter for the Microscope.

factorily studied. To apply intermittent light to the microscopical examination of ciliated organisms, the writer has devised the electrically rotated apertured disk shown in Fig. 303, which is arranged to interrupt the beam of light employed in illuminating the object to be examined.

The instrument consists of an electric motor of the simplest kind mounted on a plate having a collar fitted to the sub-stage of the microscope, as shown in Fig. 304. The shaft, which carries a simple bar armature before the poles of the magnet, also carries upon its upper extremity a disk having two or four apertures, which coincide with the apertures of the stage and sub-stage two or four times during the rev-

FIG. 304.

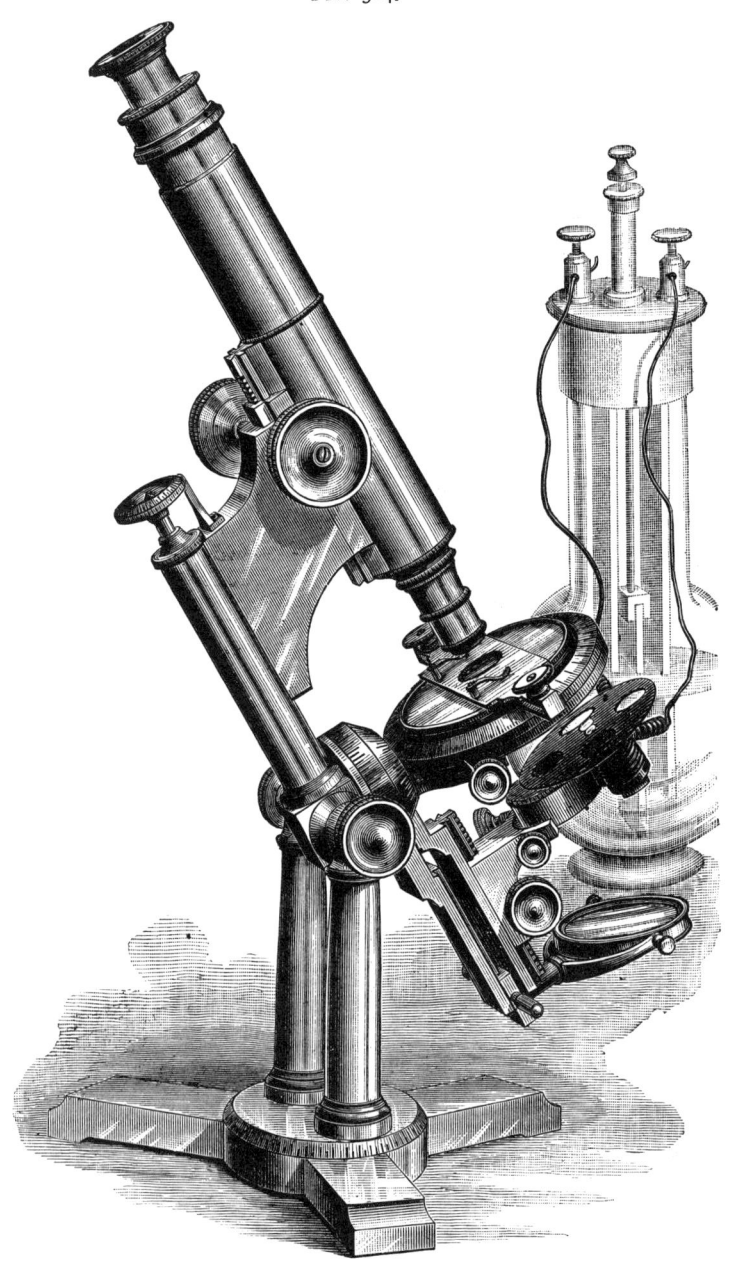

Microscopic Examination of Ciliated Organisms by Intermittent Light.

olutions of the disk. The shaft carries a commutator, and the course of the current from the battery through the instrument is through the spring touching the commutator, through the shaft and frame of the instrument to the magnet, thence out and back to the battery. There are two methods by which the speed of rotation of the apertured disk may be varied; one is by plunging the elements of the battery more or less, and the other is by applying the finger to the shaft of the motor as a brake, the motor in the latter case being started at its maximum speed, and then slowed down to the required degree by the friction of the finger.

Experiment shows that the period of darkness should be to the period of illumination about as three to one for the best effects. Closing two diametrically opposite holes in the disk represented in the cut secures about the correct proportion.

Various rotifers examined by intermittent light showed the cilia perfectly stationary. The ciliary filaments of some of the infusoria, vorticella, and the stentor, for example, when viewed by intermittent light, appeared to stand still, and their length seemed much greater than when examined by continuous light. The interrupted light brings out not only the cilia around the oral aperture, but shows to good advantage the cilia disposed along the margin of the body.

What interrupted light may reveal in the examination of flagellate or ciliated plants the writer is unable to say, as no objects of this character have been available. It is presumable, however, that something interesting will result from the examination of volvox and other motile plants, by means of this kind of illumination. Although it is necessary to interrupt the beam of light regularly, for continuous observation, the effect of intermittent light may be exhibited to some extent by an apertured disk, like that above described, twirled by the thumb and finger or revolved like a top by means of a string; or by using a larger apertured disk fitted to a rotator, and placed between the source of light and the mirror of the microscope.

CIRCULATION IN ANIMAL AND VEGETABLE TISSUES.

Among vegetable organisms in which the circulation of the sap is visible, the nitella is prominent. So, also, is the beautiful desmid colosterium.

FIG. 305.

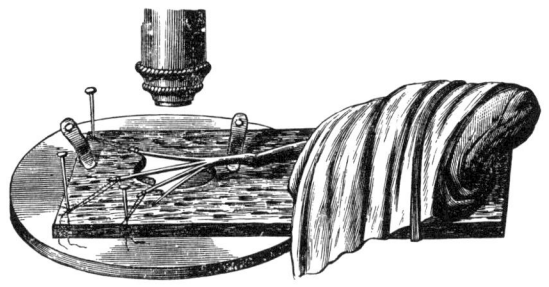

Simple Frog Plate.

Among animal organisms, the daphnia, or water flea, is extremely interesting, the minute heart being made clearly visible by the transparency of the shell of this little creature.

FIG. 306.

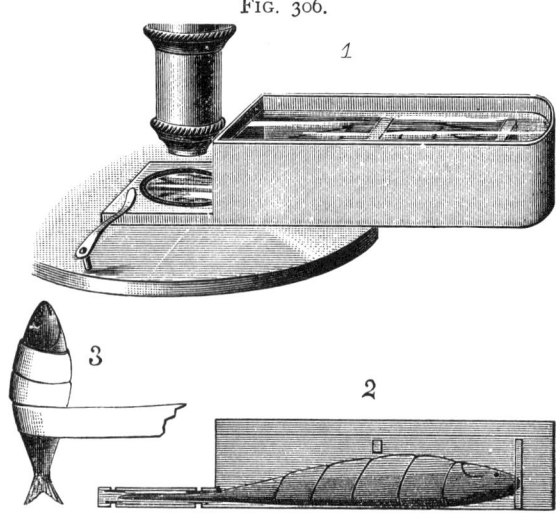

Kent's Trough for showing the Circulation of Blood in a Fish's Tail.

The circulation of blood in a frog's foot may be shown by stretching the foot so as to distend the web, as shown in Fig. 305. One form of apparatus consists of a thin, apertured piece of wood, provided with a glass slide upon which to rest the frog's foot. A piece of cork has been used for this purpose without the glass slide.

The slice of cork has a hole near one end corresponding with the hole in the stage of the microscope. The frog is wrapped in a wet cloth and held in place upon the cork by means of a small rubber band (Fig. 305). One of the frog's legs is extended. To two or three of the toes are attached threads which are held under tension by ordinary pins stuck into the cork. The foot is moistened to render the web more transparent, and the circulation is observed with a three-fourth or one inch objective.

The circulation of blood in the tail of a gold fish requires more complicated apparatus. It consists of a metallic tank provided with a thin extension, having in its upper and lower sides glass windows, formed of cover glasses set in recesses and secured by marine glue. The fish is wrapped in a strip of thin muslin, as shown at 3, to deprive it of the use of its fins, and laid upon its side in the tank, as shown at 2, in Fig. 306, with its tail between two windows, allowing the light to pass upward through the tissues from the mirror of the instrument. The tank is filled with water, and to prevent the fish from jumping, small wooden cross bars are placed in different positions in the tank. Arranged in this way, the fish may be observed for about twenty minutes. The blood is seen flowing in crimson streams in various directions through the tissues of the tail, the corpuscles being distinctly visible. A one-inch or three-quarter inch objective is ample for this purpose.

The blood of the frog is white, and the corpuscles are larger than those of the fish. As compared with the corpuscles of human blood, those of the fish are larger.

QUICK METHODS OF MOUNTING DRY OBJECTS.

There is a certain class of microscopic objects that need little or no preparation for mounting, and require no protection beyond a well secured glass cover. Many of these objects are interesting and in some degree valuable; but the microscopist considers them hardly worth the trouble of mounting. For such objects the method shown in the annexed engraving (Fig. 307) is of great utility, as it permits of inclosing the object quickly, completely, permanently, and in a presentable form, and while it seems especially adapted to such objects as are common and liable to remain unmounted, it is, of course, applicable to almost any dry object.

To carry out this method, only two articles, in addition

FIG. 307.

Quick Method of mounting Microscopic Objects.

to those usually possessed by microscopists, are required; one being the ring with an internal flange at the top and an external flange at the bottom, the other a heating tool, consisting of a ring of brass attached to a suitable handle.

The rings, of which the walls of the cells are formed, are spun or stamped from disks of Britannia metal, sheet brass, or other sheet metal, with a narrow internal flange or fillet at the top for receiving the cover glass, and a wider external flange at the bottom, for attachment to the slide. The rings vary in depth according to the depth of cell required. The under surface of each ring is coated with thick shellac varnish and allowed to dry thoroughly. When the varnish is dry and hard, a clean cover glass is dropped into each ring, and the ring is placed bottom upward on the warming stand and heated until the shellac melts and thoroughly covers the edge of the cover glass. The ring is now allowed to cool, when the cover will be ready for use. It will, of course, be understood that a quantity of rings and covers are thus prepared and held in reserve. In fact, it is to be hoped that the manufacturers of microscopists' supplies will furnish the rings and covers thus prepared, ready for instant use.

The object to be protected is attached to the slide by means of cement, in the usual way.

A ring containing a glass cover is arranged over the

FIG. 308.

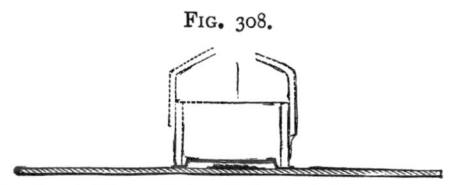

Sectional View of the Slide and Heating Tool.

object, and the heating tool is warmed and placed upon the outer flange of the ring, as shown in the sectional view, Fig. 308. By this means sufficient heat is imparted to the ring to melt the shellac upon that portion touched by the heating tool, and cause it to attach itself to the glass slide. It is the work of an instant to cover an object in this way, and the slide needs no further finish; but the operator may, if he choose, lacquer the rings to prevent them from tarnishing.

A thin ring provided with the coating of shellac may be applied to an ordinary balsam mount to increase its security.

By applying to the ring a suitable cement, a liquid cell may be made. The object to be mounted in the liquid cell is wet with the liquid and placed on the slide. The ring is then secured in the manner above described, and the liquid is afterward introduced into the cell through an aperture previously made in the side of the ring. This aperture is stopped with cement, applied with a hot wire or needle.

Dr. Stiles' wax cell is simple in construction, beautiful in appearance, and very effective for dry objects.

Sheet wax, such as is used by the makers of artificial flowers, is the material employed in the construction of this cell. Three or four sheets of different colors are pressed together by the thumb and finger to cause them to adhere, and a square of the combined sheet thus formed of sufficient size for a cell is cut out and pressed upon a glass slide. The

FIG. 309.

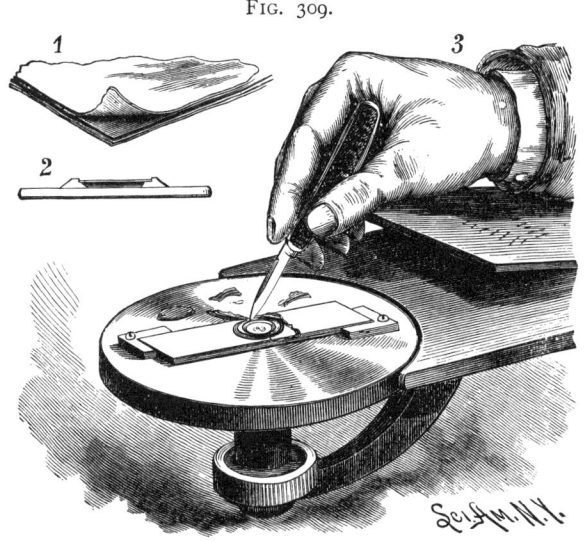

Making the Wax Cell.

slide is then placed upon a turn table, as shown at 3, Fig. 309, when, by the dextrous manipulation of an ordinary penknife, the wax is cut into a circular form, and the center is cut out to the required depth. If the cell is to contain a transparent or translucent object, the entire central portion of the wax is removed, as shown at 2; but if a ground is required for the object, one or more layers of wax are allowed to remain. A portion of the upper layer of wax is removed to form a rim for the reception of the cover glass. Where a black ground is required, a small disk of black paper is pressed upon the lower layer of wax. The final finish is given to the cell by a coating of shellac varnish, applied while the slide is on the turn table. These cells are very quickly made and have the finished appearance of a cell formed of different colored cements.

MICROSCOPICAL EXAMINATION OF THE PHENOMENON O. COLORS OF THIN PLATES.

As all works on light and on general physics treat of the phenomenon of the interference of light as exhibited in thin transparent plates or films, it will be unnecessary to go into an examination of this subject in detail; but it will doubtless prove both interesting and profitable to those interested in microscopy to take up the study of this subject with the aid of the microscope.

There is nothing more beautiful than Newton's rings, or a soap film, or extremely thin plates of mica when viewed in a microscope by properly directed light. Even the gorgeous colors of polarized light cannot be excluded in this comparison; but it is difficult with ordinary appliances to see these exquisite tints.

The writer, after some experiment, devised mounts for the ready exhibition of Newton's rings and interference phenomena, as shown by the soap film.

The device for the exhibition of Newton's rings is shown in Fig. 310, 1 showing the position of the mount on the microscope stage, 2 being a perspective view of the slide and 3 a diametrical section of the rubber cell containing the plane and convex glasses. The plane glass is a disk cut from one of the finer kind of glass slips, commonly used in mounting objects. The convex disk is cut from an ordinary biconvex spectacle lens, having a focal length of 24 inches. The cell is screw-threaded internally, and provided with a screw-threaded ring, which clamps the two glasses together. It has, in diametrically opposite sides, cavities for receiving the ends of the wire frame, which is clamped to the face of the slide by a clip and two screws. The cell containing the glasses is in this way supported adjustably so that it can be raised or lowered, or tilted at any required angle.

The position of the cell relative to the source of light is shown at 1. The cell and the source of light or the mirror should be arranged so that the image of the flame used for illumination or the broad light of the sky will be reflected up the tube. The objective (a 2 inch, with 2 inch eyepiece) may now be focused, when the rings, which about fill the field, will appear with great brilliancy. The effect may be

FIG. 310.

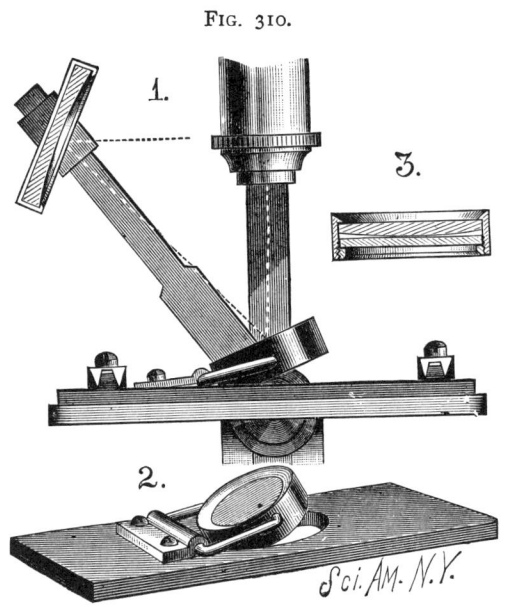

Mount of Newton's Rings for the Microscope.

somewhat varied by turning the cell at different angles, and moving the source of light accordingly. The concave mirror is used to concentrate the light; but, of course, a condenser may be used instead, or, if the light is strong enough, the beam may be received directly on the glass of the cell, and thrown up the tube.

With the unaided eye the rings appear as a very small disk, with no very noticeable beauty; but in the microscope it is not only greatly magnified, but properly illuminated.

An interesting experiment, showing the difference between the effect of pure sunlight and artificial light, consists in adjusting the mirror so as to simultaneously receive light from the sky and from a lamp or gas light. The portion of the disk illuminated by the lamp light shows the predominance of yellow, a greenish hue taking the place of the blue; the red being also modified.

Monochromatic light, such as is secured by passing light

FIG. 311.

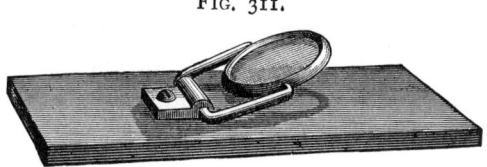

Holder for Soap Film.

through a deep red glass, for example, shows the rings as alternately red and black.

The device for exhibiting the soap film, which is shown in Fig. 311, will now need little explanation. A ring is pivoted in the same manner as the cell already described. By dipping the finger in soapy water, and passing it over the ring, a film will remain in the ring, which may be viewed

FIG. 312

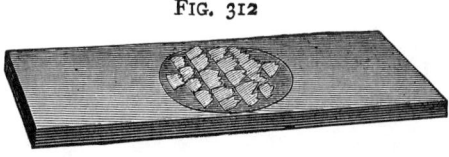

Mount of Mica Plates.

in the same manner as Newton's rings. The bands of iridescent color are very brilliant.

Thin plates of mica exhibit the same phenomenon. By tearing a very thin plate of mica, so as to leave a ragged edge, many extremely thin points will remain projecting from the torn edges; these may be cut off, and cemented in a suitable position for observation. These little points are quite difficult to handle. Probably the easiest way to manage them is to cut the piece of mica down quite small, and then take the bright point in a pair of clean forceps, and cut the larger part off, then touch the edge of the bright piece with Canada balsam, and put it in position on the slide. These little plates of mica are viewed in the same manner as the Newton's rings.

It is perhaps hardly necessary to say that having prepared a good mount of the mica plates, it is advisable to inclose it under a cover, as soon as convenient, to exclude dust.

MICROSCOPIC OBSERVATION OF VIBRATING RODS.

A metal rod fixed in a vise at one end, with a silvered glass bead attached to the other end, constitutes Sir Charles

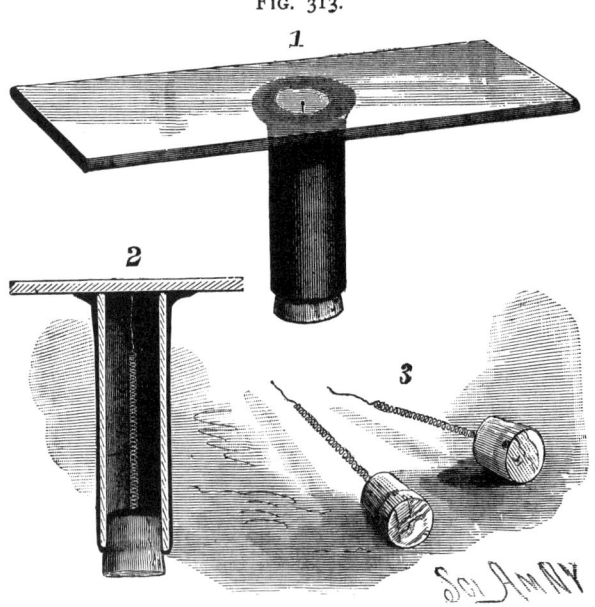

FIG. 313.

Vibrating Rod mounted for Microscopic Observation.

Wheatstone's apparatus for the study of the transverse vibrations of rods.

By vibrating a rod arranged in this way, Wheatstone was enabled to obtain an almost infinite variety of symmetrical and beautiful luminous scrolls.

It is a simple matter to repeat Wheatstone's experiment

with the apparatus alluded to, but it is not always convenient to do it.

A vibrating rod permanently mounted in a cell and arranged for observation with a microscope is shown in Fig. 313, 1 representing the mount in perspective, 2 showing it in section, 3 showing the rods detached from the mount.

To an ordinary 3×1 inch glass slip is connected a paper tube $\frac{5}{16}$ inch internal diameter and 1¼ inches long, well blackened on the inside.

The cement is applied carefully, so as to have the glass clean and clear with the tube. To a cork fitted to the open end of the tube is cemented a wire spiral formed of about 4 in. of No. 40 spring brass wire. The diameter of the spiral is $\frac{3}{32}$ inch. The end of the spiral next the glass slip terminates in a straight arm ¼ in. long, upon the end of which there is

FIG. 314.

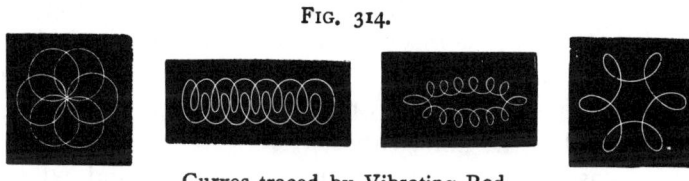

Curves traced by Vibrating Rod.

a minute bead of black glass. A smooth bead is secured by first fusing borax on the end of the wire, then touching the borax while in a fused state with a thin thread of black glass, then breaking the thread a short distance from the end of the wire, and finally fusing it by gradually pushing it forward into the flame until a perfect bead of the required size is formed.

The cork with the spiral is inserted in the paper tube with the bead arranged centrally with reference to the tube, and only a very short distance below the glass.

By placing the mount thus prepared under a 1 in. or 2 in. objective, and allowing light to fall on the bead from one direction, it will be noticed that the black glass bead is rarely at rest, the bright pencil of light reflected from it continually describing curves of various forms. Stepping on the floor of the room in which the microscope is located is generally sufficient to set the spiral into active vibration. Rapping on the table on which the microscope rests will cause the bead to describe intricate curves.

By striking the side of the paper tube with more or less force, different figures will be produced.

Illuminating the bead from two points produces parallel curves.

While this mount is perhaps not strictly a microscopic object, it may nevertheless be viewed to advantage by the microscope.

SIMPLE POLARISCOPE FOR THE MICROSCOPE.

To the draw tube of the microscope is fitted a paper tube, which is readily made by gumming writing paper and winding it around a cylindrical stick of the proper size. To the paper tube is fitted a second tube, and this last tube is cut diagonally through the center at an angle of 35° 25'. One of these pieces is inserted in the first tube, and sixteen or eighteen elliptical glass covers, such as are used for covering mounted microscopic objects, are placed on the diagonally cut end of the inner tube.

The glasses should be thoroughly cleaned, and when in position in the tube they are held by the remainder of the diagonally cut tube. The sectional view of the instrument clearly shows the position of these glasses in the draw tube.

The tube which goes under the stage is made in precisely the same way, and is supported in position for use by a short paper tube secured to a cardboard casing adapted to slide over the stage of the microscope, as shown in the engraving. Notches are formed in the rear edge of the upper part of the casing to allow it to slip by the slide-holding clips. The lower tube must be capable of turning in the short fixed tube, and it may be prevented from falling out by gluing a cardboard band or a piece of small cord around its upper end, forming a sort of flange. The hole in the upper part of the casing is made larger than the movable tube, to admit of inserting the tube from the top of the casing. The part of the attachment below the stage is the polarizer. The part in the draw tube is the analyzer.

By turning the polarizer, the light being thrown directly up the tube by the mirror, the field of the microscope will appear alternately light and dark, showing the partial extinguishment of the polarized beam twice during each revolution of the polarizer.

When the field is darkest, a piece of mica of the proper thickness inserted between the stage and objective renders

FIG. 315.

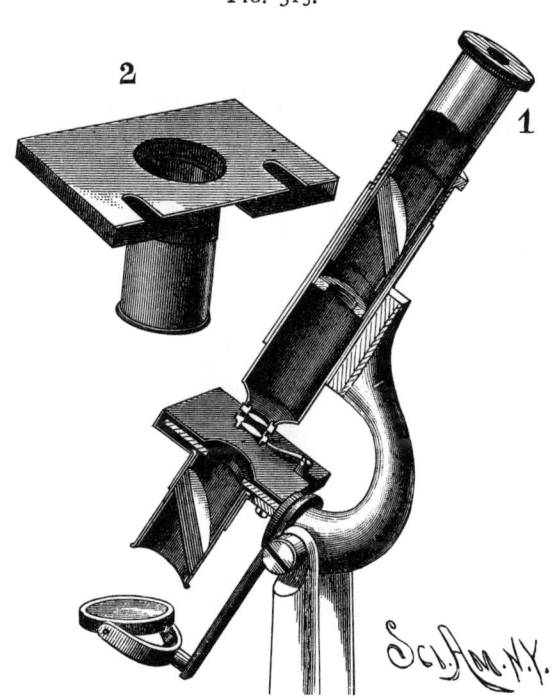

Simple Polariscope for the Microscope.

the field light, and it may, in addition to this, produce a color effect. The colors depend on the thickness of the film and upon its position in the instrument.

There are various chemical salts and animal and vegetable substances which produce brilliant color effects in the polarized beam. Salicine is a favorite. Santonine is good. Tartaric acid, boracic acid, and cane sugar are easily prepared by allowing their solutions to crystallize on the glass slip. Some of these substances, salicine for example, may be fused upon the slip and recrystallized.

The colors may be heightened by placing a film of mica behind the object during examination. Different colors will be produced by different thicknesses of mica.

Among animal substances to be examined in this way are fish scales, parings of the finger nails and of horses' hoofs, parings of corns and of horn.

Among vegetable substances, the sections of some woods, the cuticle of plants, the rush for example, form good polariscopic objects.

Many minerals show well in polarized light, but they are generally difficult of preparation. Selenite is an exception. It may be readily reduced to the proper thickness to secure brilliant effects.

The polariscope above described, although not as desirable as one provided with a pair of Nicol prisms, is nevertheloss worth having, and will give its possessor a great deal of satisfaction.

CHAPTER XIV.

THE TELESCOPE.

Some hints are here given as to the construction of a cheap and efficient telescope which will give its possessor a great deal of enjoyment, and will serve to stimulate astronomical observation and research.

Plate IV. represents the telescope, its standard, and the various parts in section and in detail. The object glass, A, shown in the engraving, is a meniscus lens $2\frac{1}{2}$ inches in diameter and 36 to 38 inch focus. It is mounted in a wooden cell, B, having an internal flange or fillet about $\frac{1}{16}$ inch wide, forming a true support for the lens and bearing against the end of the paper tube, D, which forms the body of the telescope. The lens is retained in its cell by a flat strip, E, of brass which is sprung into the cell and pushed down against the lens. The cell is fastened to the tube by common wood-screws, which pass through the collar into the paper forming the tube. It is perhaps needless to say that the cell should be made of some thoroughly seasoned hard wood, which is not liable to change under atmospheric influences. Hard maple answers a good purpose, but mahogany is preferable.

To protect the objective when not in use, a cap, F, of tin or pasteboard, neatly covered with morocco or velvet, is fitted to the cell.

The paper tube, of which the telescope body is formed, is such as is commonly used for rolling engravings for mailing. It is 3 inches external diameter and 32 inches long (about 4 inches shorter than the focal length of the objective). The exterior of the tube is covered with Java canvas attached by means of bookbinder's paste (flour paste with glue added), and varnished when dry with two or three thin coats of shellac varnish. This gives the tube an elegant and durable finish.

PLATE IV.

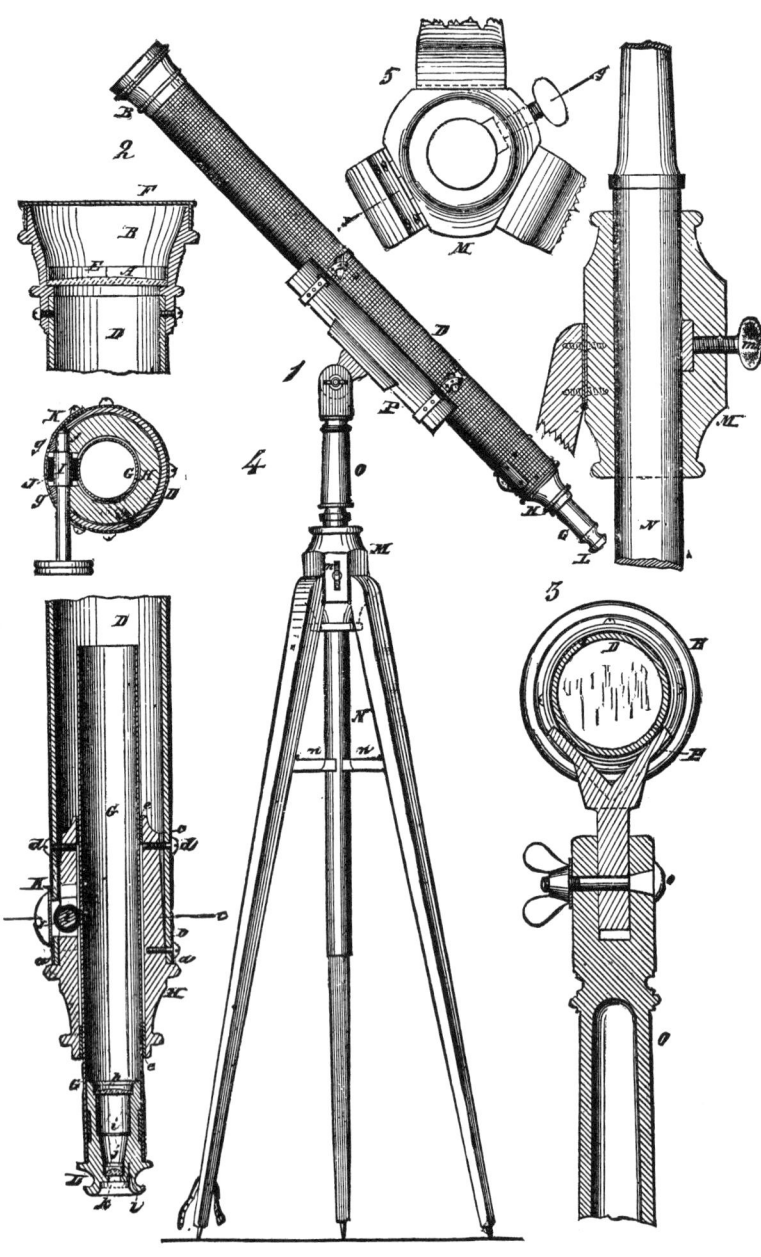

Easily Made Telescope.

The focusing tube, G, which is of brass, 1¼ inches internal diameter and 11 inches long, is guided by a turned wooden piece, H, fitted to the end of the pasteboard tube, D, and held by three or four ordinary round-headed wood-screws.

The piece, H, has a shoulder, *a*, against which the end of the pasteboard tube abuts, and only about three-quarters of an inch of the piece, H, actually fits the tube, the portion from *b* to *c* being tapered as indicated in the engraving, and near the extreme inner end, about 3½ inches from the shoulder, there are three screws, *d*, used in collimating the focusing tube, G.

The bore of the piece, H, is somewhat larger than the focusing tube, G, and is provided with a cloth lining, *e*, at each end to insure the smooth working of the tube.

A short distance from the shoulder, *a*, a mortise about three-quarters of an inch square is made through the side of the tube, D, and the piece, H, and a transverse slot, *f*, is formed to receive the wooden spindle, I, which is enlarged in the middle to receive the rubber thimble, J, and has on one end a milled head by which it may be turned. The spindle, I, is held in place by concave pieces, *g*, which in turn are retained by the curved plate, *k*, attached to the tube, D, by screws. The rubber thimble, J, is of sufficient diameter to reach to and press upon the focusing tube, and the latter has a series of transverse grooves filed in it to insure sufficient friction to move the tube, G, in and out when the spindle, I, is turned. This simple device may be used instead of the usual focusing mechanism, but a rack and pinion is preferable.

The cell, B, piece, H, and spindle, I, are blacked and polished on the outside, and the cell is left dead black on the inside. The interior of the tubes is also made dead black. Such a surface may be secured by adding lampblack to a little very thin shellac varnish, and applying it to the inside of the tube by means of a swab.

The focal lengths of the lenses of the astronomical eyepiece should be to each other as three to one; the field lens, which is nearest the object glass, having the greatest diameter and the longest focus, and the convex side of each lens should be turned toward the object glass. Their distance apart is one-half the sum of their focal lengths. These lenses are mounted in a wooden cell, L, whose exterior is fitted to the focusing tube, G, and grooved circumferentially to receive a strip of cloth, which is glued in, and insures a good fit. The cell is bored in different diameters to receive the field lens, h, the diaphragm, i, and the eye lens, j, all of which are held in place against shoulders formed in the cell, by circular springs of brass, which are sprung in, as in the case of the object glass. The eye aperture is about ¼ inch, and the aperture of the diaphragm is about the same.

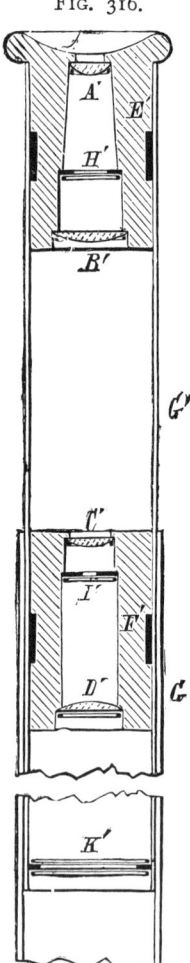

FIG. 316.

Terrestrial Eyepiece

It is well to make the diaphragm adjustable, so that it may be moved back and forth to secure the best position. It will be found, however, that if placed just beyond the focus of the eye lens, it will give the best results.

A circular recess, k, is formed in the face of the eyepiece to receive a sun glass, which is retained in place, when in use, by the short-curved spring, l. The sun glass is simply a disk of very dark glass. It must, in fact, be nearly opaque; some of the glass known as black glass answers the purpose very well.

If but one astronomical eyepiece is made, probably the most satisfactory combination would be as follows: Field lens, 1½ inches focal length; eye lens, ½ inch; distance apart, 1 inch. It is advisable, however, to have three eyepieces for different purposes—one of higher power and one of lower power than the one described.

A terrestrial eyepiece is illustrated in the sectional view,

THE TELESCOPE.

Fig. 316. It is of little use to adapt such an eyepiece to this instrument unless it is first provided with an achromatic objective. It is then a powerful telescope, which will enable one to see well for many miles. The method of mounting the lenses described in connection with the astronomical eyepieces will be followed here, therefore little more than the diameter and focus of the lenses and their distance apart need be given. There are four plano-convex lenses, A', B', C', D', mounted in two pairs in wooden cells, E', F', fitted to the tube, G', which in turn is fitted to the focusing tube, G. The cell, E', has a $\frac{1}{4}$ inch aperture for the eye and a bead which projects beyond the tube, G'. The lens, A', is about $\frac{7}{16}$ inch in diameter and 1 inch focus. The lens, B', is $\frac{3}{4}$ inch diameter and $1\frac{1}{2}$ inch focus. The lens, C', is $\frac{7}{16}$ inch diameter, $1\frac{1}{4}$ inch focus. The lens, D', is $\frac{5}{8}$ diameter and $1\frac{1}{4}$ inch focus. The plane face of A' is $1\frac{3}{4}$ inches from the plane face of B', and a stop, H', having a $\frac{7}{16}$ inch aperture is placed $1\frac{1}{4}$ inches from the face of the lens, A'. From the plane face of the lens, B', to the plane side of the lens, C', it is $3\frac{3}{8}$ inches. The distance between the plane side of the lens, C', and the plane face of the lens, D', is $1\frac{7}{8}$ inches. At a distance of $\frac{7}{16}$ inch from the face of the lens, C', there is a diaphragm, I', having a $\frac{1}{8}$ inch aperture. It will be observed that in this case the convex sides of the lenses, C' D, are turned toward each other.

At the extreme inner end of the tube, G', there is a diaphragm, K', of $\frac{15}{16}$ aperture, which is held in place by two circular springs. The interior surfaces must be well blacked to prevent reflection.

The method of mounting the lenses here shown and described is inexpensive and fairly efficient. If something better is desired the reader may, of course, make the mountings of brass, and fit the instrument up according to his taste and ability.

The arrangement of the various parts is clearly shown in the sectional view, at 2, and the focusing device is shown at 4, which is a transverse section.

In regard to collimation: by cutting off the ends of the paper tube truly in a lathe, the cell, B, and piece, H, will be measurably true. To determine whether the focusing tube, G, and cell, B, are axially in line, a truly cut cardboard disk with a pin hole exactly in the center may be placed in the cell, B. A similar disk may also be placed in each end of the focusing tube, G.

Now, by adjusting the piece, H, by means of the three screws, d, the three pin holes in the disks may be readily brought upon the same axial line; then, if the lenses have been carefully centered by the manufacturer, the telescope will be found sufficiently well collimated. If, however, it is desired to ascertain whether the lens is truly centered, it may be turned in its cell, while the telescope is in a fixed position, and directed toward some immovable object. If the image moves as the lens is turned, it shows that the centering is defective.

If there are doubts as to whether the axis of the objective coincides with the axis of the tube, the latter may be supported in V-shaped supports adapted to the truly turned ends, then by placing a candle at some distance from the face of the lens, and turning the tube in its V supports, at the same time viewing the reflection of the candle in the lens, it will at once be known by the movement of the reflection that the cell requires adjustment to render the axis of the objective and that of the tube coincident.

With a telescope of this description a large number of celestial objects may be examined with great satisfaction. The moon furnishes an unending source of delight, showing as it does a face that is ever changing throughout the lunar month. Jupiter is an interesting study of which one does not soon tire. The telescope described will show the satellites in their varying positions from night to night and the dark belt across the face of the planet.

Saturn is a grand object with the telescope. His ring may be clearly seen. The meniscus lens will show a little color, and its definition will be quite defective when directed to such bright objects as the moon, Jupiter, Saturn, Mars, or Venus, with the full aperture, therefore the aperture should be reduced by a diaphragm of black cardboard. A little experiment will determine the best sized aperture.

THE TELESCOPE.   315

For nebulæ, star groups, and double stars, the full aperture should be used. The great nebula of Orion is an interesting object; many of the star groups are very pleasing. The sun also, when the spots are visible, may be viewed with satisfaction. Of course, the sun glass will be applied before the observer attempts to view the sun, otherwise the eye may be injured or destroyed. A double or plano-convex lens, of long focus, may be used for an objective, but the meniscus is better.

If the mountings have been carefully made, the meniscus or the plano or double convex lens will soon be supplanted by a good achromatic objective, which will increase the efficiency of the instrument many fold.

As to the telescope stand, little need be said, as its construction is so clearly shown in the engraving. It cannot be made too solid. If it is very clumsy, this is no objection. If it is slender, it will shake. Every tremor has the benefit of the magnifying power of the telescope, and is amplified to a wonderful extent.

The stand represented is easily constructed and answers an excellent purpose. From the ground to the top of the hexagonal hub, M, it is four feet. Three of the alternate sides of the hub are wider than the intermediate ones, to receive the wrought iron hinges by which the legs are attached. To attach the hinges, the pin is first driven out; one-half of the hinge is then attached to the leg, and the other half to the hub, M, when the pin is replaced.

No. 5 is a top view of the hub and the upper portion of the legs; 6 is a vertical section. A 1½ inch hole is bored through the hub to receive the standard, N, which supports the telescope. To each of the legs is hinged an arm, n, which folds down against the standard, so as to spring the legs outwardly, and thus render the stand very rigid. The lower ends of the legs are provided with spikes, and a strap is attached to one of the legs to bind them all together when the instrument is not in use.

The upper end of the standard, N, is reduced in size, and made slightly conical for receiving a socket, O, to the upper end of which is jointed an arm attached to the V-shaped

trough, P, in which the telescope is secured by straps. The
form of the joint is shown at 3, which is a vertical transverse

FIG. 317.

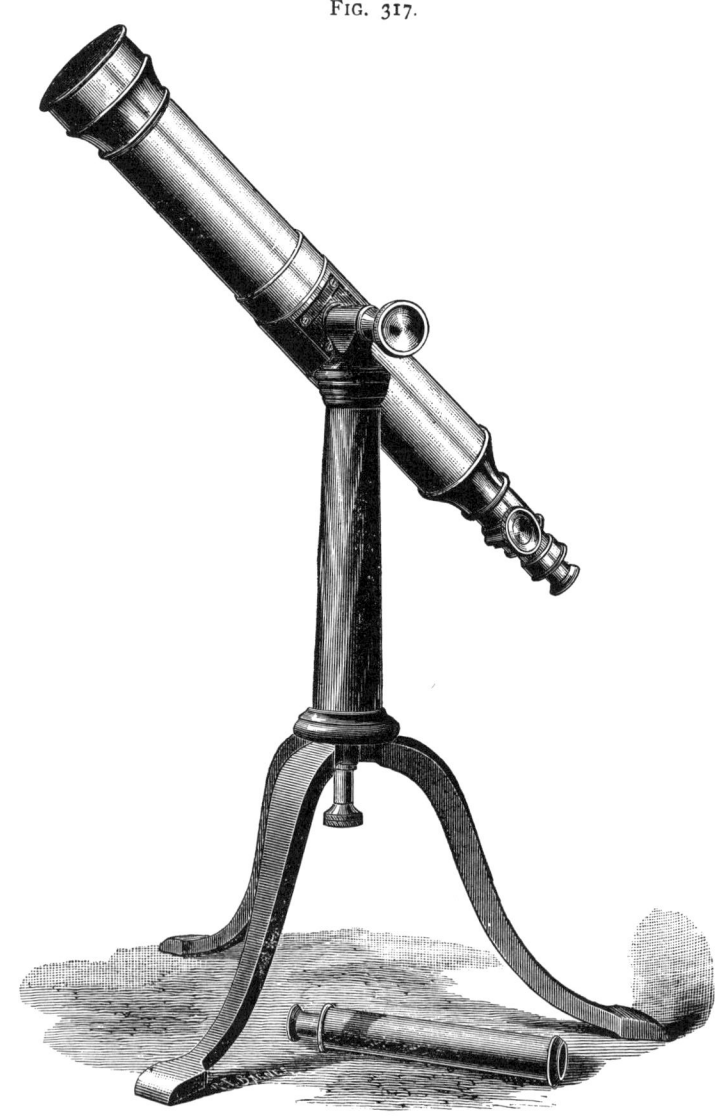

Compact Telescope—2½ inch Aperture, 24 inch Focus.

section. A strong bolt, *o*, forms the pivot of the joint
between the socket, O, and trough, P, and is provided with

a wing nut by which it may be tightened. The surfaces of the joint as well as the upper end of the standard should be coated with black lead to insure smooth working. A post set firmly in the ground, while it cannot be moved from place to place, has the advantage of being rigid. It forms one of the best of cheap stands.

COMPACT TELESCOPE.

In Fig. 317 is represented a fine telescope of $2\frac{1}{2}$ in. aperture, the optical parts of which are made after the formulæ of the late R. B. Tolles.

This telescope is suitable for either celestial or terrestrial observation. The high perfection of the objective permits of a very short focus (24 inches), which is a feature of considerable importance in portable telescopes.

Saturn with his rings and satellites, Jupiter and his moons, the nebula of Orion, and other nebulæ, the various star clusters and many of the double stars may be seen with a great deal of satisfaction with this little telescope.

CHAPTER XV.

PHOTOGRAPHY.

Probably no branch of applied science is so familiar to all classes of people as that of photography. The art is practiced by professionals and amateurs with different degrees of skill, varying from that which can produce only a recognizable shadow to that which is capable of securing results little short of perfection.

A great deal depends upon manipulative skill, and much depends upon the apparatus, and, while a camera of fair quality is indispensable, the best instrument obtainable will not compensate for carelessness nor for lack of the finer judgment required in many of the operations of photography.

Since the introduction of the dry plate, the camera and its accessories, together with a few pans and measuring glasses, constitute the outfit with which the operations are carried on.

The lens is a vital part of the outfit. It should be selected with more regard for its quality than its cost.

While photographs can be taken with a single lens, a compound achromatic lens is very desirable. There are many kinds of lenses in use; those having a wide angle and short focus, employed for photographing buildings, street views, near objects and interiors, and those of a narrower angle and longer focus, adapted for views having considerable distance. When only one lens can be purchased, a lens of the latter class is preferable. Lenses of either kind may be adapted to different conditions of use by means of stops or diaphragms with apertures of different sizes.

After acquiring a sufficient knowledge of photography to judge of the capabilities of a lens, the beginner should procure the best lens he is able to purchase. The writer has for years owned "good" lenses, and he might truthfully say "very good" lenses, but recently he has purchased some of the best lenses of recent construction and the "good" lenses have been relegated to the second-hand dealer.

The marvelous perfection of modern lenses can hardly be appreciated without the actual trial. The new lenses have great definition, flatness of field, surprising depth of focus, and rapidity equal to any demand. They are also non-astigmatic—a characteristic that cannot be overestimated.

Any good camera box will answer, provided it is light-tight. The more expensive boxes with swing backs, rising fronts and focusing mechanism are convenient and desirable. The modern plate holders are easily manipulated in the dark room, and they are not cumbersome to carry. By the use of kits, large plate holders may be adapted to small plates. A small and light tripod may be chosen, but it should have sufficient rigidity to hold the camera steadily.

The cloth used to cover the head while focusing should be light-tight, also waterproof, as in case of a storm it may be used to protect the camera and the plate holders.

The dealers furnish a great variety of plates from which to choose. Beginners will experience the greatest satisfaction in slow plates, as with these the danger of over-exposure is small. Plates must be kept in a dry place and carefully protected from the light. The boxes of plates should be opened and the plates inserted in the plate holders in a perfectly dark room, if possible. If a light is required, a ruby lamp capable of giving a dark red light may be used, but the light must be used cautiously. Probably more plates are fogged in a dark room than elsewhere by needless exposure to the ruby light. It seems hardly necessary to say that the plates should be placed in the plate holders with the film side out, that is, toward the slides. They should be carefully dusted with a fine, soft camel's hair brush before closing the slides.

The camera is pointed at the object to be taken, and adjusted so that the inverted image on the ground glass is in the desired position. The focusing cloth is then thrown over the head and over the camera, and the movable portion of the camera box is adjusted until a position is found at which the particular object appears sharp on the ground glass. If the image is too large, the camera must be moved back; if not large enough, it must, of course, be moved forward. After focusing, a suitable stop is inserted in the lens tube. This will vary with the light and with the intended exposure. It will be found that the light acts very much quicker on a July day than in December, and that the duration of exposure varies with the hour of the day as well as with the time of year, so that a larger stop must be used or a longer exposure made in winter than in summer, and in the morning and evening than at midday.

The use of a stop gives more detail in the shadows, in consequence of allowing a longer exposure; it also gives greater depth of field. After the insertion of the stops the cap is put on, the plate holder is inserted in its place and the slide withdrawn. Everything being ready for the exposure, the cap is removed and replaced. On a bright summer day, with an achromatic lens, the exposure of a slow plate with the smallest stop will require from three to five seconds, but the time cannot be given with accuracy; it must be learned by experience. With a fast plate, an exposure given by removing the cap and immediately replacing it is sufficient. With a quick-working lens this exposure would be too long. An instantaneous shutter would be required.

If it can be avoided, the camera should never be pointed toward the sun,* but if it becomes necessary, the lens must be shielded in such a way as to afford adequate protection without interfering with the field of view. The best landscapes are secured in the morning or afternoon, the shadows being longer than they are at midday. Photographing on windy days should be avoided, unless the exposures are to be instantaneous. The duration of the exposure of the plate varies greatly under different conditions. Interiors frequently require an exposure of an hour or two, often longer.

For copying from books, engravings or photographs for

---

* With recent lenses this precaution need not be observed. Some of the best effects are secured by pointing the camera toward the sun.

lantern slides or for reproductions, the ordinary camera box will usually be found too short, but a pasteboard extension may be fitted to the box. For copying, a good achromatic rectilinear lens is necessary. When the work is done by daylight, the camera should be placed with the back or side toward the window, the object to be copied being placed in front of the camera and well illuminated. In this class of work much depends on careful focusing. A magnifying glass of 8 or 10 inch focus is of great utility in this connection. By employing a kerosene or gas lamp provided with a reflector, copying may be successfully carried on at night. The exposures under these circumstances vary from ten minutes to a half hour.

Instantaneous photography is attractive and interesting, but difficult. It should be practiced only when necessary. Time exposures are always preferable when they are feasible. Excellent instantaneous pictures may be taken, however, after a little practice, but success is not always certain.

For instantaneous work, a good shutter and a quick-working lens will be required. The camera is focused in the usual way. A large stop is inserted in the lens tube and a fast plate is used. The slide is removed, and when the object is sighted, the shutter is let off.

The exposure and development of a plate are intimately related to each other; a properly exposed plate may easily be spoiled in developing, while, on the other hand, an unduly under or over exposed plate can never be made to produce a good negative by any process of developing. A perfectly dark room illuminated only by a ruby light with an orange colored glass superposed is indispensable. It should be furnished with a sink and running water, but progress may be made with no other conveniences than a pitcher of water and a washbowl. Several pans of gutta percha, glass, or porcelain are required for developing, fixing, etc., also two graduated glasses and a glass funnel are necessary. A pan should be provided for each kind of developer and one for hyposulphite of soda. The glasses, funnels and pans must be kept scrupulously clean, and the latter should always be used for the same kind of solution.

There are several developers for dry plates. The following is one of the best:

Beach Pyro-Potash Developer.

No. 1.—Pyro Solution.

Sulphite of soda (chemically pure crystals) 4 ounces.
Warm distilled or melted ice water.......4 "

When cooled to about 70° Fah., add:
Sulphurous acid water (strongest to be had)............................$3\frac{1}{2}$ ounces.
And lastly, pyrogallic acid.............1 "

No. 2.—Potash Solution.

A.

Carbonate potash (chemically pure).......3 ounces.
Water................................4 "

B.

Sulphite soda (chemically pure crystals)..2 ounces.
Water................................4 "

Make A and B separately and then combine in one solution.

For a 5×8 plate, pour into the graduated glass 1 drachm of the pyro solution and $\frac{1}{2}$ drachm of the potash; add 3 oz. water. Mix well. The plate should be lightly brushed clean with a soft camel's hair brush, and placed with the film side up in a pan containing fresh water; soak for about a minute, then pour the water off, and pour on the developer; rock the pan gently, so as to flow the developer evenly over the plate. The pan should now be brought close to the ruby light, and the plate examined. An image should begin to appear within two or three minutes. The plate should be closely watched. The high lights (sky, etc.) develop first, and appear as a darkening of the plate. The other objects follow. Development should be proceeded with until all parts of the picture show clearly by transmitted light, and until the plate turns gray, and the image seems to fade away. The outlines of the image appear on the back of the plate when it is sufficiently developed. If a plate comes up quickly, say within a minute, it is over-exposed, and should be removed to a pan containing water to which is added a small quantity of developer with the pyro solution in excess, or the plate may be placed in the developer, to which has been added a few drops of a solution containing 150 grains of potassium bromide in 2 oz. water.

In case a plate is very much over-exposed, it will not come up in a long time, and will be worthless. If a plate should not come up in a reasonable length of time, more of the potash solution should be added. An under-exposed instantaneous plate may be started by placing it in a weak solution of potash and water, then developing with an excess of alkali. Fogging is produced by too much alkali.

Over-development produces a hard negative, from which it is difficult to make a good print. Weak negatives having clear shadows, with plenty of detail, but lacking intensity in the high lights, are the result of over-exposure. Too strong high lights with weak shadows are due to under-exposure. Transparent spots (pinholes) are caused by dust, or air bubbles formed in development. If a plate during development is seen to lack detail in places, the development may be forced at such points by applying a large, soft, round camel's hair brush charged with moderately strong developer. The brush is rapidly passed over the portion of the plate to be brought out, care being taken not to touch the other parts. If negatives show too great contrast between the dark and light portions, the developer should be reduced with water.

After development the plate should be thoroughly washed with water and put in a clearing or fixing solution formed of sodium hyposulphite ("hypo") 1 oz., water 5 oz. A very small quantity of hypo mixed with the developer is sufficient to defeat all dark-room operations. Therefore, it must be isolated from everything else, and the hands must be thoroughly washed after handling it. When the hypo solution is discolored, it must be replaced with a fresh solution

The plate is left in the fixing solution for a short time after the yellow color has entirely disappeared from the film; then it is washed thoroughly until every trace of the hypo is removed. Soaking for several hours in clear cool water, frequently changed, is effective in removing the hypo. The permanence of the negative depends on this washing. The negatives should be placed on a rack to dry, in a cool place, free from dust.

The hydrochinon developer is largely used, and gives good satisfaction. With this, the development of the plate can be as easily controlled as with the pyro, and it has the advantage of not staining the fingers to any great extent. An under-exposed plate which is beyond saving with the pyro developer can be brought out by a long treatment with hydrochinon.

The formula for Carbutt's hydrochinon developer is as follows:

A.

Hydrochinon...................... 10 grains.
Crystallized sulphite soda.......... 50 "
Distilled water....................500 "

B.

Carbonate potash (pure)............ 25 grains.
Distilled water....................200 "

A and B should be mixed in equal volumes. The quantities here given make a very small amount of solution. It is advisable to make a much larger quantity. For normal exposures this developer should be reduced somewhat with water. For under-exposed plates it may be used at full strength. Development should be carried on in the same manner as with pyro; it should, however, proceed further. The developer is saved, as it may be used repeatedly, working a little slower each time. Old developer may be used for over-exposed plates and for lantern slides. Fine effects may be obtained by beginning the development with strong hydrochinon and finishing with the old, weaker solution, or this order may be reversed. The plates are washed and fixed as before described. The developer should be kept in well-corked bottles in a dark, cool place.

If a negative lacks density in the high lights, it may be intensified with bichloride of mercury and ammonia or silver cyanide; but if it is over-exposed, intensification will not help it. The negative may be intensified at any time after the final washing, even after it has been dried, but if dry, it should be soaked in water for a few minutes before intensifying.

An ounce of bichloride of mercury in a quart of water constitutes the intensifying solution. In this immerse the negative, rocking it gently until it is of a light straw color all through. It is then rinsed thoroughly.

After the negative is thoroughly washed it is placed in a solution of water and ammonia (1 drachm strong ammonia to a pint of water), or in the silver cyanide, where it is allowed to remain until it is blackened through the film. It is then washed thoroughly.

Lantern slide plates measure $3\frac{1}{4} \times 4$ inches. They are made of much thinner glass than the ordinary negative plates. If negatives are not of the proper size for lantern slides, they may be printed in the camera.

For contact prints the negative is laid in the printing frame face up, and a lantern slide plate is laid face down on this. A piece of black paper is then placed over the back of the lantern slide plate, and the back of the printing frame is put on and fastened. The exposure is made either by daylight or by artificial light, the latter being preferred. The plate is exposed by holding the printing frame about one foot from the light. Very weak negatives may be held at a distance of five or six feet, but the time of exposure must be very much increased. The time of exposure is from two or three to thirty seconds when a five-foot gas burner is used at a distance of one foot.

All lantern slide plates are slow, and admit of the use of orange light in the dark room. A good developer for this purpose is a weak hydrochinon solution. The image should begin to appear in from three to five minutes, and should be completely developed in fifteen minutes. Over-exposure is liable to veil the high lights, and while the slide may be handsome to look at, it will be worthless for projection. The high lights in a slide should be perfectly transparent. With a negative having clear shadows and a dense sky, care should be taken not to print too heavily, for while the high lights will be clear, other parts will be dark and without detail. After development is completed, the plate is rinsed thoroughly and fixed in hypo as already described. The fixing solution should be fresh and clean. When the fixing is complete, the plate is washed thoroughly and finally swabbed with wet cotton wool. After the prints are dry they are coated with thin collodion by flowing it evenly over the plate. The slides are covered with thin glass of the same size. Worthless negative and positive plates may be cleaned with very dilute hydrofluoric acid, and used as slide covers. A mat is interposed between the print and the cover, and the two glasses are bound together with adhesive paper cut into one-half or three-eighths inch strips. A small label should be placed on the lower left hand corner of the slide to serve as a guide in putting it into the lantern.

PRINTING.

Few amateurs find profit in preparing their own paper for printing. And as various good ready sensitized papers are found in the market accompanied with full directions for toning, we will confine ourselves to the gelatino-chloride paper, which is easily worked. The back of the negative is cleaned before printing, the negative is placed face up in the printing frame, and a piece of paper is placed face down upon it. The back of the frame is put on, and the paper is exposed through the negative to the sunlight. Weak negatives should be printed in the shade. A cover of tissue paper placed over the printing frame during printing preserves details. With a good negative of a landscape, for example, the printing should be continued until it is a few shades darker than required in the finished print.

It is advisable to trim all prints before toning. The trimming may be done on a glass plate, using a glass trimming form to guide the knife. Prints should be carefully kept from the light until they are toned. They should be toned within two or three days from the time of printing—the sooner, the better. The prints are thoroughly washed in eight or ten waters until the free silver is washed out and the water is clear, before they are placed in the toning solution.

Formulas of several toning baths are given below:

For Purple or Black Tones.

Chloride of gold	2 grains.
Bicarbonate of soda	8 to 16 "
Water	16 ounces.

For purple tones the smaller quantity of bicarbonate soda is used; for black tones, the larger quantity. This solution should be made up an hour before use, and not kept in stock.

For Deep Brown Tones.

Chloride of gold	1 grain.
Sodium acetate	20 "
Water	10 ounces.

Make up several hours before use.

In either case use enough solution to fill the pan. About one grain of gold is used for each 20 × 24 sheet of paper. If the prints tone too slowly, the solution must be slightly strengthened. Only a few prints are put into the bath at a time, and they are kept in motion until the red disappears and they are a little darker than they should appear when finished. Prints may have a bluish tinge, or the color may run into a purple. A print when undertoned is red. The art of toning can be learned only by practice. After toning, the prints are placed in water for a time. The solution made according to the last formula should be filtered, and kept in a dark place for future use.

The fixing bath consists of:

Water............................1 pint.
Sodium hyposulphite...............4 ounces.

The hypo should be dissolved before the toning begins. The prints may all be put into the hypo at the same time—not more than two sheets of paper to each pint of solution. They should be turned and moved about continually. The time required for fixing will be from fifteen to twenty minutes. The hypo reddens and fades prints which have been only partially toned or printed too light. The color of all prints is rendered lighter by the hypo. After fixing, the prints must be thoroughly washed for an hour in running water or in several waters and allowed to soak for a considerable time, say half an hour between the washings. The permanence of the print depends upon this washing.

The prints when completely washed, and while still wet, may be squeegeed.

A POCKET CAMERA.

No equipment for a tour or a summer's vacation is now complete without a photographic outfit for making instantaneous memoranda of scenes and objects met with upon the road, on the river or lake, or in the picturesque nooks of mountain and valley. The principal trouble with photography in these days is not with the plates and chemicals, as of old, but with the more or less cumbersome camera and accessories, which must be ever present with the artist.

If large pictures are desired, a large camera and tripod of corresponding size will, of course, be required. To these must be added a complement of plate holders if a number of pictures are to be made in a short time. Some of the recently devised cameras are very portable, and in every way desirable. The writer adds to the list an instrument which differs in some respects from others. The principal feature is the plate-changing device, which is quite simple and admits of the use of flexible bags for holding the plates before and after exposure. The bags—which hold one plate each—are made of the stout black paper known in the trade as leatherette. Each bag has a very thin covering of leather, such as is used by bookbinders on very light work, and around the mouth of the bag is glued a band of

FIG. 318.

A Pocket Camera.

thin, tough pasteboard. The bags are made over a wooden form. A dozen filled bags occupy very little more room than the plates in the original package. The light is excluded, and the plates are held in the bags by folding over the top, as shown in the engraving. Each bag is provided with a rubber band extending around it lengthwise, to prevent it from unfolding.

In the present case, the plate holder proper is made of brass and fitted to the camera box, from which it is never removed, except in case of some disarrangement of the interior parts of the camera. The holder consists of a flat sheath, made of suitable size to readily admit the plate, and provided with an opening in the front side, of the size of the field of the lens. This opening is surrounded by a flange which fits light-tight into the camera box.

Two light bowed springs, *a*, are soldered to the back of

FIG. 319.

Interior of Pocket Camera.

the sheath, and tend to press the plate forward to bring the film into the focal plane.

The end of the sheath, which projects upward above the top of the camera box, is of suitable size to be received in the stiffened ends of the bags, and a channel is formed around the end of the sheath near its upper end by soldering an angled strip of brass around the mouth of the sheath, as shown Fig. 319. Into this channel the stiffened end of the bag is inserted before it is unfolded. The channel is blackened, so that when the end of the bag is inserted in it, no light can enter. Now, by straightening the bag and shaking the camera, the plate contained by the bag will be made to fall into the holder. The bag can now be folded against the back of the holder and held there by one of the elastic bands extending over the top and under the bottom of the box. The removal of the plate from the camera is simply the reverse of what has just been described; that is, the bag is unfolded, and the camera being inverted, the plate is dropped into the bag, when the bag is again folded and removed from the holder.

The shutter of this little camera is both simple and effective. It admits of instantaneous and time exposures, and can readily be adjusted to any required speed without opening the camera box.

The shutter consists of a light metallic disk, A, provided with a central boss arranged to turn on a stud projecting from a plate secured to the inner surface of the front of the box. A stout but fine cord, b, is attached by one end to a small loop soldered to the face of the shutter and wound once around the boss of the shutter; the remaining end passes through a hole in the end of the spring, c. A screw, d, passes through the top of the camera, through a slot in the spring, c, the nut being fitted to the slot of the spring and provided with shoulders which support the spring. By turning the screw, d, the spring may be made to turn the shutter with more or less rapidity, as may be required. A cord, e, inserted in an eye on the boss of the shutter and wound in a direction opposite that of the cord, b, passes out through a hole in the box and serves to set the shutter.

The shutter is provided with two small studs, fg, the stud, f, being arranged near the periphery of the disk, in position to be engaged by the spring catch, h, when the shutter is drawn around by the cord, e, preparatory to making an instantaneous exposure. The stud, g, is placed in such a position relative to the catch, h', that its engagement with the catch will hold the shutter open, or with its opening, i, coincident with the opening of the tube, as indicated in dotted lines.

The catch, h', is provided with a wire arm, j, which

extends behind the catch, h, in such a way as to allow the catch, h', to move a short distance before releasing the catch, h. Each catch is provided with a stud which projects through the camera box and presses against the leather covering, forming two small convex projections, l, m. When an instantaneous exposure is desired, the shutter is released by pressing the projection, l. When a time exposure is to be made, the button, m, is pressed. This operation first throws the catch, h', into the path of the stud, g, then releases the stud, f, allowing the shutter to turn until the stud, g, strikes the catch, h. This will arrest the shutter in an open position. When the catch, $h,'$ is released, the shutter closes. For time exposures the camera box may be placed on any convenient support.

For instantaneous exposures, the camera may be held in the hand. One desiring to make a camera of this kind, and having the proper facilities, could substitute a toothed sector and pinion for the shutter boss and the cords used in operating it.

The camera lens is of the spherical, wide angle kind, with a fixed focus for all distances from five feet upward.

The camera box is 2 inches deep and $3\frac{1}{2}$ inches square, outside measurement. The camera was designed especially as a tourist's companion for taking lantern views, and it has served its purpose very well indeed.

SIMPLE PHOTOGRAPHIC AND PHOTO-MICROGRAPHIC APPARATUS.

While first class photographic instruments can be made only by makers having the greatest skill and large experience, an ordinary camera that will serve the purposes of the amateur may be made by the amateur himself with the expenditure of an insignificant sum for materials.

Nos. 1 to 12, Plate V., show a camera tube, box, and tripod, the materials of which cost less than a dollar. The construction is within the range of any one having a little mechanical ability. The camera is intended for 4 by 5 plates, therefore the size of the plate holder and the focal length of the tube will determine the size of the camera box. To avoid turning the camera or plate holder, the box is made square, and the inside dimensions of the plate holder are such as to permit of placing the plate either horizontally or vertically, according to the subject to be photographed. The plate holder is 5¾ inches square inside, and is provided with a wooden back of sufficient thickness to support the hooks employed for holding the plate. There are four V-shaped wire hooks, *a*, at the bottom of the holder, two for receiving the end edge of the plate, and two farther apart, and arranged higher up, for receiving the side edge of the plate; and near the top of the holder there are three Z-shaped hooks, *a*, one in the center for engaging the end edge of the plate, and one near each side of the holder for receiving the side edge of the plate. The top of the frame is slotted, and the sides and bottom are grooved to receive the slide, which covers the plate before and after exposure. To the under surface of the upper part of the frame of the plate holder is attached a looped strip of elastic black cloth, such as broadcloth or beaver, which closes over the slot of the plate holder, as shown at 10, Plate V., when the slide is withdrawn, and thus shuts out the light. The interior of the plate holder. as well as the slide, should be made dead black, by applying a varnish made by adding three or four drops of shellac varnish to one ounce of alcohol, and stirring in lampblack until the required blackness is secured.

The main frame of the camera box is made square, and is secured at right angles to the base board. The frame is provided with a narrow bead or ledge that will enter the front of the plate holder and exclude the light.

To the front of the frame are secured four trapezoidal pieces of pasteboard, of the form and size given at 6. These pieces of pasteboard are secured to each other and to the camera box frame by tape, glued on as shown. If the box is made of junk board, it may be nailed together with wire nails. In this manner a pyramidal box is formed which is strong, light, and compact. In the smaller end of the box is fitted the beveled, centrally apertured block shown at 7. The aperture of this block must be made to fit the camera tube shown at 1 and 2, after having received a lining of plush or heavy felt. The camera tube may consist of paper or

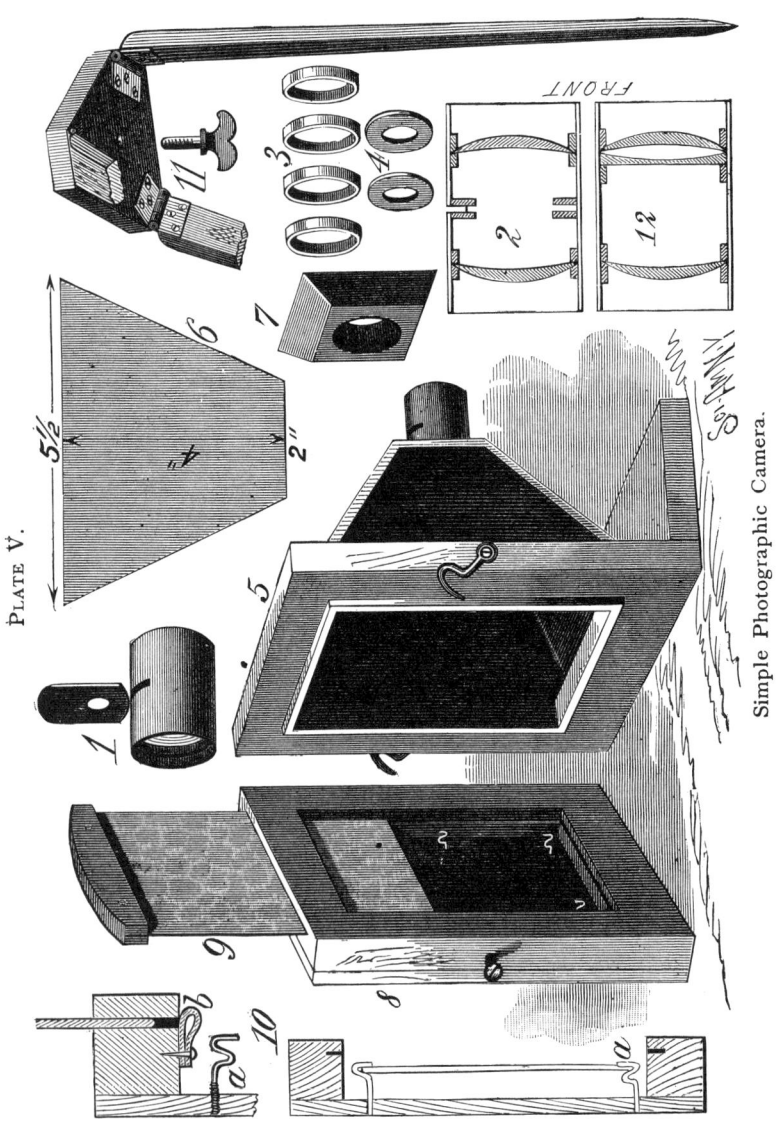

Simple Photographic Camera.

metal. Paper answers well, and costs nothing. The internal diameter of the tube is determined by the diameter of the lenses. Ordinary meniscus spectacle lenses of eight inch focus are employed. These lenses are secured in place by paper rings, shown at 3, the inner rings being glued in place, the outer ones being made removable for convenience in cleaning the lenses. The lenses are arranged with their convex sides outward; the distance between them is $1\frac{1}{4}$ inches, and in one side of the tube, half way between the lenses, is made a slot to receive the diaphragms, as shown at 1 and 2. Upon each side of the slot, within the tube, are secured flat rings, shown at 4, which together form a guide for the diaphragms, as shown at 2 Plate V.

The tube is adjusted at the proper focal distance from the plate by temporarily securing at the back of the box a piece of ground glass or tracing paper, in exactly the same plane as that occupied by the plate in the plate holder. The tube is then moved back and forth until a focus is obtained which shows the image fairly sharp throughout the field. In arranging for a fixed focus, it is perhaps best to favor the foreground rather than the distance. The tube should move with sufficient friction to prevent it from being easily displaced. By using a small diaphragm, it will be found unnecessary to focus each object separately.

At 12, Plate V., is shown a combination of cheap lenses, which is effective for portraits and for other classes of work in which focusing is admissible. It consists of two meniscus lenses, each of $8\frac{1}{2}$ inches focus, having their convex sides arranged outwardly and a plano-concave lens, 16 inches focus, arranged with its concave side against the concave side of the outer lens of the system. The plano-concave and the rear meniscus lenses are arranged $1\frac{1}{2}$ inches apart. Diaphragms may be used as in the other case, and a box about 8 inches deep will be required.

The tripod is formed of a triangular centrally apertured board, to which are hinged three tapering wooden legs, by means of ordinary butt hinges, as shown at 11, Plate V. The base of the camera box is secured to the tripod by means of an ordinary thumb screw.

This outfit will enable the amateur to cultivate his tastes, and learn much about photography. Dry plates will, of course, be used. They are procurable almost anywhere, and are inexpensive. As to the treatment of plates after exposure, and printing and toning, the reader is referred to the first article in this chapter and to the works on photography.

The amateur who possesses one of the microscopes already described may arrange it for projection, and may insert the end of the microscope tube in the camera box

FIG. 320.

Microscope and Camera arranged for Photo-Micrography.

above described, after removing the tube, and project the image of the microscopic object on the sensitive plate, and thus produce good negatives of the objects, from which prints may be made which will be interesting both to the operator and his friends.

The eyepiece of the microscope referred to is a very good objective for photo-micrography. Although special objectives are made for this purpose, almost any good objective will produce a good negative. In photographing micro-

PHOTOGRAPHY.

scopic objects, it will be necessary to employ a focusing ground glass, and to focus very carefully by the aid of a magnifier.

Slow plates are preferable for this use, as they bring out the detail much better than fast plates. The time of exposure will vary with the object, from fifteen seconds to as many minutes. In some cases the time extends to hours.

Fig. 320 shows the arrangement of the lantern, the microscope, and the camera box. It will be noticed that the annular space in the end of the camera box around the microscope tube is stopped by a black cloth wound loosely around the microscope tube. This and other precautions are necessary for preventing the light from reaching the plate except through the object and the microscope.*

DAGUERREOTYPY.

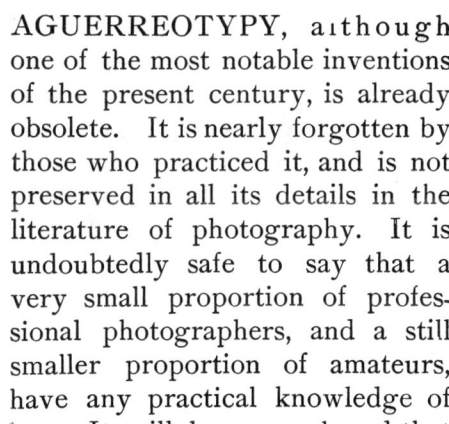

DAGUERREOTYPY, although one of the most notable inventions of the present century, is already obsolete. It is nearly forgotten by those who practiced it, and is not preserved in all its details in the literature of photography. It is undoubtedly safe to say that a very small proportion of professional photographers, and a still smaller proportion of amateurs, have any practical knowledge of the subject. It will be remembered that Niepce and Daguerre sought independently of each other for a method of producing sun pictures. Niepce at first employed plates coated with bitumen. He formed a partnership with Daguerre in 1829, but died before the invention now known as daguerreotypy was perfected.

After the death of Niepce, Daguerre improved the art

---

* For full information upon this subject, the reader is referred to "Photo-Micrographs and How to Make Them," by George M. Sternberg.

to such an extent that Niepce's son allowed it to go under its present name. Both inventors received annuities from the government for giving the invention to the public.

In this country the art was first practiced by Morse, and was improved by Draper soon after it was introduced here.

Daguerreotypy was very simple, easily understood, and easily managed, and was learned by many who found it a light business, requiring little capital and returning large profits.

The plates employed were copper faced with silver. The metal was hard-rolled, and the plates, as received from the

FIG. 321.

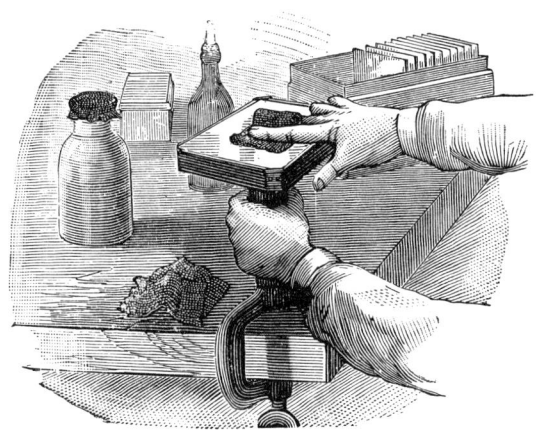

Scouring the Plate.

manufacturers, were flat and quite smooth, but not polished. The first step toward the preparation of the plate for use was to clip the corners and turn down the edges slightly, in a machine designed for the purpose, to bring the sharp edges of the plate out of reach of the buff employed in producing the necessary polish.

The plate was held, for scouring, in a block having clips on diagonally opposite corners for engaging the corners of the plate. One of the clips was made adjustable, to admit of readily changing the plates. The block was mounted pivotally on a support clamped to the table, as shown in Fig. 321.

PHOTOGRAPHY.

The scouring was effected by sprinkling on the plate the finest rottenstone from a bottle having a thin muslin cover over its mouth, and the rottenstone as well as the square of Canton flannel with which it was applied was moistened with dilute alcohol. The center of the Canton flannel square was then clasped between two of the fingers, and moved round and round with a gyratory motion until the plate acquired a fine dead-smooth surface. The last traces of rottenstone were removed by means of a clean square of flannel. The plate was then transferred to a block mounted on a swinging support, and buffed by the vigorous applica-

Fig. 322

Buffing.

tion of a straight or curved hand buff formed of a board about four inches wide and thirty inches long, padded with four or five thicknesses of Canton flannel, and covered with buckskin charged with the finest rouge. Scrupulous cleanliness was imperative in every step of the process.

The buffs were kept clean and dry, when not in use, by inclosing them in a sort of vertical tin oven (Fig. 322), which was warmed by a small spirit lamp. A careful operator would prepare a plate having a bright black polish without a visible scratch, while an incompetent or careless man would fail in this part of the process, and would prepare plates full of transverse grooves and scratches. The beauty of the picture depended very much on the careful preparation of the plate.

Occasionally, a buff would in some manner receive par-

Fig. 323.

The Rotary Buff.

ticles of matter which would cause it to scratch the plate. The remedy consisted in scraping the face of the buckskin, and brushing it thoroughly with a stiff bristle brush, gen-

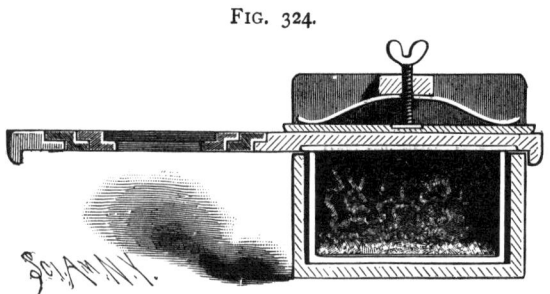

Fig. 324.

The Coating Box.

erally a hair brush devoted especially to this use. The buff was then recharged by dusting on rouge from a muslin bag.

When the rotary buff wheel was adopted, it insured rapid work, but it was otherwise no improvement over the hand buff. At first, the wheels were made cylindrical, but

PHOTOGRAPHY. 341

that incurred the necessity of an objectionable seam or joint where the leather lapped. The conical buff wheel (Fig. 323) allowed of the use of a whole skin, thereby dispensing with the seam.

After buffing, the plate was taken to the dark room to be

FIG. 325.

The Dark Room—coating the Plate.

sensitized. The room had a side window, generally covered with yellow tissue paper, for the examination of the plate during the process. The room contained two coating boxes, one for iodine, the other for bromine. The construction of these boxes is clearly shown in Fig. 324, which is a longitudinal section of one of them. The two boxes were alike except in the matter of depth; the bromine box being about twice as deep as the iodine box.

Each box contained a rectangular glass jar having ground edges. In the top of the box was fitted a slide more than twice as long as the box. In the under surface of one end of the slide was fitted a plate of glass, adapted to close the top of the jar, and in the opposite end of the slide was formed an aperture furnished with a rabbet for receiving the plate. Upon the top of the slide was arranged a spring-pressed board, which held the slide down upon the top of the jar.

On the bottom of the jar of the iodine box were strewn the scales of iodine, and in the bromine box was placed quicklime charged with bromine. The bromine was added to the lime drop by drop, and the lime occasionally shaken until it assumed a bright pink hue bordering on orange. The lime was thus prepared in a glass-stoppered jar, and transferred to the jar of the coating box as needed; one inch being about the depth required in the coating box. The polished plate was placed face downward first in the slide of the iodine box and coated by pushing in the slide so as to bring the plate over the iodine in the jar. It was there exposed to the vapor of iodine until it acquired a rich straw color, the plate being removed and examined by the light of the paper window, and replaced if necessary to deepen the color. The plate was then in a similar manner subjected to the fumes of the bromine until it became of a dark orange color. It was then returned to the iodine box and further coated until it acquired a deep brownish orange color bordering on purple. The time required for coating the plate depended upon the temperature of the dark room. The process was very rapid in a warm room and quite slow in a cool room.

The plate, rendered sensitive to the light by the thin layer of bromo-iodide of silver, was placed in a plate holder, and exposed in a camera according to the well known method. The time of exposure was much longer than that of modern photography. A great deal depended on the quality of the lenses of the camera. The exposure in the best cameras was reasonably short. The old time gallery, with its antiquated camera and fixtures, and the dark room with the appurtenances, are faithfully represented in the

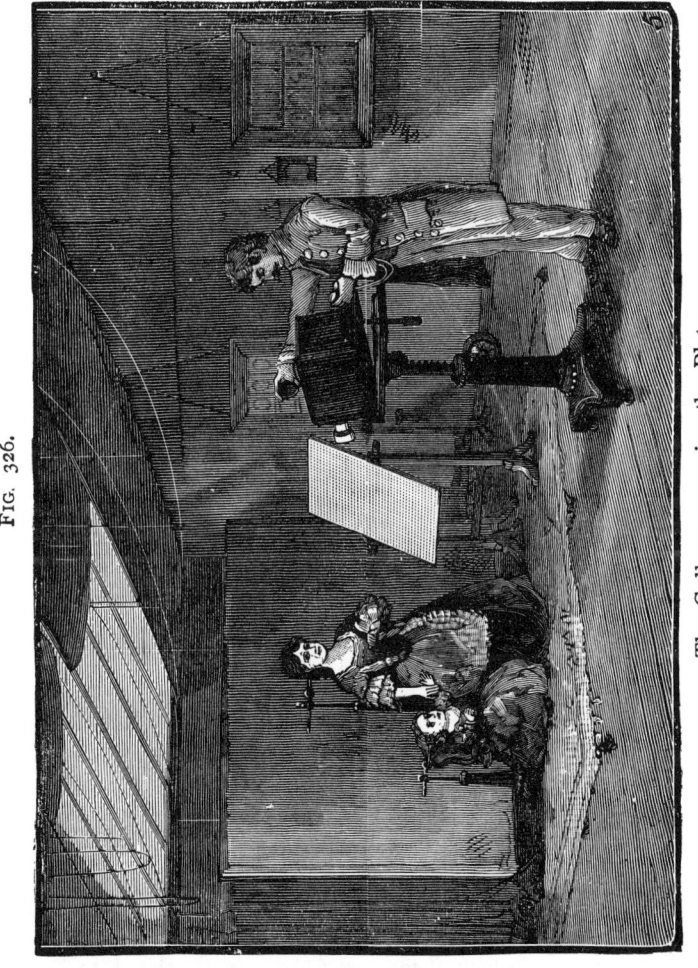

Fig. 326. The Gallery—exposing the Plate.

engravings (Figs. 325 and 326). After exposure, the plate was taken to another dark room for development. It was placed face downward over a flaring iron vessel (Fig. 327), in the bottom of which there was a small quantity of pure mercury. The mercury was maintained at a temperature of

120° to 130° Fah. by means of a small spirit lamp. The temperature was measured by a thermometer attached to the side of the vessel. The plate was raised occasionally and examined by the light of a taper, until the picture was fully brought out, when it was removed from the mercury bath and fixed.*

The fixing (Fig. 328) consisted merely in flowing over the plate repeatedly a solution of hyposulphite of soda having sufficient strength to remove in about half a minute all

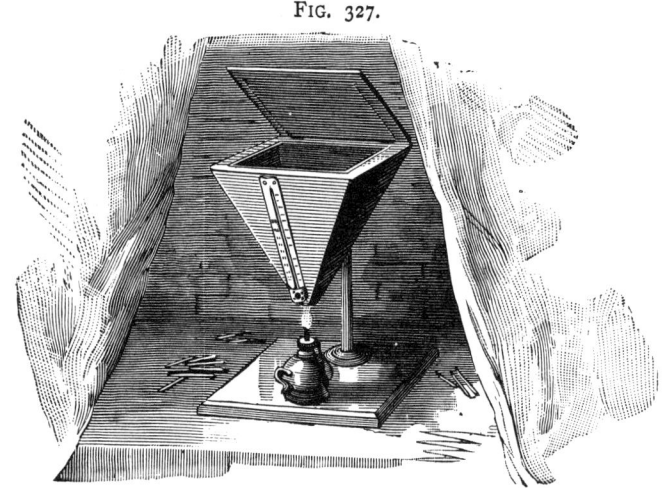

FIG. 327.

Developing the Plate.

the bromo-iodide of silver not acted upon by light. The plate was then thoroughly washed, and afterward gilded or toned by pouring upon it a weak solution of chloride of

* A fortunate accident led to the discovery of the development of the photographic impression by means of the vapor of mercury. Previous to this discovery, the image was brought out by a long-continued exposure in the camera. Daguerre on one occasion placed some under-exposed plates, which were considered useless, in a closet in which there were chemicals. Afterward, happening to look at the plates, he was astonished to find an image upon them. After taking one chemical after another from the closet until apparently all were removed, the images on his plates were still mysteriously developed. At length he discovered on the floor an overlooked dish of mercury, and the mystery was solved. He ascertained that the effects produced by mercury vapor spontaneously given off could be secured at will by suitable apparatus.

gold and heating it gently by means of a spirit lamp until a thin film of gold was deposited upon the plate and the pic-

FIG. 328.

Fixing.

ture attained the desired tone. The plate was then washed in clean water, and finally dried evenly and quickly over a spirit lamp.

FIG. 329.

Gilding or Toning.

This operation added to the strength and beauty of the picture, and also served to protect the surface of the plate to a great extent against the action of gases.

The finished picture was protected by a cover glass, and the edges of the glass and plate were securely sealed by a strip of paper attached by means of an adhesive coating.

Later on a metallic binding was added, which was called the "preserver." The pictures thus mounted were fitted to cases and frames which were more or less elaborate, varying in cost from a few cents to many dollars. Many daguerreotypes were inserted in gold lockets and charms, and occasionally they were fitted to finger rings made to receive them.

CHAPTER XVI.

MAGNETISM.

Nature furnishes permanent magnets "ready made," the lodestone being an example of such a magnet. She is able to induce magnetism in magnetic bodies, the earth itself being the great magnet by which the induction effects are secured. It is to the directive force of this great magnet that the compass owes its value.

The magnetism of the lodestone is due, doubtless, to a

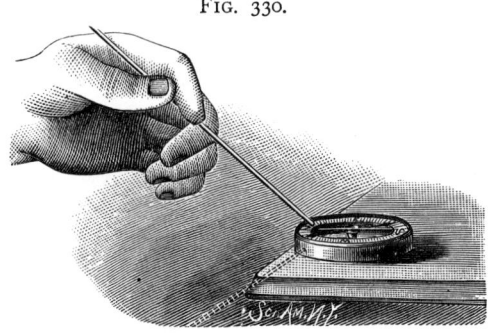

Magnetism by Induction from the Earth.

long exposure to the inductive influence of the earth's magnetism. Any body of magnetic material becomes temporarily magnetized to some extent when placed in the magnetic meridian parallel with the dipping needle, and if it be a body like soft iron, without coercive force, it loses its magnetism when arranged at right angles to this position in the same plane. This may be shown by placing a rod of well-annealed wrought iron in the magnetic meridian in an inclined position, with the lower end toward the north, as indicated in the dotted lines in Fig. 330, with its upper end in close proximity to the end of a compass needle. The needle will be instantly deflected, showing that the rod has become magnetic. When turned in the plane of the magnetic meridian to a position at right angles to its former position, it will lose its magnetism and will not repel the needle. By placing a bar of hardened steel in the magnetic meridian

FIG. 331.

Development of Magnetism by Torsion.

and striking it several blows on the end with a hammer, it becomes permanently magnetic, not strongly, but sufficiently to exhibit polarity when presented to a magnetic needle.

By twisting a rod of soft iron having one of its ends in

FIG. 332.   FIG. 333.

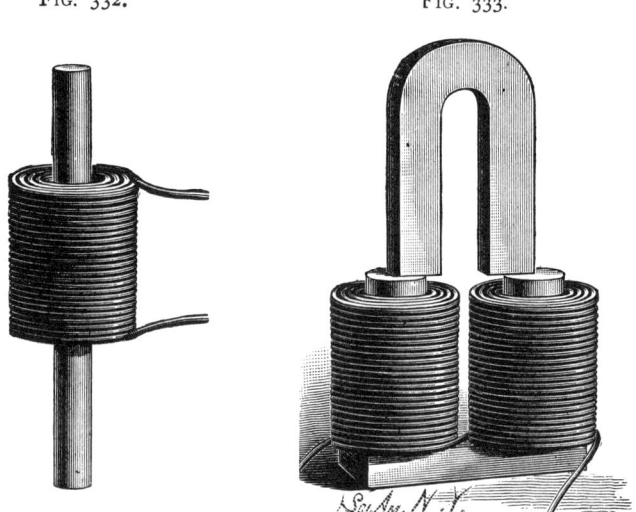

Magnetization of Bars.   Magnetization of U-Shaped Bars.

proximity to a magnetic needle, it is shown by the deflection of the needle that magnetism is developed by torsion (Fig. 331). By this and similar experiments it may be shown that stress and compression favor magnetization.

MAGNETISM. 349

Artificial magnets are produced by the contact of hardened steel with magnets or by means of the voltaic current. The latter is the more effective method, provided a strong current and a suitable helix or electro-magnet is available. For the magnetization of bars of steel a helix like that shown in Fig. 332 is needed. Its size and the amount of current required will, of course, depend upon the size of the bar to be magnetized. For all bars up to $\frac{3}{8}$ inch diameter, a helix $\frac{5}{8}$ inch in internal diameter, 2 inches external diameter, and $2\frac{1}{2}$ inches long, made of No. 16 magnet wire, is sufficient.

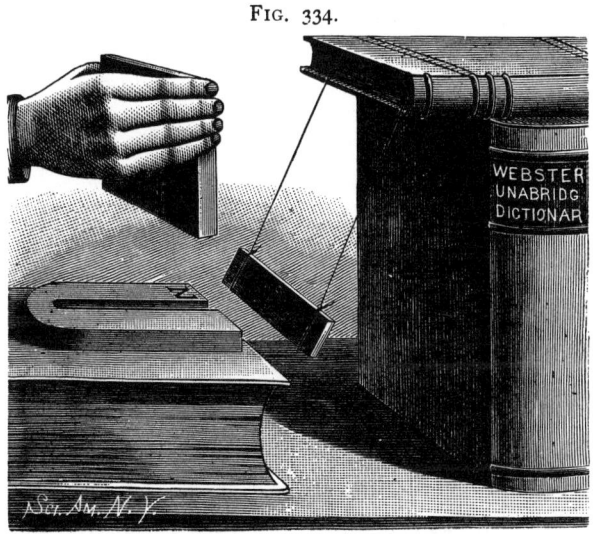

FIG. 334.

Motion produced by a Permanent Magnet.

A current from five or six cells of plunging bichromate battery is required, or in lieu thereof a similar current from a dynamo.

The bar to be magnetized is hardened at the ends and placed in the helix, the current is then applied, and the helix is moved from the center of the bar to one end, then to the opposite end and back to the center, when the current is discontinued, and the bar is removed. If several bars are to be magnetized, they may be placed end to end, and passed through the coil in succession. The magnetization of U-shaped

350 EXPERIMENTAL SCIENCE.

bars may be accomplished by means of an electro-magnet formed of two coils above described and a suitable soft iron core (Fig. 333). The U-shaped bar is placed on the poles of the electro-magnet as shown, when the current is sent through the coils for a short turn and then interrupted. Another method, which is perhaps more effectual, consists in drawing the U-shaped bar several times across the poles of the electro-magnet.

In the search for perpetual motion, vain efforts have been made to discover a substance which could be interposed between the magnet and its armature, and removed without the expenditure of power, and which would intercept the lines of force, so as to allow the armature to be alternately drawn forward and released, but no such substance has ever

FIG. 335.

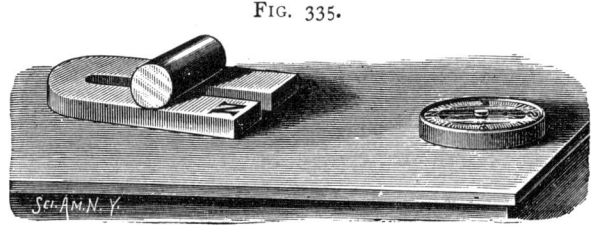

Effect of the Armature.

been discovered. The lines of force may be intercepted by a plate of soft iron placed between the magnet and its armature, but it requires more power to introduce the plate into the magnetic field, and withdraw it therefrom, than can be recovered from the armature. Fig. 334 illustrates an experiment showing how motion may be produced by the force of a permanent magnet. An armature is suspended by threads in the field of a permanent magnet. The magnet attracts the armature, slightly deflecting its suspension from a true vertical line. The introduction of a soft iron plate between the magnet and its armature intercepts the lines of force, thus releasing the armature, when it swings back under the influence of gravity. If at this instant the iron plate is withdrawn, the magnet again acts upon the armature, drawing it forward. Another introduction of the

MAGNETISM. 351

iron plate into the field again releases the armature, when it swings back, this time a little farther than before. By moving the iron plate in this manner synchronously with the oscillations of the armature, this may be made to swing through a large arc.

When a piece of soft iron is placed in direct contact with the poles of a permanent magnet, the magnetic force is nearly all concentrated upon the soft iron, so that there is very little free magnetism in the vicinity of the poles of the magnet. This may be readily shown by arranging a U-mag-

FIG 336.

Permanent Magnet and Bar magnetized by Induction.

net parallel with the magnetic meridian, placing in front of and near the poles of the magnet a compass so adjusted with reference to the poles as to cause the needle to rest at right angles to the magnetic meridian, then applying to the poles of the magnet a massive armature. It will be found that the needle, under these conditions, immediately tends to assume its normal position, showing that the power of the magnet over the needle has been, to a great extent, neutralized. By rolling a cylindrical armature along the arms of the U-magnet, as shown in Fig. 335, it is found that as the armature recedes from the poles of the magnet the influence of the magnet upon the compass needle is increased, while the movement of the armature in the opposite direction diminishes the power of the magnet over the needle.

In Fig. 336 is illustrated an example of temporary magnetization by induction, and of the effect of a permanent magnet on the iron so magnetized, showing that the iron bar inductively magnetized acts like a permanently magnetized needle. The soft iron bar is freely suspended, and receives its magnetism from the fixed magnet. The end of the suspended bar adjacent to the N pole of the magnet becomes S, as may be shown by presenting to it the S pole of another permanent magnet. The S end of the swinging bar will be

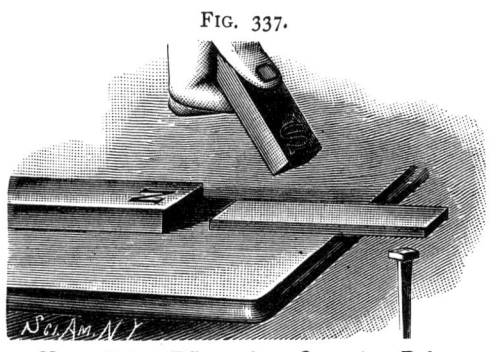

Fig. 337.

Neutralizing Effect of an Opposing Pole.

immediately repelled. If the S end of the permanent magnet be presented to the opposite end of the suspended bar, the reverse of what has been described will take place, *i. e.*, that end of the bar will be attracted, showing that its polarity is N.

In Fig. 337 is illustrated an experiment showing the neutral effect produced by induction from two dissimilar magnetic poles. A bar of soft iron is arranged near, but not in contact with, the pole (say the N pole) of a magnet, so that it becomes magnetized by induction to such an extent as to support a nail. The N pole of the magnet produces S polarity in the end of the soft iron bar adjacent to it and N polarity in the opposite end. The S end of another permanent magnet presented to the same end of the iron bar will produce exactly the opposite effect in the bar, and will, therefore, neutralize the magnetism induced in the bar by the first magnet, and cause the nail to drop.

A similar effect is produced when the iron bar is in actual

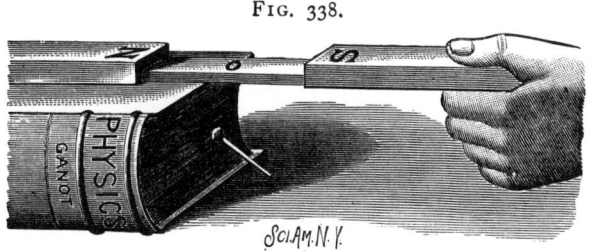

Neutral Point between Unlike Poles.

contact with the N pole of a magnet and the S pole of another magnet is brought into contact with the opposite end of the bar, as shown in Fig. 338. The nail will adhere to the bar when either magnet alone is in contact with the bar; but when dissimilar poles are brought into contact with oppo-

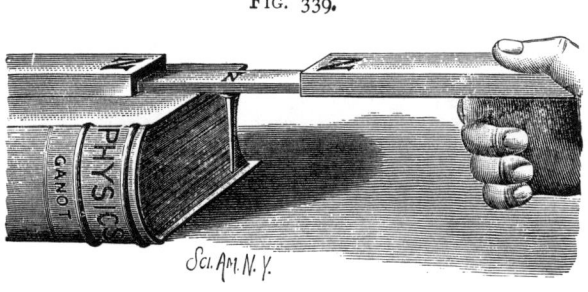

Consequent Pole.

site ends of the bar, its middle portion becomes neutral, and is no longer able to support the nail.

When like magnetic poles are presented to the ends of the iron bar, as in Fig. 339, a strong consequent pole is developed in the center of the bar, which is of the same name as that of the ends of the magnets touching the bar.

MAGNETIC CURVES.

A great deal may be learned about the properties of magnets by causing them to delineate their own characteristics. The common method of doing this is to form magnetic curves by dusting iron filings on a glass plate, then jarring the plate to cause the particles to arrange them-

FIG. 340.

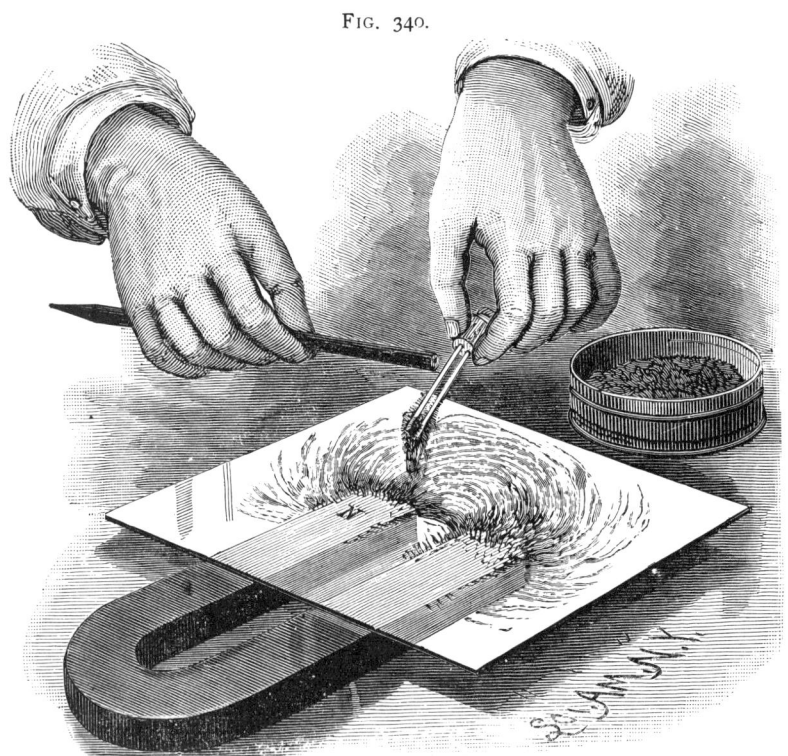

The Formation of Magnetic Curves.

selves parallel with the lines of force extending from the magnetic poles. The figures thus formed are not, of course, entirely autographic; and as they tend to develop in lines, they convey the erroneous idea that the lines of force, as spoken of in connection with magnets, are really separate lines or streams of force.

There is no way of exactly representing the magnetic

field of force by forms or figures, but the annexed engravings serve to illustrate a method of forming and fixing curves which has some advantages over the method referred to above. The magnetic particles fall in the position in which they are to remain, and no jarring is required.

To make a flat plate for lantern projection or individual use, a plate of glass flowed with spirit varnish is laid upon the magnet, and iron dust reduced from the sulphate, or fine filings, or dust from a lathe or planer, is applied by means of a small magnet in the manner indicated in Fig. 340. The small magnet in this case consists of two magnetized carpet needles inserted in a cork, with unlike projecting poles arranged about one-quarter of an inch apart. A little of the iron dust is taken up on the small magnet, and the slightly

Fig. 341.

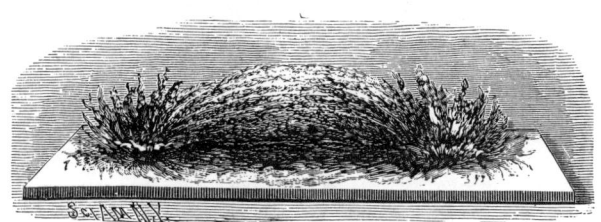

Magnetic Curves in Relief.

adhering particles are shaken off. The remaining portion is then disengaged from the small magnet by rapping the magnet with a pencil, the small magnet being held above the poles of the larger one. The particles having been polarized by the small magnet, arrange themselves in the proper position while falling. Several applications of the iron dust will be required to complete the figure. Of course the iron must be applied before the varnish dries, and the plate should be allowed to remain on the magnet until dry.

To make the curves in relief, as shown in Fig. 341, a slightly different method is employed. The glass plate is warmed, coated with paraffine, and allowed to cool. It is then placed on the magnet, and proceeded with as in the other case. With care the curves can be built up to a considerable height, especially if the larger magnet be a strong one. Iron filings or turnings of medium fineness are required in this case.

When the curves have assumed the desired proportions, a few very fine shreds of paraffine, scraped from a paraffine block or candle, are deposited very gently on the curves, and melted by holding above them a hot shovel. More shreds are then added and the hot shovel is again applied, and so on until the mass of iron filings is saturated with paraffine, when it is allowed to cool. The plate to which

Fig. 342.

Arborescent Magnetic Figures.

the filings are now attached may be removed from the magnet after having applied the armature, if it be a permanent magnet, or after interrupting the current, if it be an electromagnet, when the curves will retain their position.

The arborescent figures shown in Fig. 342 are built upon a cap of brass, which incloses the poles of the magnet separately. The magnet in this case is arranged with its poles downward. The fixing of these curves is somewhat difficult, on account of being obliged to work under the plate, but it can be accomplished by proceeding in the manner described. Instead of the hot shovel, an alcohol lamp or

Bunsen burner is used in this case for melting the paraffine, but considerable care is required to prevent the iron dust from burning. The figure when cool may be removed from the magnet and preserved.

FLOATING MAGNETS.

The ordinary magnetic fish, ducks, geese, boats, etc., are examples of floating magnets, which show in a very pleasing way the attraction and repulsion of the magnet. The little bar magnet accompanying these toys serves as a wand for assembling or dispersing the floating figures; or it may serve, in the hands of the juvenile experimenter, as a baited fish hook.

FIG. 343.

Floating Magnetic Figures.

FIG. 344.

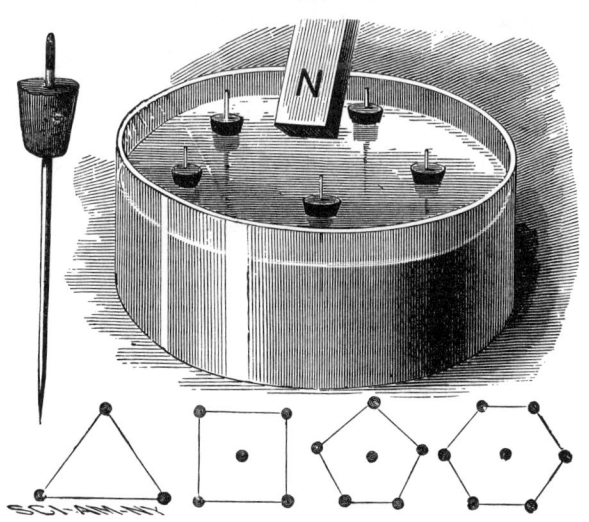

Mayer's Floating Needles.

Prof. A. M. Mayer has devised an arrangement of floating magnetic needles which beautifully exhibits the mutual repulsion of similarly magnetized bodies. A number of strongly magnetized carpet needles are inserted in small corks, as shown in Fig. 344.

When floated, these needles arrange themselves in symmetrical groups, the forms of the groups varying with the number of needles.

One pole on a bar magnet held over the center of a vessel containing the floating needles will disperse the needles, while the other pole will draw them together.

ROLLING ARMATURE AND MAGNETIC TOP.

The rolling armature applied to a long

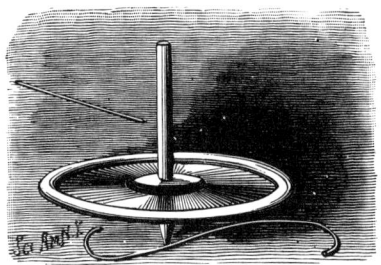

Fig. 345.

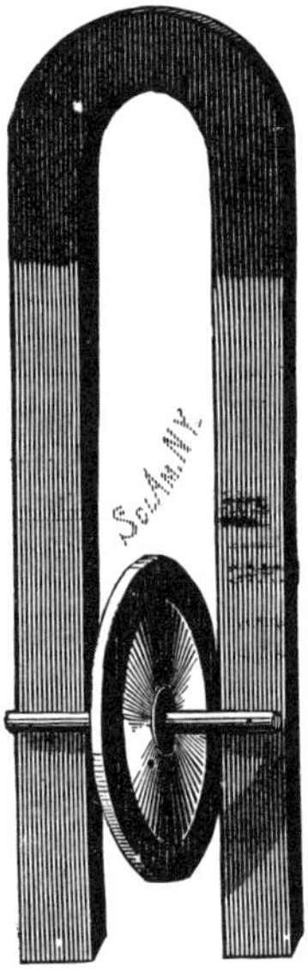

Magnet and Rolling Armature.

Fig. 346.

Magnetic Top.

U-magnet exhibits the persistency with which an armature adheres to a magnet. The wheel on the cylindric armature acquires momentum in rolling down the arms of the magnet which carries it across the polar extremities and up the other side (Fig. 345).

A very pretty modification of this toy has recently been devised. It consists of a top with a magnetic spindle and straight and curved iron wires (Fig. 346). The top is spun by the thumb and fingers in the usual way, and one of the wires is placed against the side of the point of the spindle. The friction of the spindle causes the wire to shoot back and forth with a very curious shuttle motion. The point of the top rolls first along one side of the wire and then along the other side.

CHAPTER XVII.

FRICTIONAL ELECTRICITY.

Many different views have been entertained regarding the nature of electricity, but notwithstanding the multiplicity of electrical inventions and discoveries and their numerous practical applications, the problem of the real nature of electricity remains unsolved. Recent experiments, however, have shown quite conclusively that electricity, like light, heat, and sound, is a phenomenon of wave motion. Laws

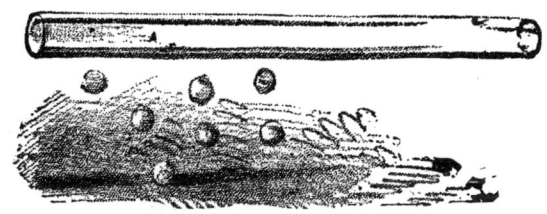

FIG. 347.

Attraction and Repulsion of Pith Balls by an Electrified Rod.

governing its various manifestations have been discovered, so that, knowing the conditions of its production and use, results can be determined with certainty.

Electricity is evoked from bodies by friction, pressure, chemical action, and other causes. A glass rod or stick of sealing wax rubbed with dry silk or flannel becomes electrified, so that when it is held over bits of paper or small pith balls, as shown in Fig. 347, these will leap at once to the glass or sealing wax, and after a brief contact they will be repelled, to be again attracted and repelled, and so on.

It is a matter of indifference whether the rod be of glass or sealing wax; the result is the same. It is easy to determine by a very simple experiment that the electrification of the glass rod differs from that of the sealing wax. A pith ball is suspended by a silk thread from an insulating standard, and when an electrified glass rod is brought near the pith ball, the latter is immediately attracted, and after a brief contact is repelled.

The attraction of the pith ball by the electrified glass is due to the electrification of the ball in the opposite sense by induction from the glass rod.

Bodies oppositely electrified mutually attract each other. When the pith ball touches the glass rod, its former charge of electricity becomes neutralized, and it receives a charge by conduction which is like that of the glass rod. The two

FIG. 348.

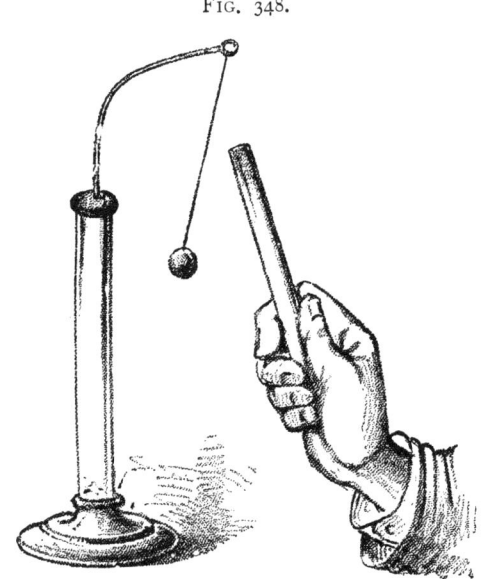

Electric Pendulum.

bodies being now similarly electrified, the pith ball is repelled. Bodies having like charges of electricity mutually repel each other. Now, while the pith ball is charged with electricity received by conduction from the glass, if an electrified stick of sealing wax be brought near the pith ball, the latter will be at once attracted by the former, thus showing the electrification of the two bodies to be different.

Two glass rods delicately suspended by silk threads, and electrified, will repel each other.

Two sticks of sealing wax treated in like manner will act

toward each other in the same way; but if one of the electrified glass rods be brought near one of the electrified sticks of sealing wax, there will be mutual attraction.

These two manifestations of electricity were originally called *vitreous* and *resinous* electricity, in consequence of being developed respectively upon glass and resin. Now, however, that which is evoked from glass is known as positive electricity, and that from resin as negative electricity, but these are merely convenient conventional names given to opposite phases of the same thing.

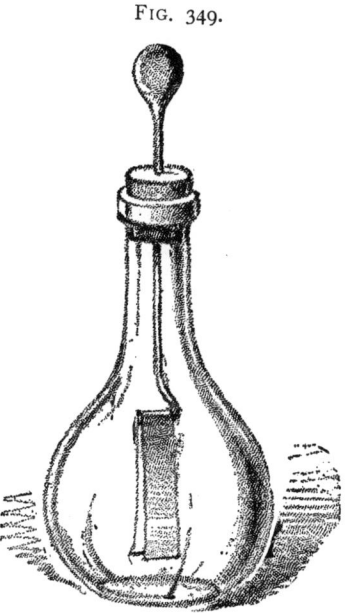

FIG. 349.

Electroscope.

An electroscope is an instrument for determining the presence and kind of electricity.

The electroscope in its simplest form is shown in Fig. 349. It is far more sensitive than the electrical pendulum, and may be used in many instructive experiments.

It consists of a small flask or bottle, through the stopper of which is inserted a brass wire having at its upper end a metal ball and at its lower end a hook bent out horizontally to receive two strips of very thin metal leaf, either Dutch-metal leaf, silver or gold leaf, or aluminum leaf, the latter on account of its extreme lightness being preferable. The strips, which are three-eighths inch wide and two inches long, are fastened to the top of the wire hook by means of gum or even saliva alone.

To determine when a body is electrified, present it to the ball. If the leaves mutually repel each other and diverge, electricity is present. A slight touch of a glass rod, a rubber comb or ruler, or a wooden ruler, upon the clothing or carpet, or even upon a wooden surface, develops electricity in sufficient quantities to affect the electroscope. Very little friction is required to evoke a perceptible amount of electricity. One movement of the clothes brush upon the clothes or carpet affects the electroscope from a long distance. A feather duster brushed once over a varnished chair will cause the leaves of the electroscope to diverge at a distance of eight to ten feet, the effect in this case being produced by electrical induction, more fully described later on. An ordinary elastic rubber band drawn across the edge of the desk develops sufficient electricity to widely diverge

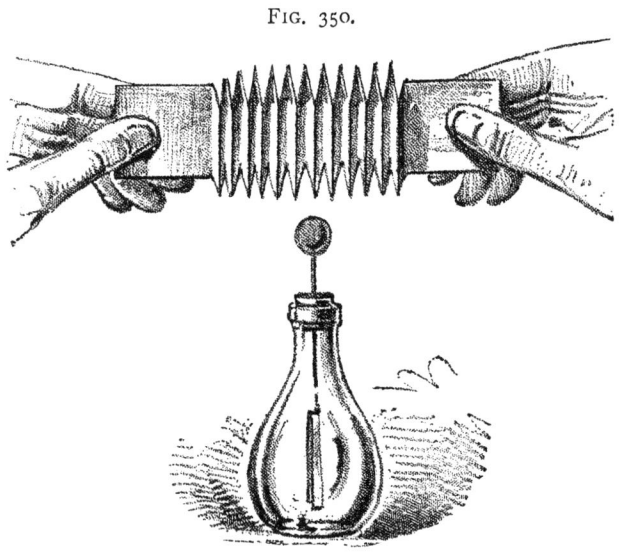

Experiment with Electroscope.

the leaves. The rubber band affords a curious example of the distribution of electricity on an extensible surface. If after electrification the rubber band is held over the electroscope, and alternately elongated and allowed to contract, the leaves of the electroscope will be seen to converge when the band is stretched, and to diverge when the band contracts.

If a piece of paper, folded like a fan and well dried, is struck several times with a dry silk handkerchief or woolen cloth, and afterward alternately closed and opened over the

FRICTIONAL ELECTRICITY. 363

electroscope, as shown in Fig. 350, the reverse of what occurred in the case of the rubber band will happen. That is, when the paper is stretched out the leaves will diverge, and when it is closed up they will fall together, showing that in the latter case the electricity is masked.

There are many other interesting experiments that may be tried with the electroscope in connection with simple objects that may be found anywhere.

A toy exhibiting some of the phenomena of frictional electricity is shown in Fig. 351. It has received the name of Ano-Kato. It is a flaring box lined with tin foil, covered

FIG. 351.

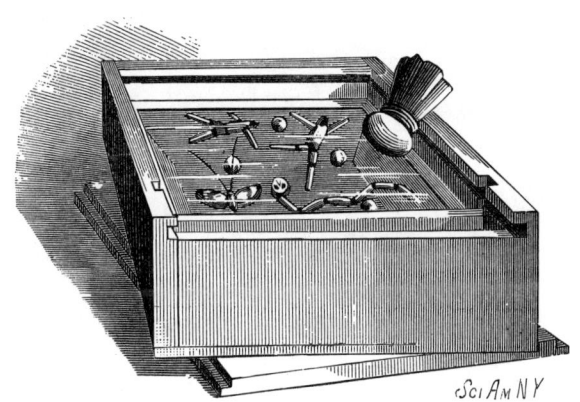

Ano-Kato.

with a piece of ordinary window glass, and containing figures made of pith.

By rubbing the glass with a leather pad charged with bisulphide of tin, the electrical equilibrium is disturbed, and the figures are attracted and repelled, and made to go through all sorts of gymnastics.

An interesting example of the mutual repulsion of similarly electrified bodies is shown in Fig. 352.

For the experiment illustrated, the rubber strips were seventeen feet long.

A manufacturer in handling some of the rubber threads used in making suspenders and other elastic webs noticed that the threads at times repelled each other. The repulsion was naturally attributed to electrification, and the experiment illustrated was at once suggested. The elastic rubber strips used in the experiment were suspended from the ceiling in one of the apartments of the *Scientific Ameri-*

Fig. 352.—Mutual Repulsion of Electrified Threads.

can office, and were electrified by simply brushing them over with a feather duster. The threads became more and more divergent as the electrification proceeded, until it finally became impossible to approach the threads without becoming entangled in them.

FRICTIONAL ELECTRICITY.

Upon gathering all of the free ends of the threads together, the repulsion of the threads at their mid-length caused them to separate widely. When once electrified, in a dry day, the threads retain the charge for hours. They are discharged by connecting them with the ground through the body, and drawing them through the hand.

When the mercury in a barometer tube is agitated, the friction of the mercury on the glass generates electricity and produces effects which are visible in the dark.

FIG. 353.

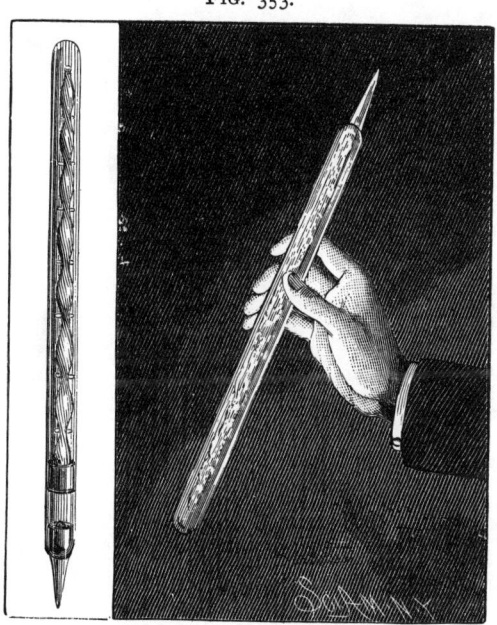

Self-exciting Geissler Tube.

The self-exciting vacuum tube, shown in Fig. 353, operates in the same manner. The electrical effect is produced by the friction of mercury on the inner surfaces of the vacuous glass tube, as the tube is inverted or shaken. The tube is ingeniously contrived to prevent breakage by the falling of the mercury against the end of the tube, and at the same time to increase the effectiveness of the device by arranging two tubes concentrically, the inner tube being beaded, and provided with little knobs for breaking the fall of the mer-

Fig. 354.

Self-luminous Buoy.

FRICTIONAL ELECTRICITY.

cury. The inner tube is sealed to the outer tube near one end, and in the inner tube, a short distance above this sealing, is formed an aperture which determines the amount of mercury to be retained between the inner and outer tubes when the tube is inverted preparatory to use, as all of the mercury between the two tubes and above the aperture will run through the aperture into the lower end of the tube. In this manner the mercury is equally divided, so that when the tube is reversed, one-half of the mercury flows through the inner tube, and the other half flows downward between the inner and outer tubes.

The full effect is realized only when the mercury is allowed to flow quickly from one end of the tube to the other, but any agitation of the mercury in the tube produces some phosphorescent light. This tube is a beautiful object in a dark room.

Fig. 354 illustrates illuminating apparatus designed as an auxiliary to bell buoys and whistling buoys. It is based upon the generation of electricity by the agitation of mercury in a high vacuum or in an attenuated gas. It involves the same principle as the self-exciting vacuum tube just described. The buoy represented in the cut is adapted to ring the bell by the rolling motion imparted to it by the waves. Advantage is taken of this motion to agitate mercury in the annular tubes placed in the upper portion of the frame of the buoy. The tubes are made very heavy and strong, and each contains barriers for causing friction of the mercury against the sides of the tubes.

To insure the action of one or more of the tubes at all times, they are inclined at different angles. A slight motion of the buoy causes the mercury to travel circularly in the tubes and generate sufficient electricity to render the tubes luminous.

Among devices tried for rendering buoys luminous are lamps arranged to burn for a long time, phosphorescent mixtures, electric illuminators supplied with the current from the shore by means of a cable, and the more recent luminous paint, which absorbs light by day and gives it out at night. Compressed gas has been employed with great success, some of the buoys having been designed to carry six months' supply of gas and to serve as lightships.

ELECTRICAL MACHINES.

The simplest machine for supplying electricity in small quantities is the electrophorus, invented by Volta. It consists of two parts, one being a vulcanite disk secured to a metallic sole plate, the other a metallic cover plate provided with a handle of hard rubber or other insulating material.

FIG. 355.

Electrophorus.

To secure the best results, the vulcanite disk and metal cover plate should be warmed, dried, and freed from dust. The vulcanite disk is rubbed with a piece of warm flannel or a cat skin, when it becomes charged with negative electricity. The cover plate is then placed on the vulcanite disk. The negative electricity of the vulcanite disk acts inductively upon the cover, positive electricity being attracted to the lower side of the cover, while negative electricity is repelled to the upper side. By touching the upper side of the cover while it is still in contact with the vulcanite disk, the negative electricity will pass from the cover through the body of the operator to the ground, and only positive electricity will be retained by the cover. If now the cover be raised from the vulcanite disk by means of the insulating handle, a spark may be drawn from it. This is due to the combination of the positive charge on the cover with the negative induced in the hand by this charge.

The cover may be replaced upon the vulcanite disk, and the operation may be repeated indefinitely, when the conditions are favorable, without further excitation. Instead of a vulcanite disk, a cake formed of resin, shellac, and a small

FIG. 356.

Winter's Electrical Machine.

proportion of Venice turpentine may be used. The materials are melted, thoroughly mixed, and poured into a circular tin pan. The cake thus formed is allowed to remain in the pan, and is used in the same manner as the vulcanite disk.

Winter of Vienna devised a simple frictional electrical machine, an inexpensive form of which is shown in Fig. 356. In the top of the cast iron standard is journaled a shaft having at one end a crank by which it may be turned, and furnished at the opposite end with a pair of collars between which is clamped a vulcanite disk. In a socket at the base of the standard is inserted a forked bar of wood which extends upwardly and embraces the vulcanite disk, each arm of the fork being provided on its inner face with a silk or woolen cushion charged with bisulphide of tin and arranged to press on the disk. Upon a vulcanite column rising from the base board near the edge of the disk is supported a metallic ball,

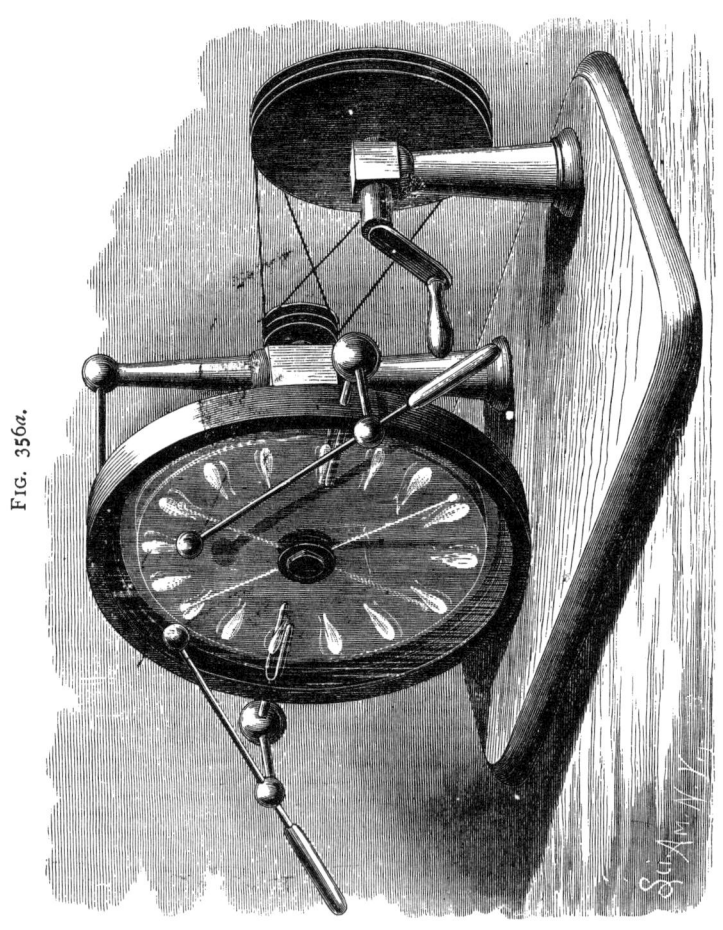

FIG. 356*a*.

Modified Wimshurst Induction Machine.

to which are attached metallic rings arranged on opposite sides of the vulcanite disk, and provided with a number of short points projecting inwardly. To the forked wooden bar is attached one edge of a segmental silk case, which incloses a portion of the disk between the cushions and the collector.

FRICTIONAL ELECTRICITY. 371

When the machine is turned in a right-handed direction, electricity is generated by the friction between the cushions and the disk, and the negative electricity is carried along to the collecting points, where it is drawn off and accumulated upon the ball. The positive electricity escapes to the ground through the rubbers and the base. If the rubbers were insulated, positive electricity could be taken from them by connecting the insulated ball with the ground. The machine will yield a spark having a length equal to about one-sixth of the diameter of the disk.

The Wimshurst electrical machine is the most recent, and on some accounts it is the best that has been devised. It is less affected by atmospheric conditions, and may be relied on in all weathers for results of some kind, while the frictional machines and the induction machines of Holtz and Toepler generally fail in a damp atmosphere.

The Wimshurst machine here shown differs from the ordinary type, mainly in having the rotary disks inclosed by a hoop, and glass cover disks to exclude dust and moisture, the stationary disks being provided with brushes which are connected electrically by strips of tin foil secured to the inner faces of the outer disks by means of shellac.

This machine is shown in perspective in Fig. 356a. Fig. 357 is a vertical section taken through the center of the disks, and Fig. 358 is an enlarged horizontal section taken on the line of the collectors.

The column supporting the revolving disks is provided with a hollow arm in which is journaled a tubular shaft, upon one end of which is mounted a disk of common window glass between two collars, the glass being centrally apertured* to receive the shaft, the outer collar being screwed on.

The opposite end of the tubular shaft is provided with a grooved pulley. A solid shaft placed within the tubular shaft, and projecting beyond the ends thereof, carries upon one end a glass disk, and upon the other a grooved pulley, as in the first case. The glass disks are separated from each other about $\frac{1}{8}$ inch. They are both coated with shellac var-

* For hints on perforating glass, see chapter on mechanical operations.

nish and allowed to dry. To each glass disk near its periphery are secured sixteen radial sector plates of tin foil or thin brass, arranged at equal angular distances apart. These sectors are coated on one side with shellac varnish and allowed to dry, when they are placed in position on the varnished glass disks, varnished side down, and secured by rubbing each one quickly with a warm, smooth iron.

A drawing should be made of a glass disk with the sectors to be placed under the disks as a guide in locating the

FIGS. 357 AND 358.

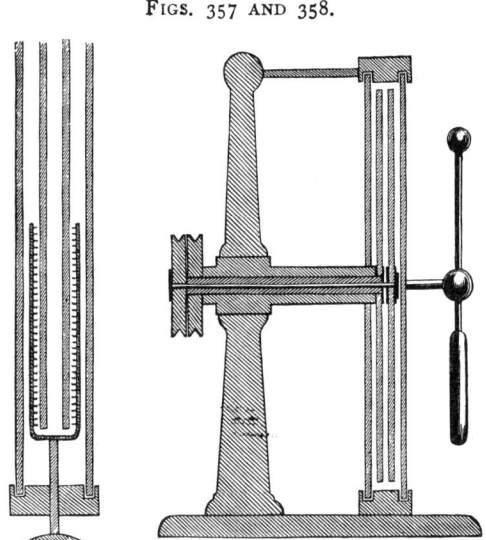

Sectional Views of Modified Wimshurst Machine.

sectors. Brass sectors are preferable on account of their superior wearing qualities.

The glass disks are placed on their respective shafts with the sectors outward. A ring of vulcanite surrounds the glass disks and is grooved internally to receive the stationary glass disks, which inclose the rotary ones. The vulcanite ring is divided at the top and bottom to allow of applying it to the stationary plates. The rear plate is centrally apertured to admit the tubular support of the shafts. The vulcanite ring is provided, at the top and bottom, where it is divided, with vulcanite dowels, and is supported by attachment at the bottom to the base board, and at the top to a wooden rod projecting from the upper end of the column.

In diametrically opposite sides of the vulcanite ring, and on a level with the axis of the disks, are inserted brass rods, provided on their inner ends with metallic forks, the arms of which extend along the outer surfaces of the rotary disks and are provided with collecting points, as shown in Fig. 358. The outer ends of the brass rods are furnished with knobs into which are inserted the supports of the discharge rods or conductors. The latter are provided with vulcanite handles by which they may be moved in these supports as may be required.

Fig. 359.

Attachment of the Leyden Jar.

The stationary glass disks are each provided on their inner faces at diametrically opposite points with small metallic sockets, attached to the glass with cement, and containing brushes of tinsel or very fine brass wire, which touch the rotary disks lightly. The brushes of each pair are connected by a narrow strip of tin foil attached to the glass. The stationary glass disks may be turned in the vulcanite ring to adjust the brushes at the required angle, which is about 45° with the plane of the collecting forks.

One of the rotary disks is driven by a straight belt, the other by a crossed belt, both belts being carried by a doubly grooved wheel fixed to a shaft journaled in a standard attached to the base. This shaft is furnished with a crank, by which it is turned.

To secure good results small Leyden jars or condensers must be connected with the conductors, as shown in Fig. 359. To the bottom of each jar is attached a small chain. These chains are brought into contact when a detonating discharge is desired, and separated for a silent discharge.

The machine is self-exciting and yields sparks varying in length from one-fourth to nearly one-half of the radius of the rotary disks, according to the state of the atmosphere and the condition of the machine.

The machine illustrated has 12-inch rotary disks and 14-inch stationary disks.

Mr. Wimshurst has constructed the diagram (Fig. 360) which shows the distribution of the electricity upon the plate surfaces when the machine is fully excited. The inner circle of signs corresponds with the electricity upon the front surface of the disk. The two circles of signs between the two black rings refer to the electricity between the disks, while the outer circle of signs corresponds with the electricity upon the outer surface of the back disk. The inventor found by experiment that when two disks made of a flexible material were driven in one direction, they close together at the top and the bottom, while in the horizontal diameter they are repelled. When driven in the reverse direction, the opposite action takes place.

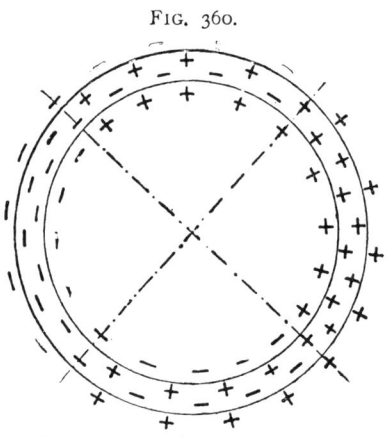

FIG. 360.

Distribution of Electricity upon the Plates.

EXPERIMENTS WITH THE INDUCTION MACHINE.

The appearance of the spark when the two conductors are separated only a short distance is shown in Fig. 361. It leaps in a straight line from one electrode to another. When the distance between the electrodes is greater, the spark takes a

FRICTIONAL ELECTRICITY. 375

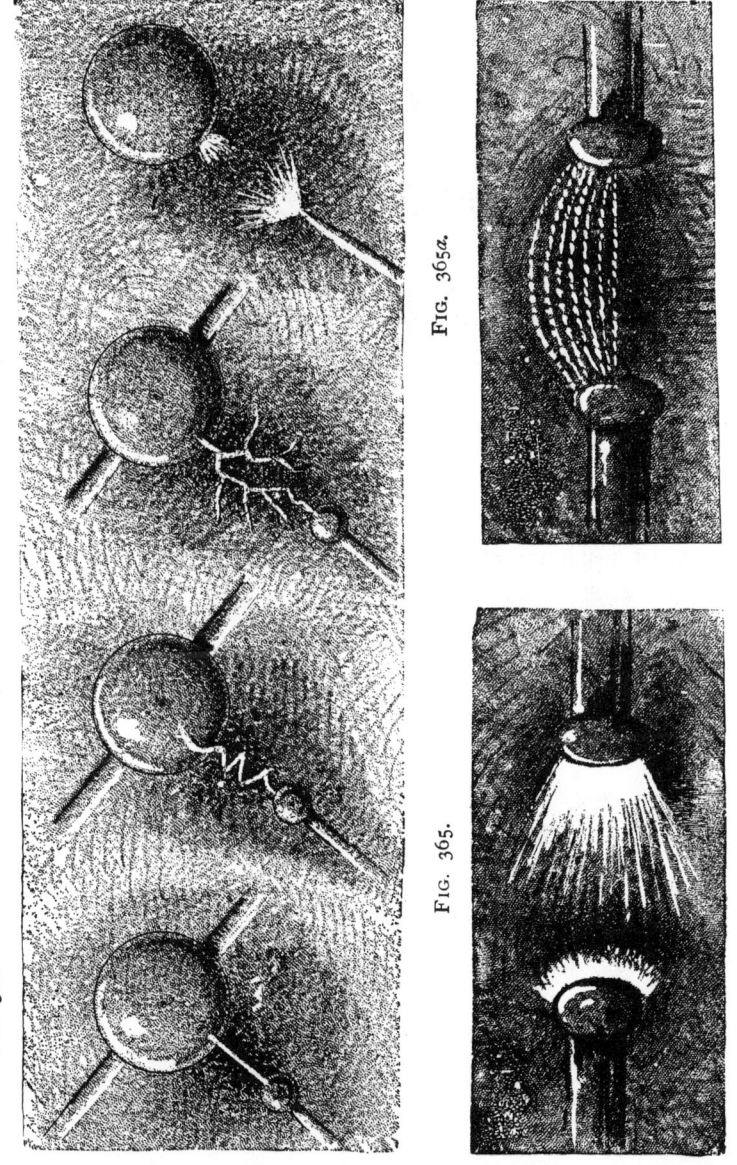

Fig. 364. Fig. 363. Fig. 362. Fig. 361. Fig. 365a. Fig. 365.

Various Phases of the Electric Discharge.

zigzag course, as shown in Fig. 362; and when a very long space separates the electrodes, the appearance of the dis-

FIG. 366.

Lengthening the Spark.

charge is as illustrated in Fig. 363. In Fig. 364 the discharge of positive electricity to a point is exhibited, and

FIG. 367.

Discharge over Finely Divided Metal.

in Fig. 365 the ends of the discharge rods are shown as they appear when a considerable distance apart, the machine being

FRICTIONAL ELECTRICITY. 377

arranged for the silent discharge. The multiple appearance of the small spark of the silent discharge, when the discharge rods are near together, is shown in Fig. 365a.

The report of the discharge is increased when a rubber plate is held between the rods, as in Fig. 366, the spark jumping over the edge of the rubber through an increased distance. Thick cardboard placed in this position is readily perforated, and the spark will pass through a pamphlet one-fourth inch thick.

Fig. 367 shows a glass plate eight inches square, furnished with a coating of finely divided metal. It is covered with a

FIG. 368.

Diversion of the Discharge by Moisture.

coat of thick shellac varnish or other suitable cement, and is thickly sprinkled with brass or iron filings before the varnish begins to dry. When the varnish is thoroughly dry, a band of tin foil is pasted across opposite ends of the glass. When opposite ends of this plate are connected with the conductors of the machine by a wire or otherwise, the discharge takes various courses over the filings, and when the machine is arranged for the silent discharge, the brilliancy of the spark is diminished, while the rapidity of the discharge is greatly increased.

The support shown in Fig. 367 is convenient for exhibiting this class of experiments. It consists of a thick plate of glass supported in a slightly inclined position by two wooden feet. Two knobs furnished with large flanges are cemented to the glass near its lower edge. The

FIG. 369.

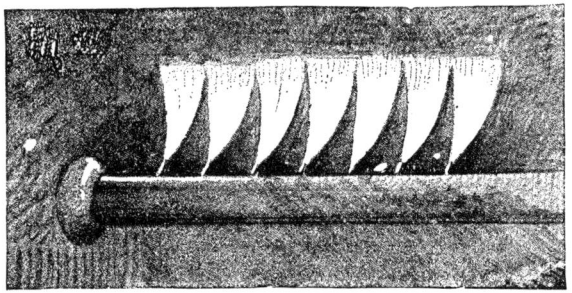

Glow at the Positive Collector.

knobs are sufficiently long to receive a tube or anything of that nature which it is desired to exhibit.

To conveniently connect the luminous panes with the machine, two U-shaped springs may be clasped on the edges

FIG. 370.

Glow at the Negative Collector.

of the glass, and connected with the machine by large wires. Unless chains with soldered links can be procured, wires or rods with rounded ends are preferable for making electrical connections, as chains afford numerous points for escape of electricity.

FRICTIONAL ELECTRICITY. 379

A case of the diversion of the electric discharge by exceedingly slight causes is illustrated by Fig. 368. The end of a vulcanite plate is moistened and placed against the ends of the conductors, and moved along so as to make a tracing of the moisture along the surface of the rubber. The discharge will follow these lines of moisture, however slight they may be, in preference to traversing the shorter route between the two conductors.

As to experiments possible with the induction machine, they are endless. The machine itself presents a weird and

FIG. 371.

Effect of the Hand on the Discharge.

interesting appearance in the dark. From the positive collector a luminous brush extends from each point, as shown in Fig. 369, while on the points of the negative collector only stars or luminous points are seen, as represented in Fig. 370. Besides these effects the inductors glow with a shimmering light, like the aurora. The brushes of the cross-arms are luminous, and all conducting points near the machine are aglow with the lambent light.

When the machine is at rest, if one hand is placed upon the negative conductor and the other hand is held a short distance above the positive conductor, as shown in Fig. 371, and if an assistant turns the machine, beams of soft purple light will radiate from the knob at the end of the discharge rod toward the hand. In this experiment the jars must be disconnected. No shock will be experienced during this experiment if it is carefully conducted.

Geissler tubes are best exhibited by placing them between the jars, allowing them to nearly touch the outer

FIG. 372.

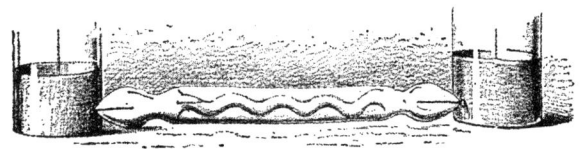

Discharge through a Geissler Tube.

coating of the jars. (Fig. 372.) The conductors should be placed one-fourth inch apart, and the machine adjusted for the silent discharge. Care should be taken in the use of tubes having long, sinuous passages, such as twisted or spiral tubes and the like, as they are very liable to be ruptured by the spark. When such tubes are used, the rods must be as near together as possible without destroying the effect.

Another method of exhibiting Geissler tubes is to hold them in the hand parallel with the face of the revolving

FIG. 373.

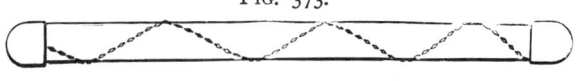

Tube with Interrupted Conductor.

plate, and about three or four inches from the large balls through which the discharge rods pass.

When the electric discharge is over an interrupted conductor, a bright spark appears at every interruption. Fig. 373 shows a tube wound spirally with a narrow strip of tin foil, cemented to it with starch paste, stratena, or shellac varnish. After the cement is thoroughly hard, the tin foil is separated at short intervals, say one-quarter inch, with a knife or file, leaving a narrow space of about one thirty-second inch between the sections. This tube may be from twelve to eighteen inches long, and for the sake of protection may be inclosed in another glass tube. A strip of foil should extend from the extremities of the interrupted strip over the ends of the outer tube. The inner tube may then be stopped with a cork at each end, which is allowed to project a short distance. These corks are rounded at their outer ends, and covered with rather thick tin foil, which is allowed to extend a short distance over the end of the outer tube. This tube, held by one end in the hand and presented by its other end to one of the conductors of the machine, exhibits a brilliant luminous spiral. The brilliancy of the

FIG. 374.

Franklin's Plate.

sparks may be increased by connecting the conductors with the ends of the tube.

By means of a condenser, large quantities of electricity may be condensed upon a small surface.

The various forms of condensers are alike in principle. They consist essentially of two insulated conductors separated by a non-conductor.

The Franklin plate or fulminating pane, shown in Fig. 374, is the simplest form of condenser. It is made by attaching sheets of tin foil to opposite sides of a pane of window glass, leaving a space of two inches all around. It will be found convenient to support the glass upon two wooden feet, as shown in the engraving. This plate is charged by connecting the tin foil on one side with the ground, and that upon the other side with one of the conductors of the machine. It is discharged by touching opposite sides with a discharger. By connecting opposite sides of the plate with the opposite conductors of the machine, the plate may be charged so that it will discharge over its edges with a loud report.

The Leyden jar, shown in Fig. 375, is nothing more than a fulminating pane rolled up. It may be made by covering a jar over the bottom and about half way up its sides with tin foil, and stopping the mouth of the bottle with a well varnished cork or wooden stopper, through which runs a one-eighth inch wire, having a knob on its upper end, and a piece of chain on its lower end resting on the tin foil lining. The uncovered portions of the glass jar should be coated with shellac varnish. The jar may be made in various sizes, and when the size is so small that it is inconvenient to apply tin foil to the inside, a little shellac varnish may be poured into the bottle, and the bottle coated half way up its sides with the varnish by turning it down upon the side and revolving it. Before the varnish begins to dry, a quantity of metal filings are poured into the bottle and shaken about. They attach themselves to the varnish and form a metallic coating that answers a very good purpose. When the varnish dries, the surplus filings may be poured out and the bottle may be coated with foil on the outside.

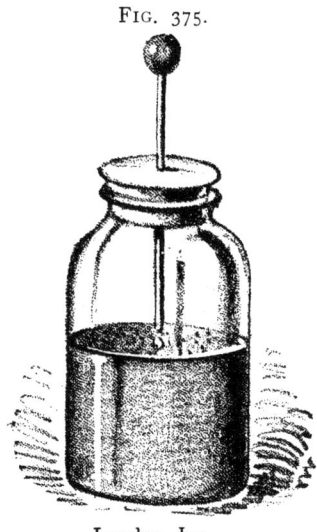

FIG. 375.

Leyden Jar.

The jar is charged by connecting the outer coating with the ground or with one of the conductors of the machine, and connecting the ball with the other conductor; and it is discharged by touching the ball and the outer coating of the jar with opposite ends of a jointed discharger. The measuring jar, shown in Fig. 376, is similar to the jar just described, the only difference being the addition of a curved wire having a ball on its lower end, and a support for the wire attached to the vertical discharge wire of the jar. The ball of the additional wire may be placed a greater or less distance from the outer coating of the jar. It is obvious that the jar can never be charged to give a spark longer than the distance between its outer coating and the ball.

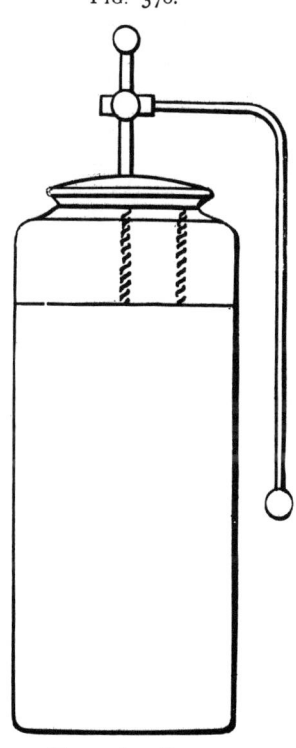

Fig. 376.

The disruptive effect of the spark can be readily exhibited by partly filling a glass bottle (Fig. 377) with kerosene, olive, or lard oil, and inserting through the cork a curved wire pointed at its lower end and provided with a ball at its upper end. The pointed end of the wire should be very near the inner surface of the glass, and the ball at the top should be connected with one of the conductors of the machine. The other conductor should be placed opposite the point of the wire and near the side of the bottle.

When the machine is turned, the sparks will perforate the glass, and will continue to pass through until the pointed wire is turned to

Measuring Jar.

a new place in the bottle, when another hole will be made. The holes made by the spark are so small that the oil will pass through very slowly, if at all.

Fig. 378 shows a chime of bells operated by the electric discharge. The three bells are suspended from a wire cross arm, which is attached to one of the conductors of the machine or to an insulated support connected with the machine. The two outer bells are suspended with chains,

FIG. 377.

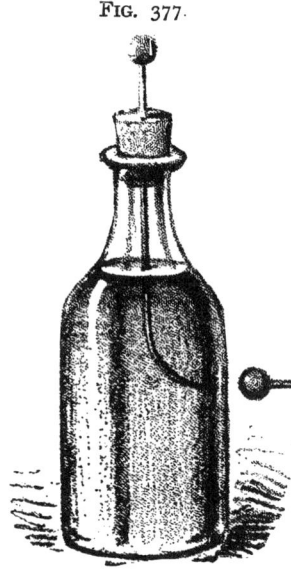

Disruptive Effect of the Discharge.

the middle one with a silk cord. Two small metal buttons are suspended by silk threads half way between the outer bells and the middle one, and the middle bell is provided with a chain which rests on the table.

When the machine is turned, the suspended buttons are attracted to the outer bells, and after becoming charged with electricity are repelled by the outer bells and attracted toward the middle one. After parting with their charge they are again attracted by the outer bells, again repelled, and so on. If the bells are connected with the ball of a Leyden jar, and the chain from the middle bell is connected with the outer coating of the jar, a slow discharge of the jar will take place. The time occupied in the discharge may be prolonged by fastening up one of

FIG. 378.

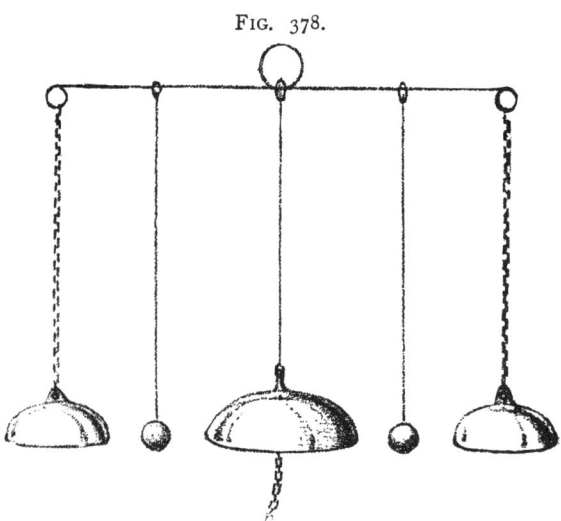

Electric Chime.

FRICTIONAL ELECTRICITY. 385

the buttons so that it will not swing. The electric fly, shown in Fig. 379, illustrates the effect of the electric discharge from points. The fly consists of a piece of metal having a slight depression in the center to receive the pivotal point on which it turns, and having a number of wire arms, pointed at their outer ends and all bent in the same direction. When the pivot of the fly is connected with the machine, the fly revolves in a direction opposite to that of the points. The motion is owing to a repulsion between the electricity of the points and the electricity imparted to the adjacent air by conduction.

FIG. 379.

The Electric Fly.

Fig. 380 shows a fly mounted on a horizontal axis, the latter being placed on two inclined wires having feet resting on a pane of glass. On connecting the incline with the machine, the fly will revolve and ascend the inclined plane in opposition to gravity. When electricity escapes from a point, the electrified air is repelled so strongly as to blow out a candle.

FIG. 380.

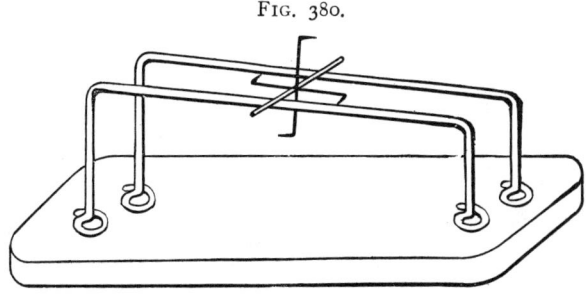

Fly and Inclined Plane.

For various experiments with the electrical machine and with Leyden jars a jointed discharger is required. A simple and inexpensive one is shown in Fig. 381. It consists of two wires bent one around the other to form a joint, and bent out nearly parallel in one direction to receive vulcanite handles, while the opposite extremities are curved and provided with balls at the ends.

In many experiments in static electricity the wires must

FIG. 381.

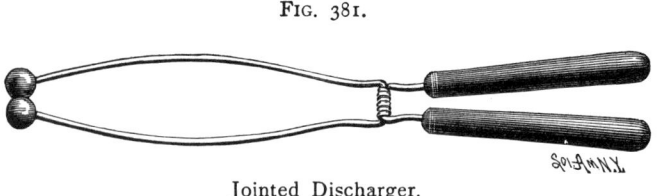

Jointed Discharger.

terminate in balls, to prevent escape and to secure the desired form of spark. It is a matter of considerable labor to make a large number of metal balls on the lathe. Balls which will answer every purpose may be cast directly upon the wires by using an old-fashioned bullet mould with a hole

FIG. 382.

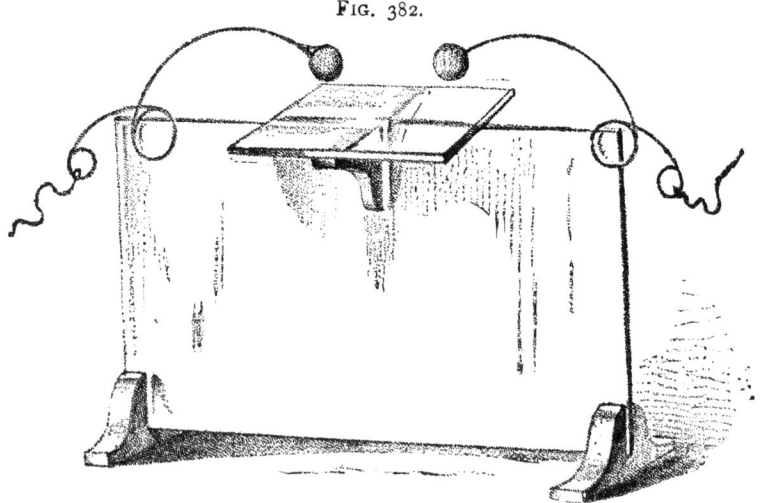

Universal Discharger.

drilled in the bottom to receive the wire upon which the ball is cast. Type metal is excellent for this purpose, but lead will answer very well. An alloy of tin and antimony makes a very fine ball, having the appearance of silver.

In a certain class of experiments the universal discharger

FRICTIONAL ELECTRICITY. 387

is convenient if not absolutely necessary. A cheap and simple form of this instrument is shown in Fig. 382. It consists of a pane of glass a foot long and six inches high, supported on wooden feet. Upon the upper edge in the center there is a glass table supported by two wooden brackets cemented to both glasses. Upon opposite corners of the upright glass there are two curved wires bent into the form of a spring in the middle to clasp the glass, and having at their upper ends balls and at their lower ends rings to receive the conductors which connect the discharger with the machine. By means of this instrmuent the electric discharge may be made to pass through or over any substance placed on the glass table.

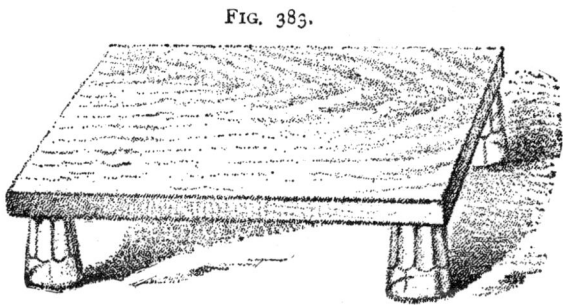

FIG. 383.

Insulating Stool.

The simplest method of making an insulating stool for supporting a person while being charged with electricity is shown in Fig. 383. It consists of a board resting on four common tumblers. An insulated spherical conductor is shown in Fig. 384. It may be made of any thin metal or it may consist of a pasteboard or wooden ball covered smoothly with tin foil. This sphere is provided with lateral arms terminating in knobs, and is supported upon a glass standard inserted in a wooden base. Fig. 385 shows a cylindrical conductor about four inches in diameter and twenty inches long. It has rounded ends, and is supported on a glass standard at the same height as the spherical conductor. With these two conductors the phenomena of static induction may be exhibited. In each end of the cylindrical conductor is inserted a standard from which two pith balls are suspended by silk threads. A pair of pith balls may also be suspended at the center of the conductor. Now, by charging the spherical conductor with positive electricity, and bringing it within a few inches of one end of the cylindrical conductor, the pith balls at the ends of the latter will diverge, while those at the center will remain quiet. By testing the charges of the conductor, it will be found that the electricity of the end of the conductor nearest the sphere is negative, while that of the remote end is positive.

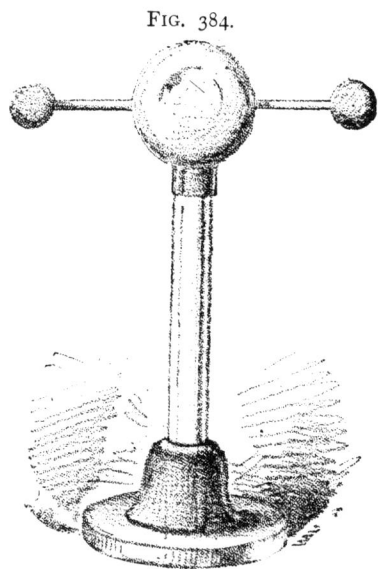

Fig. 384.

Insulated Sphere.

The positive charge of the sphere attracts the negative electricity of the cylinder and repels the positive, thus disturbing the equilibrium which existed before the approach of the positively charged sphere. On testing the middle portion of the cylinder by means of an electroscope, it is found to be neutral.

In Fig. 386 is shown a gas pistol, consisting of a metallic tube permanently stopped at one end with insulating material, and having a wire inserted in the stopper so that it nearly touches one side of the tube. The tube is filled with a mixture of illuminating gas and air, and lightly stopped with a cork.

An electric discharge through the wire and tube explodes the charge of gas. Fig. 387 shows a somewhat

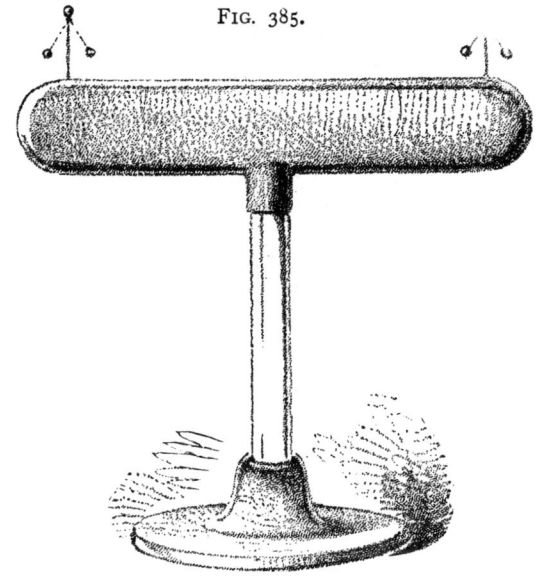

Cylindrical Conductor.

similar device for exploding gunpowder. It consists of a block of wood having a central cavity into which are inserted two wires nearly touching. The powder is placed in the cavity, and the spark sent through the wires, and in

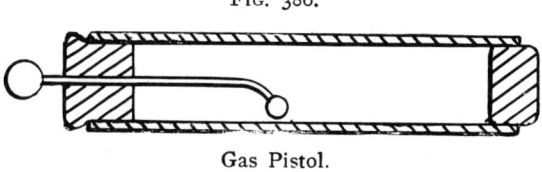

Gas Pistol.

leaping the space between their inner ends, ignites the powder.

Fig. 388 represents a simple apparatus for exhibiting the alternate attraction and repulsion of pith balls when placed between two metallic plates connected with opposite conductors of the machine. To prevent the pith balls from flying in all directions, they are confined by the glass jar. Four pieces of window glass forming a hollow square may replace the jar in this experiment.

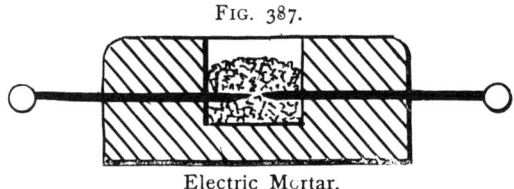

Fig. 387.

Electric Mortar.

Gassiot's cascade, shown in Fig. 389, is a beautiful experiment, but requires an air pump. A goblet coated with tin foil in the manner of a Leyden jar is placed under the tubulated bell of an air pump; a rod extends through the bell into the goblet, and when the electric discharge takes place (the rod and plate of the air pump being in communication

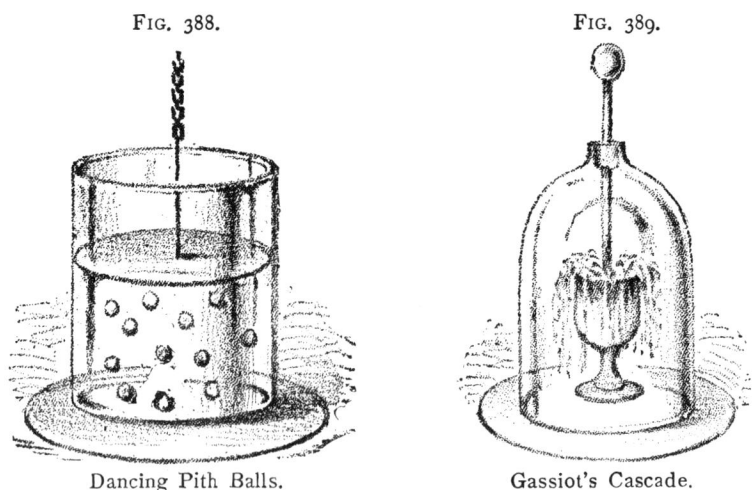

Fig. 388.

Dancing Pith Balls.

Fig. 389.

Gassiot's Cascade.

with the machine), a cascade of wavy light overflows the goblet like a fountain.

The pith ball electroscope, shown in Fig. 390, consists of a rod having at its upper end a ball and at its side a scale, from the angle of which is suspended a pith ball on a filament of whalebone. The upper end of the whalebone is formed

FRICTIONAL ELECTRICITY. 391

into a loop which hangs on a delicate pivot projecting from the scale. This instrument placed on a body receiving an electric charge will indicate roughly the extent of the charge. What has been said covers a very small proportion of the experiments possible in static electricity; but

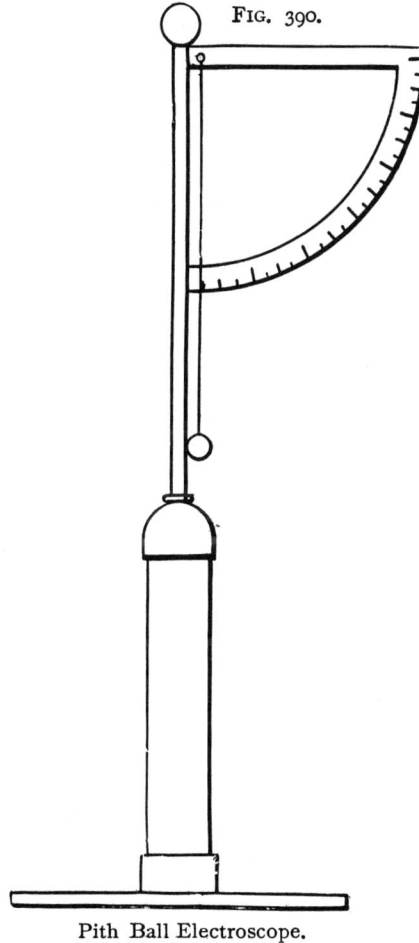

Fig. 390.

Pith Ball Electroscope.

it is hoped that some of the hints given in regard to the construction of the electrical machine, and some of the apparatus to be used in connection with it, will enable the student of electricity to at least begin a course of experiments which will prove of interest.

CHAPTER XVIII.

DYNAMIC ELECTRICITY.

GENERATION OF THE ELECTRIC CURRENT.

When two dissimilar metals, such as pure copper and pure zinc, are placed in contact in acidulated water, evidences of activity immediately appear in the form of a cloud of microscopic bubbles constantly rising to the surface of the water. If the metals are individually capable of resisting the action of the acid solution, it will be noticed that on separating them the action ceases, but it will commence again as soon as the metals are brought into contact. The same action is noticed if the two metals are brought into contact or connected by a wire above the surface of the acidulated water.

The bubbles are hydrogen resulting from the decomposition of the water. They escape from the copper, while the oxygen resulting from the analysis unites with the zinc, forming zinc oxide.

The copper is scarcely attacked, while the zinc slowly wastes away. If the wire connecting the zinc and copper be cut and the two ends placed on the tongue, a slight but peculiar biting sensation is experienced, which will not be felt when the wires are disconnected from the metals.

A piece of paper moistened with a solution of iodide of potassium and starch placed between the ends of the wires turns brown at this spot, showing that there is here a species of energy capable of effecting chemical decomposition. If a wire joining the copper and zinc is placed parallel with and near a delicately suspended magnetic needle, it will be found that it is endowed with properties capable of affecting the needle in such a manner as to cause it to swing and tend to take a position at right angles to the wire. This form of energy is dynamic or current electricity, generated in this case by chemical action and confined to and following a continuous conductor, of which the two metallic elements and the acid solution form a part, the whole comprising a complete electric circuit.

For the purpose of studying the generation and behavior of dynamic electricity, the elements referred to may be formed into an electric generator or battery, and the magnetic needle and conducting wire may be combined to form an electrical indicator or galvanometer, as illustrated in Plate VI., which shows convenient apparatus for making the primary experiments in dynamic electricity. The glass tank or cell is built with special reference to projecting the visible manifestations of the phenomena exhibited in the cell upon a screen, by means of the lantern, to enable a number of persons to observe simultaneously.

The cell consists of two plates of transparent glass 4 by 6 inches, separated by a half inch square strip of soft rubber, which is cemented to both glasses by means of a cement composed of equal parts of pitch and gutta percha. The cell is nearly filled with the exciting liquid, consisting of dilute sulphuric acid (acid 1 part, water 15 parts), in which are placed two plates, one consisting of a strip of zinc about one-sixteenth of an inch thick, the other plate being a strip of copper.

As commercial zinc is so impure as to be violently attacked by the exciting liquid, it is well to dip the zinc strip into the solution, and then apply to it a drop or so of mercury, which amalgamates the surface and prevents local action.

When these two plates are brought into contact with each other in the exciting liquid, hydrogen gas is given off copiously at the copper plate, while the action at the zinc plate is almost unnoticeable. If the plates are connected together by a conductor outside of the solution, the same phenomenon is observed.

The plane flat surfaces of the cell offer facilities for the examination of the plates by means of the microscope, and if so examined it will be found that so long as there is no metallic connection between the plates, they will remain unaltered, and no action is discoverable; but when the cir-

394 EXPERIMENTAL SCIENCE.

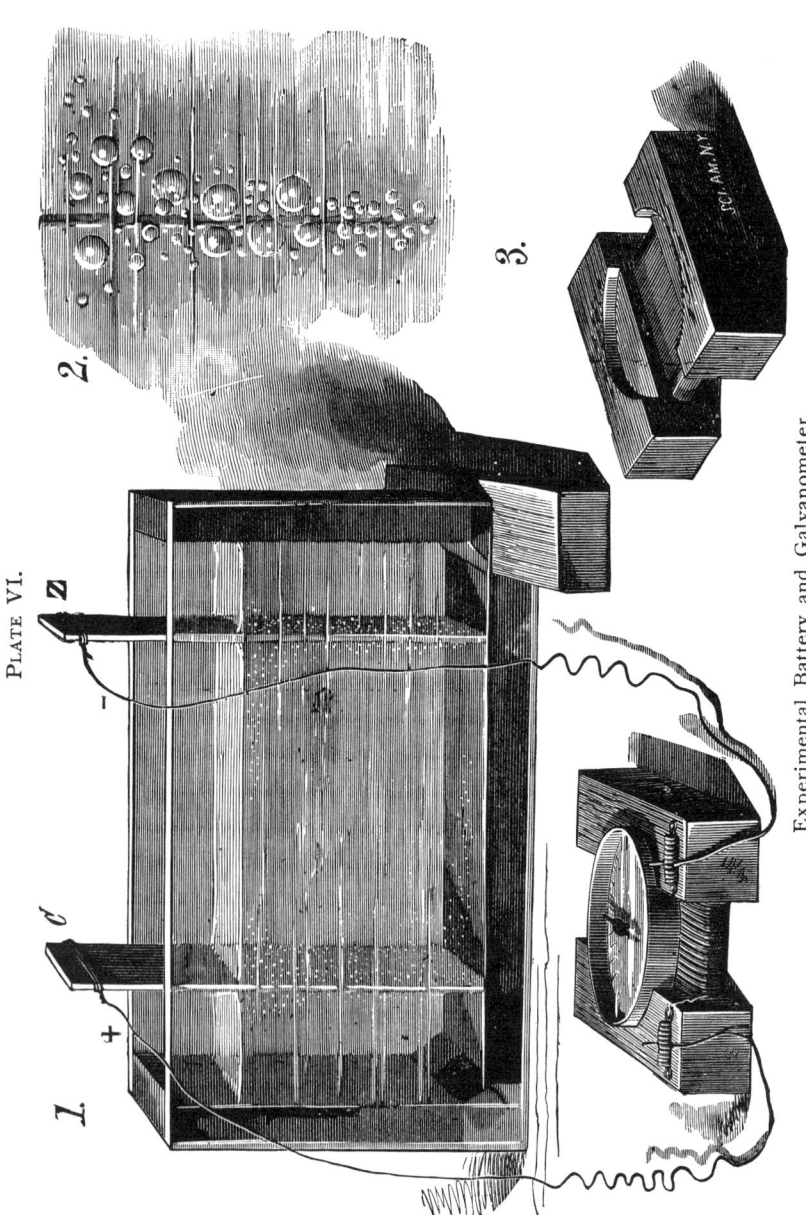

Plate VI. — Experimental Battery and Galvanometer.

cuit is completed, the first visible indication of action is the sudden whitening of the copper plate as if it were frost-covered; the next indication of action is the formation over the entire surface of the plate of myriads of minute silvery bubbles, which grow until they become detached, when they rise to the surface and escape into the air. These bubbles may be discharged into the mouth of a small test tube, and when a sufficient quantity of gas has accumulated it may be ignited, showing that it is hydrogen.

The appearance of the copper plate when the cell is in action is shown at 2 greatly magnified. The gas bubbles formed on the surface of the copper are at first very minute, but they rapidly increase in size and begin to merge one into another, taking an upward course. When a large bubble has absorbed a large number of the smaller bubbles and becomes sufficiently buoyant to overcome its adhesion to the plate, it rises to the surface and is lost in the air.

The bubbles of hydrogen are very bright, appearing and acting much like globules of mercury. Often an equatorial belt of very small bubbles will be seen surrounding a larger one.

The accumulation of hydrogen on the copper plate seriously affects the strength of the current. To ascertain to what extent and at what time this happens, a simple galvanometer, like that shown at 1, will be required. This instrument consists of a common pocket compass, a wooden frame or spool, and about 20 feet of No. 32 silk-covered copper wire. The wood spool (3) has a recess cut in the top at either end to receive the compass, which is placed a short distance from the flat body of the spool, and the wire is wound evenly around the body back and forth until the spool is full. Then the terminals of the wire are connected with two spiral springs fastened to the ends of the spool and forming "binding posts" for receiving the wires from the battery.

In regard to the adjustment of the compass, it should be arranged with the line marked N S parallel with the wires of the coil, and the instrument should be turned until the N S line is exactly under the needle, then a weak current from a constant battery should be sent through the coil and the deflection noted. The current should then be sent in the opposite direction, when the needle will be deflected in the opposite direction. If the amount of deflection is the same in both cases, the galvanometer is in condition for use; but if the deflections differ in degree, the compass must be turned in its socket until the proper adjustment is secured. The only precaution necessary in the construction of this instrument is to select a compass whose needle is delicately poised and vibrates freely.

By connecting the galvanometer with the cell as indicated in the engraving, it will be noticed that after a little time the galvanometer needle begins to fall back toward 0°, a point which it ultimately reaches if the circuit is kept closed; and the shorter the circuit, the sooner the cessation of the current. This enfeeblement of the current is principally due to three causes, one of which has already been noticed, that is, the accumulation of hydrogen on the copper plate. The film of hydrogen not only prevents contact between the exciting solution and the plate, but it actually renders the surface to a certain degree like the zinc. Another cause of enfeeblement of the current is the reduction on the copper, by the hydrogen, of a portion of the zinc sulphate accumulating in the liquid. This increases the similarity of the two plates, and consequently assists in diminishing the current. The reduction of the strength of the exciting liquid of the cell by mixture with zinc sulphate contributes still further toward the diminution of the current. All this results in making the two plates similar in their action, and in a consequent weakening of the current; but this chemical action cannot be avoided, as to secure any action in a galvanic cell the exciting fluid must be capable of decomposition. The production of local currents, the accumulation of hydrogen on the copper plate, and the weakening of the exciting solution are the three great causes of inconstancy in batteries. The first may be remedied in a great measure by amalgamation; the remedy for the last is obviously the strengthening of the solution, and the second, the accumulation of hydrogen on the copper plate, or the polarization of the plate can only be remedied by mixing with the exciting liquid some substance, such as nitric or chromic acid, which will oxidize the hydrogen as fast as it is liberated by oxidation of the zinc, or by brushing it while in the solution, or by violently agitating the exciting solution. The galvanometer needle faithfully indicates the result of either treatment. The polarization of the electrode may be strikingly exhibited by allowing the copper plate to become polarized and then replacing the zinc with a clean copper strip like the one already polarized. The galvanometer needle will be deflected in the opposite direction, showing that the polarized copper plate acts in the same manner as the zinc. Now, by removing the polarized copper plate and wiping and replacing it, the deflection of the needle will be much less, and it will not fall back to 0° until the very slight coating of zinc which has been deposited on the copper is removed from the polarized plate by means of emery paper or otherwise. Precisely the same effect is noticed when a newly amalgamated zinc plate is opposed to an oxidized zinc plate. The oxidized plate in this case will act as if it were copper.

This method of showing the effect of the polarization of the copper plate is conclusive. The phenomenon attributed to the polarized plate manifests itself in an unmistakable manner in polarizable batteries under the conditions of actual use.

While the entire office performed by the mercury in amalgamation is not known with certainty, one of its purposes is to present to the liquid a surface made up of zinc and mercury, and these two only. The acid acts on the zinc, which is at the same moment not in contact with any of the impurities, such as particles of carbon, iron, etc., that are diffused throughout the commercial zinc. Local currents are thus almost entirely avoided. The object of amalgamation is to prevent local currents as much as possible, and to present clean zinc to the liquid for oxidation. Yet, in spite of the mercury, local currents exist to some extent, and they are often quite as important as other causes in decreasing the effective value of the battery.

All batteries are more or less defective in operation, and require a great deal of care and attention. Many of the large uses to which batteries were applied a few years since now depend entirely upon dynamos for current. Nevertheless, batteries have many uses to which the dynamo cannot be conveniently or economically applied; such for example as working the smaller lines of telegraph, ringing call bells, operating indicators, annunciators, etc., and all closed circuit work where a comparatively small current is used. In telephone transmitters, and in open circuit work where a current is required only at long intervals, the dynamo cannot be substituted for batteries.

Terms such as "electric current," "electric fluid," "flow of the current," are based on the assumption that the action of dynamic electricity is analogous to that of fluids; but as nothing is positively known of the nature of electricity, these expressions are to be considered as purely conventional.

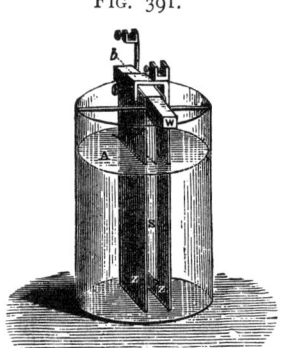

Fig. 391.

Smee's Battery.

SINGLE-FLUID BATTERIES.

Several of the batteries employing only a single exciting fluid are very useful in experimental work, and a number of them are of great value commercially. One of the oldest of these batteries is Smee's, which is illustrated in Fig. 391. A wooden strip, W, which rests upon the jar, A, supports the platinized silver plate, S. The zincs, Z, are clamped to the sides of the strip, W, by a clamp, b, which is provided with a binding post for receiving a wire. A binding post is also connected with the silver plate, S.

The wooden strip, W, is paraffined, and the zinc plates are amalgamated. The liquid generally used to charge the cell is 1 part of sulphuric acid to 10 of water.

The electro-motive force of the Smee battery is 1·09 volts when not in action, when in action it is 0·482 volt. Its internal resistance is about 1 ohm. The depolarization of this

DYNAMIC ELECTRICITY. 399

battery is due to the facility with which the hydrogen is detached from the rough surface of the platinized silver plate. The Smee battery has been used largely in telegraphy and electro-metallurgy.

The Grenet battery (Fig. 392) is a very good form of experimental battery where constancy of current is not required, as, for example, in the laboratory and mechanical work rooms. The cell is in the form of a bottle, and contains a solution formed by adding one part of sulphuric acid to five parts of a saturated solution of bichromate of potash in water. The top is provided with a brass frame, to which is fastened a vulcanite cover; to this are attached two carbon plates, that dip permanently into the fluid; and between them a zinc plate is suspended by a rod, by means of which it may be plunged into the fluid or withdrawn at pleasure. When the zinc is withdrawn, the action ceases. This battery gives a powerful current for a short time, but it rapidly polarizes. The length of time during which the fluid will retain its power depends on the use that is made of the battery.

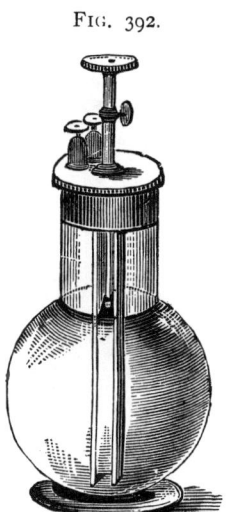

FIG. 392.

Grenet Battery, Bottle Form.

Fig. 393 represents an inexpensive and easily made plunge battery, which is very convenient for temporary use.

Ten tumblers, arranged in two rows of five, are held in place by an apertured board supported a short distance above the base board by the round standards. To these is fitted a board which is split from the standards outward, and provided with two bolts with wing nuts, by which the board may be clamped at any desired height. To opposite edges of this movable board are clamped six plates of carbon, $1\frac{1}{4}$ inches wide, $\frac{1}{4}$ inch thick, and 6 or 8 inches long. The upper ends of these plates are heated and saturated with wax or paraffine, and a copper wire is interposed between the carbon plate and the edge of the board. The strips of wood by which the carbons are clamped are ⅜ inch thick. To these wooden strips are secured zinc plates of the same dimensions as the carbon plates, by means of ordinary wood-screws passing through holes in the zinc into the wood. The wires connected with the carbons are bent over and inserted between the zinc plates and the wood, as shown in the engraving. That is, the carbon of one pair is connected with the zinc of the next pair in order, and so on throughout the series, and the terminal plates are connected with the binding posts.

Fig. 393.

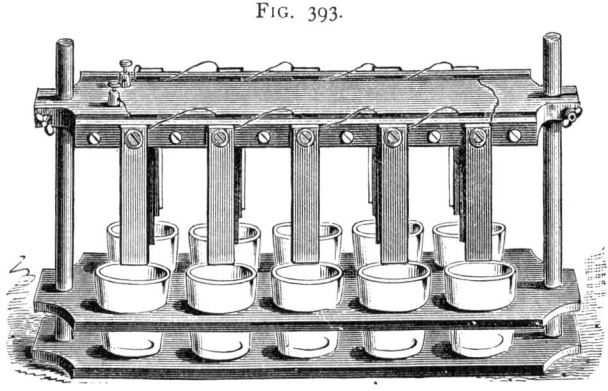

Simple Plunge Battery.

The zincs are amalgamated, and the tumblers are nearly filled with the bichromate solution.

To maintain the amalgamation of the zincs, a small quantity of bisulphate of mercury is added to the bichromate solution, say ⅛ ounce to every quart of solution.

The tumblers should be as large as can be conveniently obtained. Those holding one pint are not too large.

The plunging battery shown in Fig. 394 is a very powerful one, designed for running an electric motor or for supplying a current to three or four small incandescent lamps. The battery consists of eight elements, each formed of two 6×10 inch carbon plates ¼ inch thick, and one zinc plate of the same size, suspended in a cell $3\frac{1}{2} \times 7\frac{1}{2}$ inches and 9 inches deep.

DYNAMIC ELECTRICITY. 401

The upper ends of the carbon plates are paraffined, as shown in Fig. 395, by heating the ends only and rubbing on paraffine, allowing it to melt and soak into the pores of the plate until a strip about 1½ inches wide across the end of the

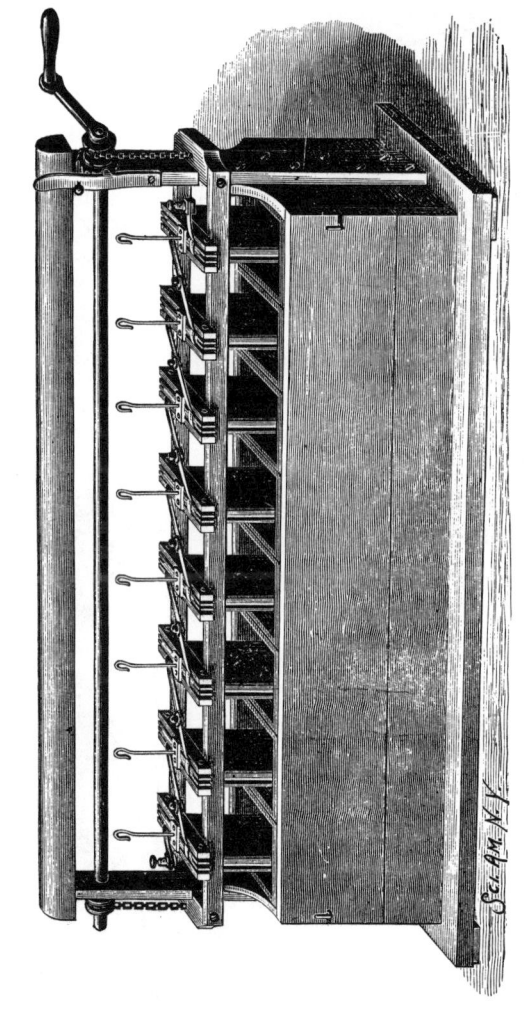

FIG. 394.

Large Plunge Battery.

plate is well filled with paraffine. This treatment prevents the solution from ascending by capillarity and destroying the connections.

The plates are arranged as shown in Fig. 396, the zinc

plate being located between two carbon plates and separated from them by strips of paraffined wood ¼ inch thick, 1¼ inches wide, and 8 inches long. The plates and separating strips are clamped together by thick strips of paraffined wood arranged upon the outer side of the carbon plates, and bolts, preferably of brass, passing through the ends of all of the strips. The electrical connection with the zinc plate is made by inserting a copper strip, *a*, between the plate and the wood strip. The connection with the carbon plates is made in a similar way, the strip, *b*, being looped so as to form a contact with both plates without touching the zinc.

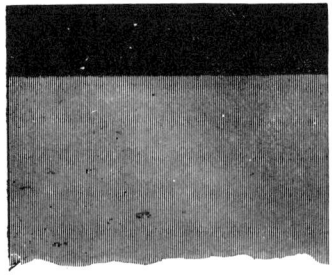

FIG. 395.

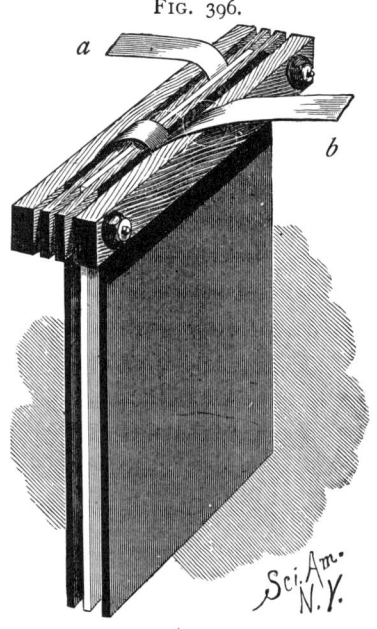

FIG. 396.

Before the elements are put together, the zinc plates should be carefully amalgamated. This is done by dipping each plate into a jar of dilute sulphuric acid (acid 1 part, water 10 parts), containing mercury at the bottom. As soon as the lower end of the plate is coated with mercury it may be lifted from the solution, inverted, and allowed to stand until the entire surface of the plate is perfectly covered with mercury. If there are portions which do not receive the mercury, they are scraped or sandpapered and returned to the acid solution, when mercury is applied locally.

If the amalgamation is perfect, the plates will not require

re-amalgamation. An amalgamating solution is made by dissolving mercury in nitric acid, then adding water so as to make a 10 per cent. solution of the mercury nitrate. A zinc plate immersed in the solution becomes amalgamated, but the operation requires frequent repetition. The cells consist of pine boxes of the size mentioned lined with gutta percha. The operation of lining is quite simple, and the cell, if well made, is durable. A wooden form is made which is the thickness of the gutta percha smaller than the boxes. Around the sides and end of this form is wrapped a sheet of gutta percha, which is ¾ inch wider than the form, the edges of the sheet being allowed to project beyond the form, as shown in Fig. 397.

FIG. 397.

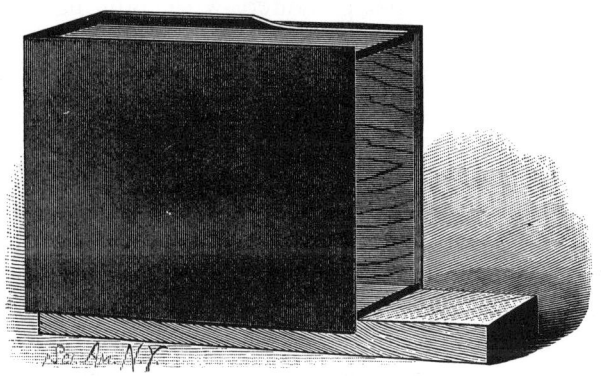

Forming the Gutta Percha Lining.

A piece of gutta percha of suitable width and length is placed upon the form within the projecting edges of the sheet already in position. The edges are then warmed sufficiently to render them adhesive, by means of a lamp flame or by holding a hot iron near enough to soften the gutta percha. The edge is then turned over in the manner illustrated. The fingers should be moistened to prevent the gutta percha from adhering to them. When the lining is complete, it is placed in the wooden box and expanded to fit by filling it with warm water. The upper edges of the lining should be turned over upon the edge of the box and made to adhere by heating. The box should be thoroughly coated with shellac varnish inside and outside, and allowed to dry before introducing the lining. Eight of these cells are placed in a box having removable sides and a frame extending over the top. To the vertical standard of the frame is loosely fitted a horizontal frame which supports the plates of the battery. In the upper part of the frame is journaled a shaft provided at opposite ends with drums, to which are attached chains for lifting the horizontal frame and plates supported thereby. The shaft is provided with a crank by which it may be turned, and with a ratchet which is engaged by a spring pawl attached to one of the standards.

The copper strips connected with the zinc plates are clamped to the strips extending from the carbon plates, and the terminal strips are provided with binding posts for receiving conductors. Each set of plates is provided with a hook, attached to the clamping strips by means of a crossbar of vulcanite or vulcanized fiber. These hooks are designed to be placed on the shaft when it is desired to use only a part of the cells, the unused plates being detached from the others and suspended out of contact with the solution. On account of the difficulty of removing the hard and almost insoluble crystals of chrome-alum formed in batteries employing a solution of bichromate of potash, a bichromate of soda solution is substituted. The crystals forming in the bichromate of soda solution are readily removed from the cell.

This solution is made by dissolving bichromate of soda in warm water to saturation, allowing it to cool, then slowly adding commercial sulphuric acid to the amount of one-fifth of the volume of the bichromate solution. As the gutta percha lining of the cells melts at a low temperature, the solution should be allowed to cool before pouring it into the cells.

The plates should not be plunged into the solution to a greater depth than is necessary for the production of the desired current, and they should always be withdrawn after use. The electro-motive force of this battery is 16·0 volts, and the maximum current is 4 amperes.

De la Rue's chloride of silver battery is well adapted for electrical testing. Its electro-motive force remains practically constant under various conditions. It is shown about half size in the sectional view (Fig. 398).

The top of the tube, A, is closed by a cork, D. The negative pole, C, consists of a cylindrical rod of chemically pure zinc supported by the cork stopper, which is perforated to receive it. The zinc rod has a hole in the top to allow the silver, connecting wire or electrode which goes to the next element to be soldered in.

The positive pole consists of a cylinder of silver chloride, B, having a silver wire or electrode, b, cast into it. This chloride rod is usually inclosed in a hollow cylindrical diaphragm of fine parchment paper. The zinc rod is amalgamated.

FIG. 398.

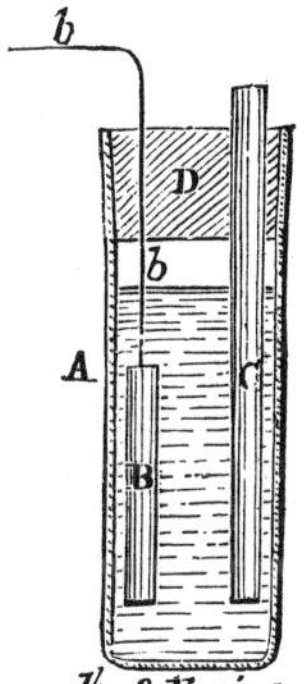

Chloride of Silver Cell.

The solution for charging the cell is made by dissolving 1 ounce of pure sal-ammoniac (ammonium chloride) in one quart of water.

The electro-motive force of each element is about 1·10 volts, and the internal resistance is about 8 ohms.

In the action of the cell, pure silver is reduced and deposited on the bottom of the cell. To prevent short-circuiting, the zinc rod is raised about three-eighths of an inch above the bottom of the cell. This pure silver deposit can be readily converted into chloride of silver, which is melted and recast into rods for use, or if preferred the pure silver may be sold.

This battery is largely used in electro-medical apparatus.

The Leclanche battery is one of the best for open circuit work. It is, in fact, a distinctively open circuit* battery. So long as the circuit is open there is no action in the cell, and as a consequence there is no loss.

This battery is shown in its improved form in Figs. 399 and 400. The carbon plate, which is suspended from the cover of the jar, supports two prisms clamped to the plate

---

* An open circuit is one which is normally without a current, and in which the current flows only while the circuit is in use.

by elastic rubber bands, as represented in Fig. 400, which shows the elements removed from the jar. The cover of the jar is perforated to receive the amalgamated zinc rod which extends down into the solution.

The prisms consist of 40 parts of granulated black oxide of manganese, 52 parts of granulated carbon, 5 parts of gum shellac, and 3 parts of potassium bisulphate. These ingredients are mixed, heated to 212° Fahr., and compressed in

FIGS. 399 AND 400.

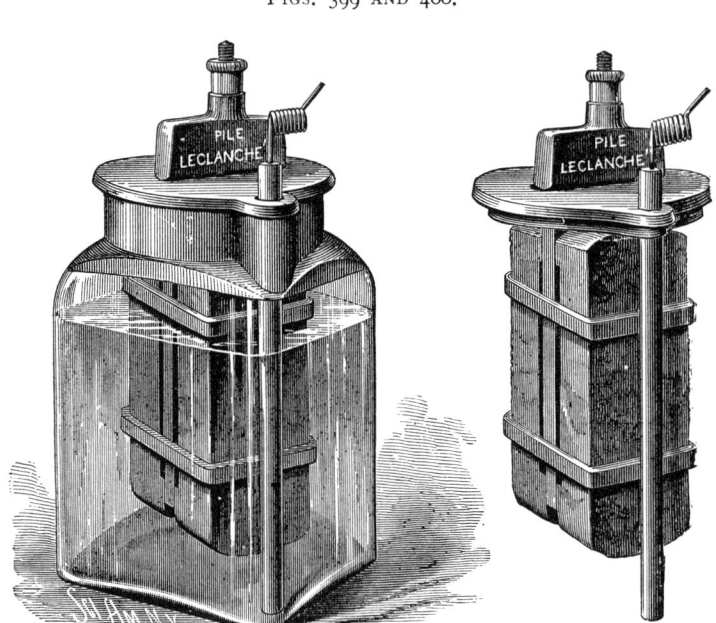

Leclanche Battery.

moulds under a pressure of two tons. A saturated solution of sal-ammoniac forms the exciting solution. In the Leclanche battery the hydrogen of the decomposed water unites with the oxygen of the manganese.

If the solution becomes too much reduced, zinc oxide is formed, and the solution becomes milky. When this occurs, more sal-ammoniac should be added. This cell has a resistance of 5 to 6 ohms, and an electro-motive force of 1·47 volts.

Dr. Carl Gassner's patent dry battery is much the same in principle as the Leclanche, but the exciting fluid is contained in a paste, and the zinc element forms the containing vessel. Two forms of the battery are made, one being cylindrical, as shown in Fig. 401, the other elliptical, as shown in Fig. 402.

The carbon rod or plate occupies about one-half of the space in the cell, and the space between the carbon and the cell is filled with the following mixture:

"Oxide of zinc, 1 part, by weight; sal-ammoniac, 1 part,

FIG. 401.   FIG. 402.

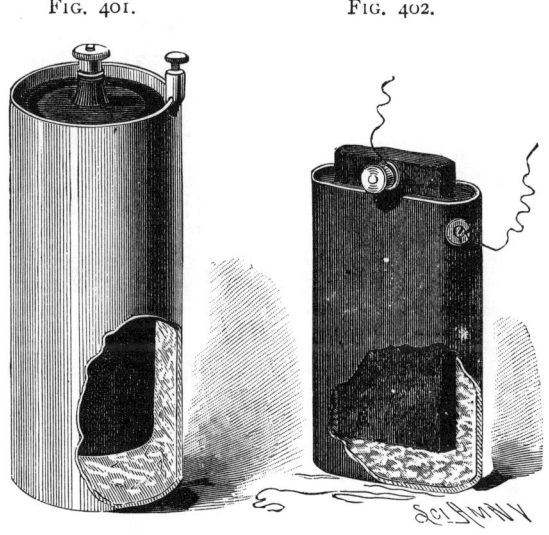

Dr. Gassner's Dry Battery.

by weight; plaster, 3 parts, by weight; chloride of zinc, 1 part, by weight; water, 2 parts, by weight. The oxide of zinc in this composition loosens and makes it porous, and the greater porosity thus obtained facilitates the interchange of the gases and diminishes the tendency to the polarization of the electrodes."

The battery works well on an open circuit, and is cleanly and portable.

The caustic potash battery represented in two forms in Figs. 403 and 404 is of comparatively recent invention. It is adapted to either open or closed circuit work, and will operate for several months without replenishing. It has been used successfully in electro-plating and in electric lighting on a small scale.

The cell is made of cast iron and serves as one of the plates of the battery. It is much heavier than a glass cell, but this is compensated for by its non-liability to breakage.

In the small pattern the iron cell, V, is closed by a rubber stopper, G, through which passes a brass rod, K, provided at its upper end with a binding post, F, and carrying

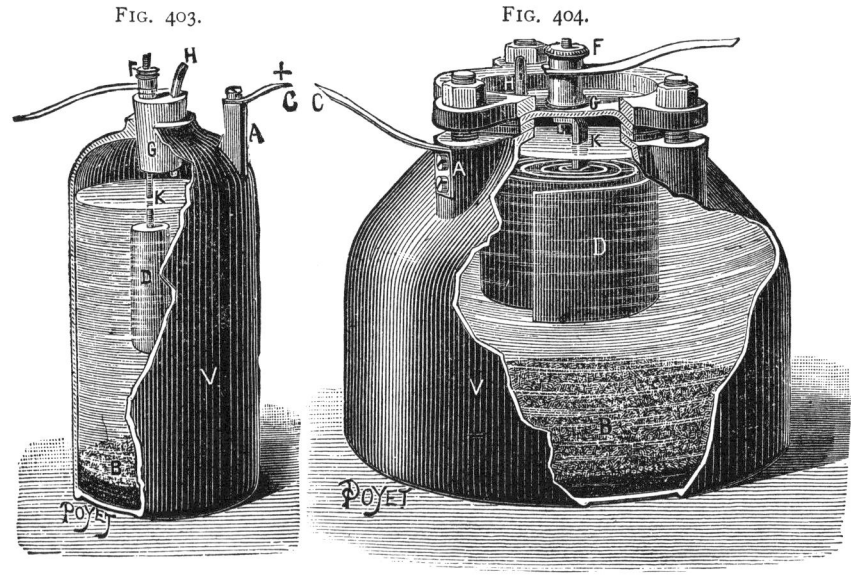

Caustic Potash Battery.

at its lower end the zinc cylinder, D. A lug, A, on the cell is provided with a binding screw for clamping the conductor, C. The cell is filled with a saturated solution of caustic potash, and upon the bottom of the cell is distributed a quantity of black oxide of copper.

A valve, H, formed of a piece of rubber tubing, is inserted in the stopper to admit of the escape of gas.

The large pattern shown in Fig. 404 is 9 inches in diameter. It is similar in its construction to the smaller cell. The zinc element in this case is formed of a plate bent spi-

DYNAMIC ELECTRICITY. 409

rally. It is not necessary to amalgamate the zincs in this battery. It is stated that the small cell yields a current of 2 amperes, while the larger one is capable of yielding 8 amperes. The E. M. F. is one volt.

TWO-FLUID BATTERIES.

The Daniell battery, shown in Fig. 305, is scentless and does not evolve any poisonous or disagreeable vapors.

In this battery, and in several cells derived from it, the

FIG. 405.

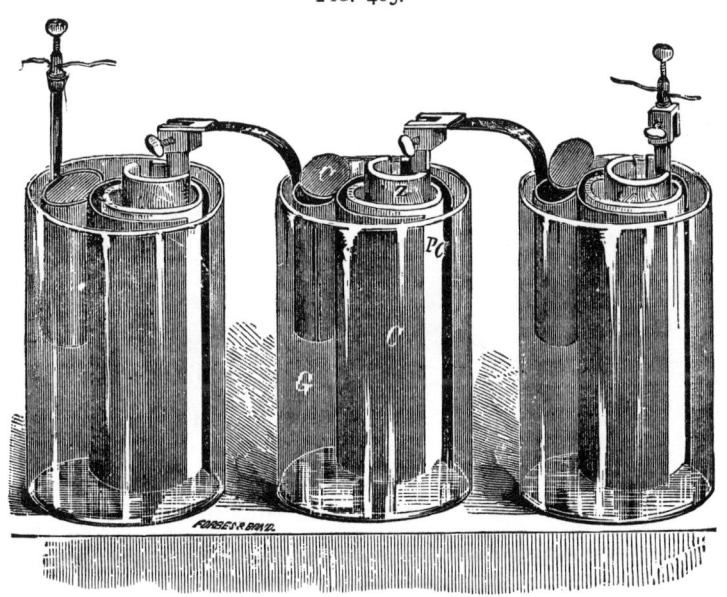

Daniell Battery.

two liquids are separated by a porous cell of unglazed clay. The glass vessel, G, is filled with a solution of copper sulphate. The porous cell, P C, contains the zinc, Z, which is not amalgamated. The curved sheet of copper, C, has attached to it a perforated pocket, c, for containing crystals of copper sulphate. The porous cell may be filled with a solution of common salt or water slightly acidulated.

This battery is especially adapted for closed circuits; it is less suitable for open circuits. It has an electro-motive force of about 1·079 volts. This amount varies somewhat with the density of the copper sulphate solution. The internal resistance of this battery varies considerably with the construction.

In a battery like that shown in the engraving, the resistance is about ½ ohm, but this may run up as high as 8 or 10 ohms in some forms.

In this battery, as well as in the gravity battery, described below, an example of the most perfect depolarizing action is found. Here the hydrogen resulting from the action of the dilute acid on the zinc is liberated on the surface of the copper plate, where it reduces the sulphate of copper, forming sulphuric acid and metallic copper, the latter being deposited on the surface of the copper plate. So long as sulphate of copper is present in the battery this action continues, and the current from the battery remains practically constant.

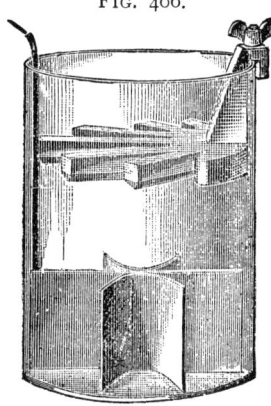

Fig. 406.

Gravity Battery.

The gravity battery, which is shown in its simplest form in Fig. 406, consists of a glass jar about 8 in. high and 6 in. diameter, having a zinc casting suspended near the top, and at the bottom three copper plates which are riveted together, the side plates being bent away from the central one as shown. One of the plates is provided with a gutta percha covered wire leading out of the jar. About two pounds of sulphate of copper are placed on the bottom of the jar, and enough water is poured in to cover the zinc about 1 inch. After standing 24 to 36 hours, the battery is in working condition. As the name of this battery indicates, its action is dependent on the separation of the zinc sulphate, which is formed at the top of the jar, and the copper sulphate solution, which gravitates toward the bottom of the jar. When the two solutions have properly separated, the fluid in the lower part of the jar will be blue, and that in the upper part will

DYNAMIC ELECTRICITY. 411

be colorless and transparent. The zinc should always be surrounded by the colorless fluid, and as the blue fluid decreases in volume, some of the zinc sulphate solution is removed and replaced by water.

When the water in the upper portion of the jar becomes saturated with zinc sulphate, the sulphate crystallizes upon the zinc plate, stopping the action of the battery. The conducting power of a solution of zinc sulphate is greater when diluted. Part of the solution, therefore, should be from time to time removed, and replaced by water. Undissolved crystals of sulphate of copper should always remain in the bottom of the jar. Any disturbance of the jars when

FIG. 407.

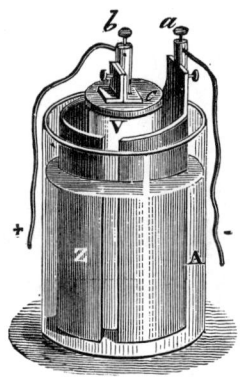 

Grove Battery.

in use causes the solutions to mix, thus seriously affecting the working of the battery. The water requires replenishing occasionally, to compensate for evaporation. The action of this battery is the same as that of the Daniell. The resistance varies from two to four ohms. Its electro-motive force is the same as that of the Daniell cell. It is used largely in telegraphy, and its electro-motive force is so nearly one volt, that it is used in making ordinary electrical measurements.

In Grove's battery the sulphate of copper solution used in the Daniell is replaced by nitric acid, and the copper by platinum. By this change greater electro-motive force is obtained. Fig. 407 represents one form of this battery.

The glass vessel, A, is partly filled with dilute sulphuric acid (1 part of acid to about 10 or 12 parts of water). In this vessel is placed an amalgamated zinc cylinder, Z, which is open at both ends and slit down one side. In this cylinder is placed the porous cell, V, containing ordinary nitric acid. A plate, P, of platinum, which is bent in the form of an S, is fixed to the porous cell cover, and is immersed in the nitric acid. The platinum is connected with the binding screw, *b*, and there is a similar binding screw, *a*, on the zinc.

In this battery the hydrogen which would be disengaged

FIG. 408.

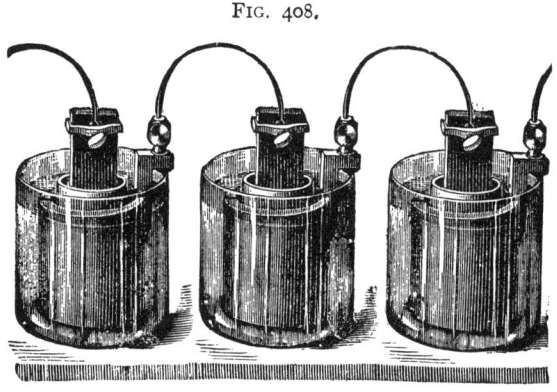

Chromic Acid or Carbon Battery.

on the platinum decomposes the nitric acid, forming hyponitrous acid, which is dissolved or is disengaged as nitrous fumes.

The resistance of the Grove cell is about half ohm. Its electro-motive force is 1·956 volts. The action of this battery is constant.

The chromic acid battery, shown in Fig. 408, is a modification of the Bunsen and is similar to the Grove in form. In this battery an amalgamated zinc cylinder surrounds the porous cup, and a rod of carbon replaces the platinum foil in the Grove. The jar is filled with saturated solution of common salt, or with sulphuric acid diluted with 12 parts of water.

The porous cell is filled with the bichromate of potash or the bichromate of soda solution previously described.

When the bichromate of potash solution is used in the porous cell, and a saturated aqueous solution of common salt is placed in the jar, the action is as follows: The chlorine of the salt unites with the zinc, forming zinc chloride, and at the carbon plate the sodium replaces the hydrogen of the sulphuric acid, forming sodium sulphate. The nascent hydrogen reduces the chromic acid of the solution, producing chromium sesquioxide.

The Bunsen battery differs from the chromic acid in employing nitric acid in the porous cell and dilute sulphuric acid in the jar.

FIG. 409.

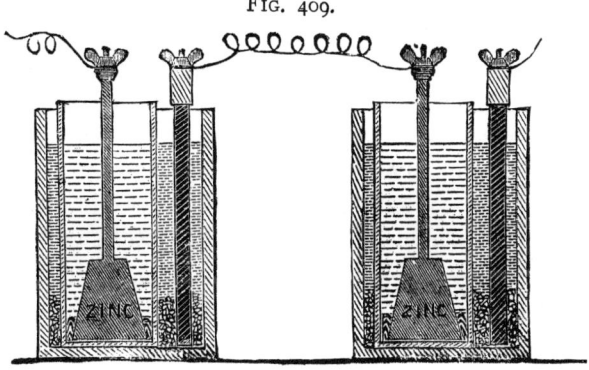

The Fuller Cell.

The electro-motive force and resistance of these batteries are about the same as in the Grove.

In the Fuller battery (Fig. 409), the zincs, so long as they last, are permanently amalgamated. In the accompanying figure two cells are shown. The carbon plate is placed in the outer vessel in the bichromate of soda solution. The zinc element, which is of the shape shown in the figure, is placed in a porous cell, into which an ounce of mercury is poured, and which is then filled up with water only. The addition of this mercury is the essential feature of the battery. The zinc plate is in this way kept permanently amalgamated so long as it lasts; the consequence is that not only is the internal resistance of the battery largely diminished, but its constancy is to a great extent insured. The action, after the battery is charged and the elements are connected with each other, commences almost immediately, and reaches a maximum in the course of a few hours.

The rod connected with the zinc element requires a protecting covering of gutta percha.

This is an excellent battery for open circuit work. It has an electro-motive force of nearly two volts, and an internal resistance of about two ohms.

MECHANICAL DEPOLARIZATION OF ELECTRODES.

In all single-fluid batteries polarization necessarily takes place to some extent, whatever precautions may be adopted for its prevention. The means of depolarizing single-fluid batteries are mechanical, and consist in the agitation of the exciting fluid by gravity, as in the fountain battery, by air jets, as practiced by Grenet and others, by stirring the fluid by mechanical means, by rotating or swinging the electrodes, and by roughening the electrode, as in the case of Smee's battery, in which the platinum plate is covered with a deposit of finely divided platinum.

In single-fluid batteries polarization may be greatly retarded by enlarging the plate on which the hydrogen tends to collect, so as to afford a great surface for its dissipation. In two-fluid batteries the depolarization is effected by chemical means, and perhaps more perfectly in the sulphate of copper batteries than any other.

In all single-fluid batteries the oxidation of the zinc liberates hydrogen, and this rapidly reduces the power of the battery in the manner explained in the former paper. In Smee's battery the microscopic points formed by the roughened platinum surface facilitate the escape of hydrogen, and in this way may tend to maintain the power of the element.

In the Grenet battery the carbon plate quickly polarizes, rendering the battery unfit for uses of more than a few minutes' duration. However, the agitation of the exciting fluid by the withdrawal and replacement of the zinc restores the battery to its normal strength. Grenet agitated the exciting fluid by means of air blown in through glass tubes, as shown in Fig. 410. This prevents polarization to a great extent, and renders the battery very active. Dr. Byrne, of Brooklyn, adopted this plan of depolarization in his battery with remarkable results.

Figs. 411, 412, and 413 show a purely mechanical agitator, consisting of spring-actuated stirrers, controlled by an electro-magnet of high resistance in a shunt around the battery. The magnet absorbs but a very small proportion of the current, and has only sufficient power to move the lever controlling the spring motor.

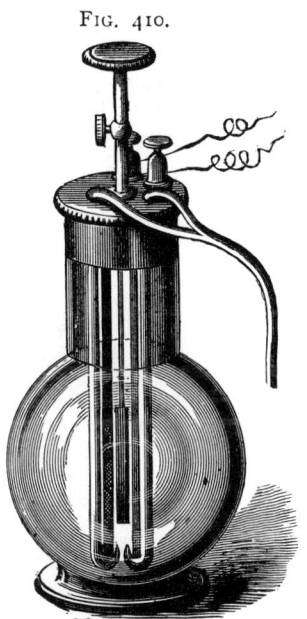

FIG. 410.

Grenet Battery, with Air Tubes

This motor, which may be of the cheaper class, is mounted on a base, A, secured to two parallel bars, B, carrying the zinc and carbon plates, $z\ c$, of the battery. These plates are placed flat against the bars, B, and secured by screws and washers. The zinc of one element is connected with the carbon of the next by a wire passing diagonally through the bar, and the first zinc and last carbon are connected with the binding posts at the ends of the bars, B.

The second shaft in the train of gearing is provided with a crank connected by a rod, C, with the lever, D, which is fastened to a rock shaft and connected with the bar, E, extending the whole length of the battery between the zinc and carbon of each element, and carries a series of vertical rods, F, of vulcanite, one such rod being located between the zinc and carbon plates of each element. The zinc in one of the elements is broken away in the engraving to show this rod, and the small horizontal sections at the top of Fig. 411 show the

416   EXPERIMENTAL SCIENCE.

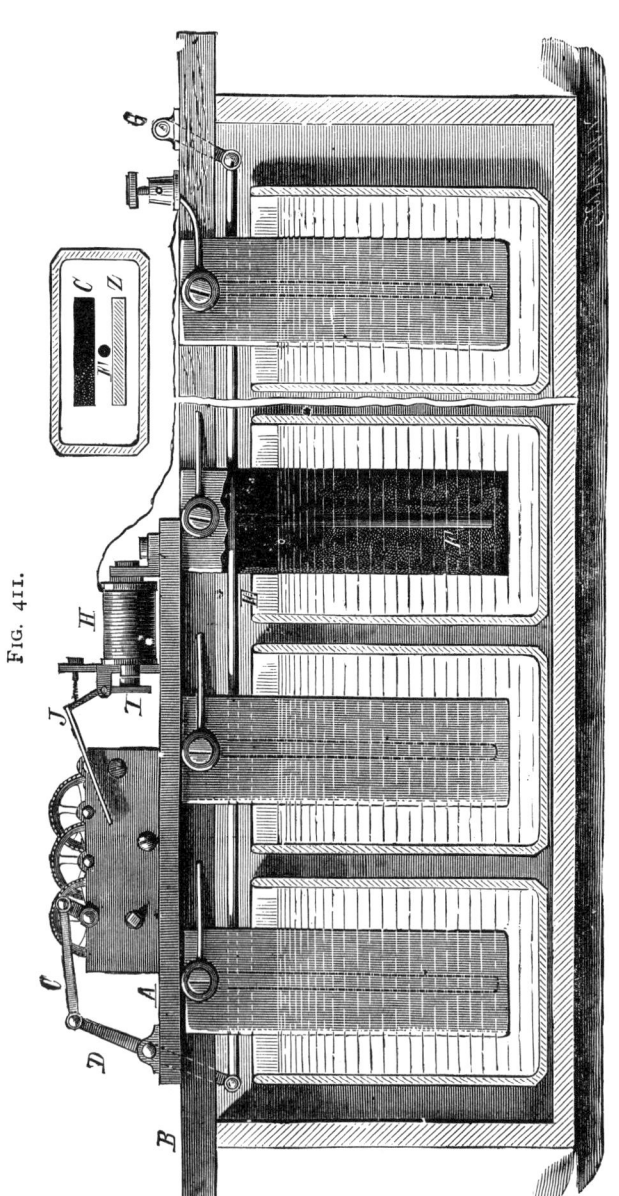

Fig. 411.

Depolarization of Electrodes by Mechanical Agitation.

DYNAMIC ELECTRICITY. 417

position of the rod relative to the plates. A swinging arm, G, supports the extremity of the rod, E. A high resistance magnet, H, mounted on the base, A, is connected with the two binding posts of the battery, so as to receive a small portion of the current. The armature attached to the lever, I, when drawn against the poles of the magnet, brings the lever, I, into engagement with the fan, J, which is the last element in the train of gearing composing the spring motor. A light retractile spring draws the lever, I, away from the fan, J, and removes the armature from the magnet when the power of the battery is reduced to a certain limit. The spring motor, being free to act, oscillates the rods, F, and by stirring the exciting liquid disengages the hydrogen from the plates, and brings fresh liquid into contact with the zinc and carbon and restores the strength of the battery, when the armature of the magnet, H, will be acted upon, bringing the lever, I, into engagement with the fan, J, and stopping the action of the spring motor until the current is again weakened, when the operation just described will be repeated.

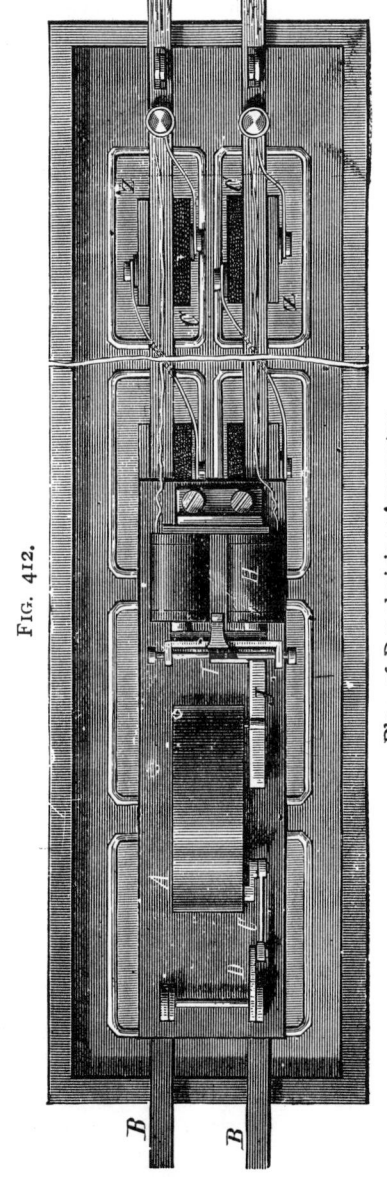

FIG. 412.

Plan of Depolarizing Apparatus.

In this way the strength of the battery will be maintained within certain limits, until the liquid is exhausted. Of course this system may be extended sidewise or lengthwise as much as may be desired.

All batteries employing mechanical means of depolarization, with, perhaps, the exception of Smee's, are only adapted to uses requiring a very strong current for a limited time.

SECONDARY BATTERY.

Probably no secondary battery can be more readily made or more easily managed than the one invented by

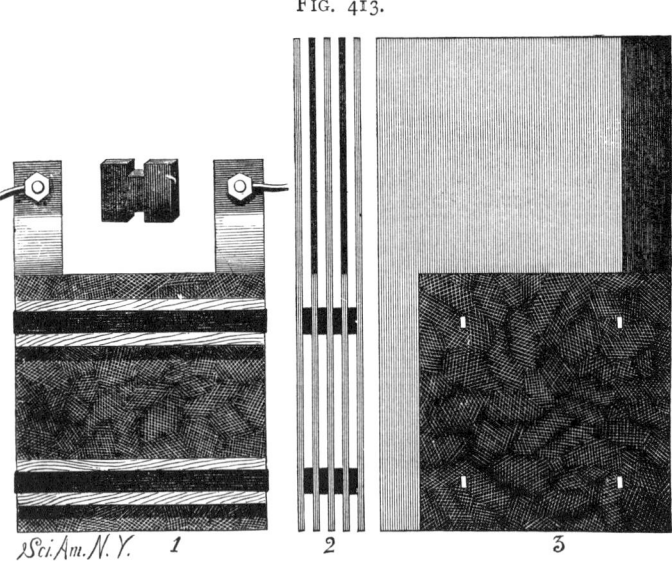

FIG. 413.

Plates of Secondary Battery.

Plante. It is, therefore, especially adapted to the wants of the amateur who makes his own apparatus. It takes a longer time to form a Plante battery than is required for the formation of some of the batteries having plates to which the active material has been applied in the form of a paste, and its capacity is not quite equal to that of more recent batteries, but it has the advantage of not being so

DYNAMIC ELECTRICITY.   419

liable to injury in unskilled hands and of allowing a more rapid discharge without affecting the active matter.

Each cell of the battery consists of 16 lead plates, each 6×7 inches and $\frac{3}{32}$ inch thick, placed in a glass jar 6×9 inches, with a depth of $7\frac{1}{2}$ inches. Each plate is provided with an arm $1\frac{1}{2}$ inches wide and of sufficient length to form the electrical connections. The plates are cut from sheet lead in the manner indicated at 3, in Fig. 413, *i. e.*, two plates are cut from a sheet of lead $8\frac{1}{2} \times 14$ inches. This method of cutting effects a saving of material. The plates after being cut and flattened are roughened. One way of doing this is shown in Fig. 413*a*. The plate is laid on a heavy soft-wood plank, and a piece of a double-cut file of

FIG 413*a*

Roughening the Plate.

medium fineness is driven into the surface of the lead by means of a mallet. To avoid breaking the file, its temper is drawn to a purple. After the plate is roughened on one side, it is reversed and treated in the same way upon the opposite side. If a knurl is available, the roughening may be accomplished in less time, and with less effort, by rolling the knurl over the plate. Half of the plates are provided with four oblong perforations into which are inserted H-shaped distance pieces of soft rubber, which project about $\frac{1}{8}$ inch on each side of the plate. The perforated and imperforate plates are arranged in alternation, with all of the arms of the perforated plates extending upward at one end of the element and all of the arms of the imperforate plates similarly arranged at the opposite end of the element.

The plates are clamped together by means of wooden strips —previously boiled in paraffine—and rubber bands. The strips are placed on opposite sides of the series of plates at the top and bottom, and the rubber bands extend lengthwise of the strips.

The arms of each series of plates are bent so as to bring them together about 3 or 4 inches above the upper edges of the plates. They are perforated to receive brass bolts, each of which is provided with two nuts, one for bending the arms, the other for clamping the conductor.

FIG. 414.

Plates Connected.

This element is placed in a glass cell, on paraffined triangular wood supports, and the formation is proceeded with.

To hasten the process, the cell is filled with dilute nitric acid (nitric acid and water equal parts by measure), which is allowed to remain for twenty-four hours. This preliminary treatment modifies the surface of the lead, rendering it somewhat porous, and, in connection with the roughening, reduces the time of formation from four or five weeks down to one week. The nitric acid is removed, the plates and cell are thoroughly washed, and the cell is filled with a solution formed of sulphuric acid 1 part, water 9 parts.

The desired number of cells having been thus prepared,

DYNAMIC ELECTRICITY.

are connected in series, and the poles of each cell are marked so that they may be always connected up in the same way. The charging current, from whatever source, should deliver a current of ten amperes, with an electromotive force ten per cent. above that of the accumulator.

Each cell of this battery has an electro-motive force of two volts, and the voltage of the series of cells would be the number of cells × 2. It is a simple matter to determine the amount of current required to charge a given number of cells. For example, a battery is required for supplying a series of incandescent lamps. It has been found uneconomical to use lamps of a lower voltage than 60. It will, therefore, require a battery having an E. M. F. of 60 volts to operate even a single lamp. This being the case, at least 30 cells of battery must be provided, and on account of a slight lowering of the E. M. F. in use, two extra cells should be added. It will, therefore, require 32 cells for a small installation, and the machine for charging such a battery should be able to furnish a current of ten amperes, with an E. M. F. of 75 volts.

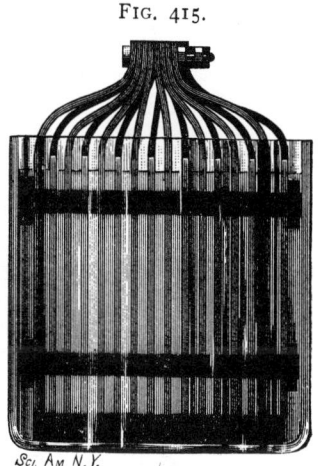

FIG. 415.

Complete Cell.

To form the battery, it is placed in the circuit of the dynamo and kept there for thirty hours continuously, or for shorter periods aggregating thirty hours. It is then discharged through a resistance of 20 or 30 ohms, and again recharged, the connections with the dynamos being reversed so as to send the current through the battery in the opposite direction. The battery is again discharged through the resistance, and again recharged in a reverse direction. These operations are repeated four or five times, when the formation is complete. It will require from five to seven hours to charge the battery after it is thoroughly formed. It must always be connected with the dynamo as connected last in charging. Although amateurs may find pleasure in constructing and forming a secondary battery, there is no economy in securing a battery in this way. It is less expensive and less vexatious to purchase from reliable makers.

THERMO-ELECTRIC CURRENT.

Professor Seebeck, of Berlin, discovered in 1821 that an

FIG. 416.

Thermo-Electric Series.

electric current could be produced by the direct application of heat to a conductor consisting of two metals soldered together, the heat being applied to the junction of the two parts of the circuit.

A simple thermopile for illustrating this phenomenon is shown in Fig. 416. It consists of a series of brass and German silver bars, alternating in position and separated by strips of mica, except at a short interval at one end of each pair, at which point the bars are connected by soldering. The soldering occurs alternately at opposite ends, as indicated in the plan view in the lower part of the cut. The battery is thus formed of a continuous conductor of dissimilar metals. The terminals of the series being connected with a galvanometer of low resistance, heat applied to one end of the series will cause a current to flow. This will be indicated by a deflection of the galvanometer needle. The current will continue to flow so long as a difference of temperature of the ends of the series is maintained.

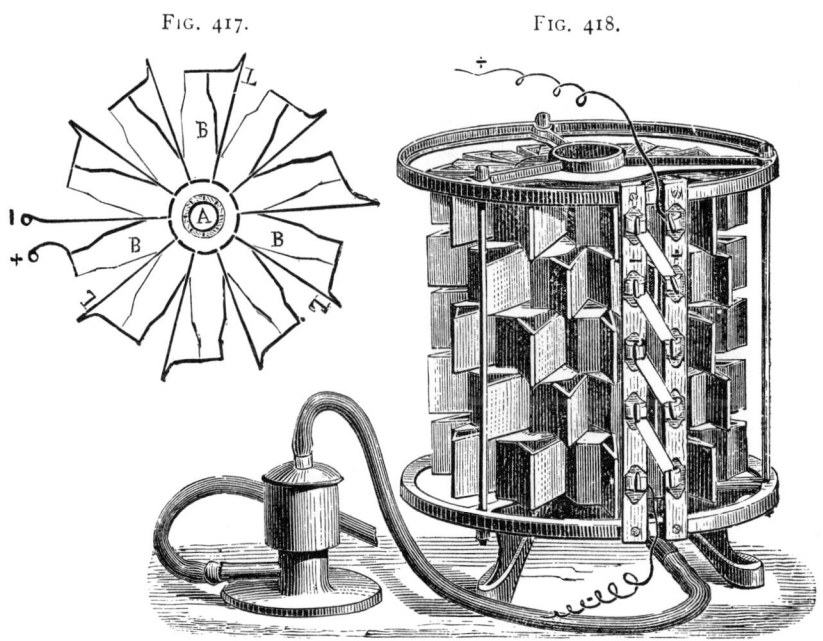

Clamond's Thermo-Electric Battery.

Nobili's thermopile, constructed on this principle from a large number of small bars of bismuth and antimony, used in connection with a delicate galvanometer, constitutes one of the most sensitive indicators of change of temperature known.

Clamond's thermo-electric battery, which is shown in plan in Fig. 417, in perspective in Fig. 418, and vertical section in Fig. 419, has been used for telegraphic purposes and for electro-plating. In this battery one element consists of an alloy of two parts of antimony and one of zinc, cast in a flat spindle-shaped bar, B, from 2 to 3 inches in length by $\frac{3}{8}$ inch in thickness. The other element is a thin strip, L, of tin plate, which enters a notch in the inner end of one antimony-zinc element and is connected in a similar way with the outer end of the next element. These are joined in a circle, as shown in Fig. 417, and are kept in position by a paste of asbestos and soluble glass. Flat rings, V, of this composition are also made and placed between the series of

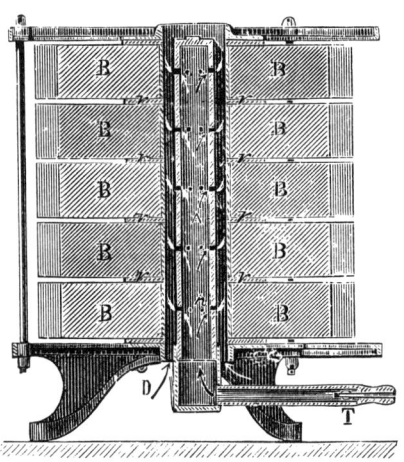

FIG. 419.

Vertical Section of Clamond's Battery.

elements, which are piled one over the other, as shown in Figs. 418 and 419. The connection between the several series is made by soldering together positive terminals of one series with the negative of the next, as shown in Fig. 417. When the battery is complete the interior presents the appearance of a perfect cylinder.

The heating is effected by means of coal gas, admitted through an earthenware tube, A, perforated with numerous small holes. The temperature should not exceed about 200° F.

A battery of sixty such elements has an electro-motive

force of three volts and an internal resistance of 1½ ohms. This battery has been used in telegraphy, in electro-metallurgy, and in charging secondary batteries.

ELECTRICAL UNITS.

Potential is a term used to express various degrees of electrical energy or power of doing work, and is used with respect to electricity in much the same way as pressure is applied to steam. The earth, so far as potential is concerned, is said to be at zero. The zero point forms a basis from which to measure the relative electrical condition of bodies which may have higher or lower potential than that of the earth.

For the sake of convenience, electricity is treated as a fluid. Any substance through which it flows is called a conductor, and the flow of the fluid over the conductor is known as a current. Any substance over which electricity will not pass is called an insulator.

The difference of potential between two points connected by a conductor causes a passage of electricity from one point to the other until an equilibrium is established, when there can be no further transfer of the current. When a current is passing, it shows that there is a difference of potential.

Electro-motive force (for convenience usually written E. M. F.) is that force which tends to move electricity from one point to another. It is proportional to the difference of the potential of the two points. There may be a difference of potential at two points without a current. When the two points are connected by a conductor, the current will be established by virtue of the electro-motive force.

All substances offer more or less resistance to the electric current. Most metals are called good conductors, because they offer but little resistance to the passage of a current. Other materials, such as wood, stone, glass, are practically non-conductors, and are therefore called insulators.

Electricity being invisible and imponderable, it is impossible to measure it as ponderable matter is measured, therefore special units have been devised for the measurement of electricity, which are of two kinds, known as absolute units and practical units, the ratio between the two being some power of ten.

In these measurements, length, mass, and time are measured in centimeters, grammes and seconds, respectively. This is known as the centimeter-gramme-second method. The abbreviation for this method is C. G. S.

The absolute units of this system are not adapted to practical use, as they involve figures of inconvenient length, but in order to show the basis of electrical measurements, the following examples are given:

The dyne or absolute unit of force is that force which, acting for one second on a mass of one gramme, imparts to it a velocity of one centimeter per second. The weight of one gramme according to this explanation is equivalent to a force of $1 \times 980.2 = 980.2$ dynes at New York, lat. 40° 41′ N. (A gramme is equal to 15,432 grains, and a centimeter to 0.3937 of an inch.) The velocity acquired by a falling body in one second is 32.16 feet, or 980.2 centimeters, at New York.

The erg or absolute unit of work is the work required to move a body one centimeter against the force of one dyne. The weight of one gramme being equal to 980 dynes, the work of raising one gramme through one centimeter against the force of gravity is 980 ergs. An erg is equal to $\frac{1}{13,560,000}$ of a foot pound. A foot pound is work done in raising one pound one foot high.

A magnetic pole of unit strength is such that, when placed at unit distance (one centimeter) from a similar pole, the two will act upon each other with unit force (one dyne).

A unit line of force is of such strength as to act on a pole of unit strength with unit force (one dyne). A magnetic field of unit intensity is one in which each square centimeter of area is occupied by one unit line of force.

A current of unit strength is such that when flowing around an arc one centimeter long on a circle of one centimeter radius, it exerts a force of one dyne on a unit pole placed at the center of the circle.

A conductor is of unit resistance when the work done in

DYNAMIC ELECTRICITY.

a second by a current of unit strength passing through it equals one erg.

The unit difference of potential or electro-motive force is that necessary to impel a current of unit strength through unit resistance.

Unit quantity of electricity is that conveyed by a unit current in one second.

The practical units in most frequent use are the volt, the ohm, and the ampere.

The volt (equal to 10^8 absolute units) or unit measure of electro-motive force, or of difference of potential, is equal approximately to the electro-motive force possessed by one Daniell cell; accurately, it is 0·95 of the E. M. F. of this cell.

The ohm (equal to 10^9 absolute units) or unit measure of resistance is approximately equal to the resistance of 250 feet of copper wire $\frac{1}{20}$ of an inch in diameter, or $\frac{1}{10}$ of a mile of No. 9 telegraph wire.

The ampere ($= \frac{1}{10}$ absolute unit) is the unit measure of current strength. If an electro-motive force of one volt be applied to send a current through a resistance of one ohm, the strength of the current produced will be one ampere; that is to say the strength of a current in amperes varies directly as the electro-motive force applied to produce it, and inversely as the resistance of the circuit. This is expressed by the formula known as Ohm's law:

$$C = \frac{E}{R} \text{ where}$$

C is strength of current in amperes,
E is electro-motive force in volts,
R is the resistance in ohms.

The coulomb ($\frac{1}{10}$ absolute unit) is the unit of quantity, and represents the amount of electricity conveyed by one ampere of current acting for one second. This is represented by the formula:

$$C = \frac{Q}{t} \text{ or } Q = Ct, \text{ where}$$

C is the current in amperes,

Q is the quantity of electricity in coulombs,
t is the time in seconds.

For example, if a current of a strength of 5 amperes flows for ten seconds, the amount of electricity which passes during that period will be 50 coulombs.

The farad (10^{-9} absolute units) is the measure of capacity, and is such that a condenser of one farad of capacity could be raised to the potential of one volt by a charge of one coulomb of electricity, or in other words, by a current strength of one ampere acting for one second.

As a condenser of the capacity of one farad would be inconveniently large, the microfarad, or one-millionth part of a farad, is the unit generally used.

Since it is frequently necessary to measure quantities millions of times greater or less than the practical units, the prefix *mega* has been adopted to represent one million times, *micro* one millionth part, and *milli* one thousandth part. In this way the megohm signifies one million ohms, and milliampere one thousandth part of an ampere.

The gramme-degree (or calorie) the C. G. S. unit of heat is the amount required to raise one gramme of water one degree centigrade, and is equal to the work of 42 million ergs or $3\frac{1}{10}$ foot pounds. The work required to raise one pound of water one degree Fahrenheit is equivalent to about 772 foot pounds.

The heat developed in a circuit depends upon the strength of the current, the time that it acts, and the resistance of the conductor, and is calculated by the following formula, called Joule's law:

$$H = \frac{C^2 R^t}{4 \cdot 2} \text{ where}$$

C is the current in amperes,
R is the resistance in ohms,
t is the time in seconds.
H is heat in calories, or gramme degrees centigrade, as above.

The joule or practical unit of heat is the amount of heat

caused by a current of one ampere acting through a resistance of one ohm in one second, and the heat may be calculated by the formula:

$$J = C^2 R, \text{ where}$$

C is the current in amperes,
R is the resistance in ohms,
J is heat in joules.

The watt or practical unit of the rate of doing work is equal to ten million ergs (10^7 absolute C. G. S. units) per second, or to the work produced in that time by one ampere of current of an electro-motive force of one volt acting through a resistance of one ohm.

The horse power is the unit of rate of work commonly used by engineers.

An actual horse power is equivalent to 33,000 pounds raised one foot in one minute, or 550 foot pounds per second.

The electrical horse power is equal to 746 watts. The work expended in a circuit in producing a current of a certain strength and of known electro-motive force, or against a known resistance, can be calculated by the following formula, which, however, only represents the work expended in the circuit itself, and does not make allowance for that wasted in the generator and in the prime motor:

$$W = C E \text{ or } W = C^2 R \text{ or}$$

$$H P = \frac{C E}{746} \text{ or } \frac{C^2 R}{746} \text{ where}$$

C is the current in amperes,
E is the electro-motive force in volts,
R is the resistance in ohms,
W is the work in watts,
H P is the actual horse power *

ARRANGEMENT OF BATTERY CELLS.

To secure the greatest efficiency in a battery, the elements must be arranged so as to adapt the electro-motive

* These concise definitions are taken from "Practical Electric Lighting," by A. Bromley Holmes.

force and the internal resistance to the resistance of the external circuit. To accomplish this the batteries are connected up in different ways, so as to yield currents of high voltage and low amperage, or the reverse.

To facilitate the explanation of the method of connecting batteries, it will be necessary to describe the conventional sign by which the element is designated. Fig. 420 represents the symbol or conventional sign for a single cell of any battery. The short, thick line represents the zinc, and consequently the negative pole of the battery, while the longer, thin line stands for the platinum, copper, or carbon plate, and the positive pole. The minus sign (−) is used to designate the negative pole, while the plus sign (+) is used to designate the positive pole.

FIG. 420.

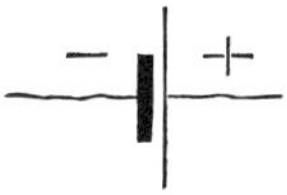

When a number of cells are connected together, as shown in Fig. 421, that is, with the positive pole of one cell connected with the negative of the adjoining cell, with the terminal cells connected with the conductors, the battery is connected up in series; and when so connected it yields the highest electro-motive force of which it is capable; that is to say, it yields the electro-motive force of a single cell multiplied by the number of cells in series.

A current of this kind is adapted to overcome high resistances. If a single cell of battery has an electro-motive force of one volt, then 12 cells of a battery connected in series would have an electro-motive force of 12 volts.

FIG. 421.

Now, to secure the best effects with a battery, the external resistance through which the current must work should be equal to the internal resistance of the battery. In this case, if each cell of battery has a resistance of 5 .ohms, the total resistance of the battery would be 60 ohms; therefore, a battery arranged in this way is best adapted to an external circuit having a resistance of 60 ohms.

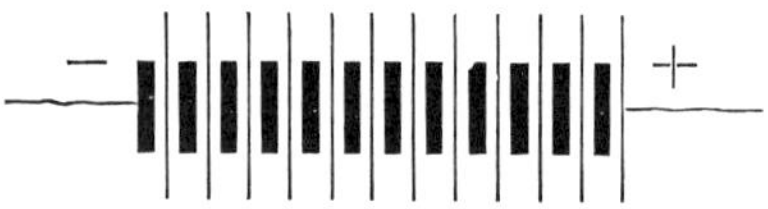

As the current is equal to the electro-motive force divided

DYNAMIC ELECTRICITY. 431

by the resistance $\left(C = \dfrac{E}{R}\right)$ in this case the electro-motive force being 12 volts and the total resistance of the circuit being 120 ohms, $C = \dfrac{12}{120} = 0\cdot 1$ ampere. We have then a current with the strength of 0·1 ampere, having an electro-motive force of 12 volts.

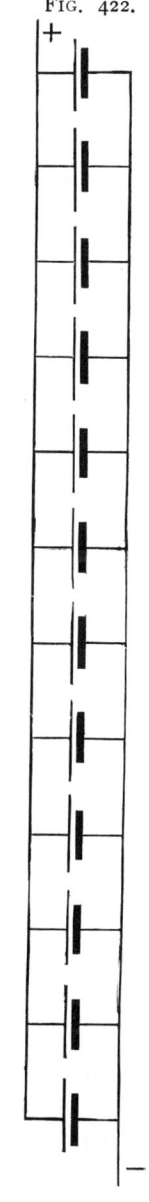

FIG. 422.

Perhaps the difference resulting from the methods of connecting up batteries cannot be better shown than by taking the opposite extreme. The 12 cells of battery are connected up in parallel circuit; that is to say, all the positive poles are connected with one conductor, and all the negative poles are connected with another conductor, as shown in Fig. 422. In this case, each cell of battery having a resistance of 5 ohms, the total resistance of the 12 cells connected in parallel will be $\tfrac{1}{12}$ of 5 ohms, which is a little more than 0·41 of an ohm, and the electro-motive force of a battery thus connected will be only that of a single cell; then, making the external resistance equal to the internal resistance of the battery, the total resistance of the circuit will be 0·82 ohm. Now, by Ohm's law, $C = \dfrac{E}{R}$ we will have $\dfrac{1}{0\cdot 82} = 1\cdot 219$ amperes.

Where the cells are connected three in series, with four such series parallel, as shown in Fig. 423, the electro-motive force will be three volts (this quantity remaining the same for any number of series of three connected parallel). The resistance is inversely as the number of series; assuming the resistance to be 5 ohms per cell, the resistance of one series would be 15 ohms, and that of four series connected parallel would be $\frac{15}{4} = 3.75$. Now, making the external resistance of the circuit equal to the resistance of the battery, the

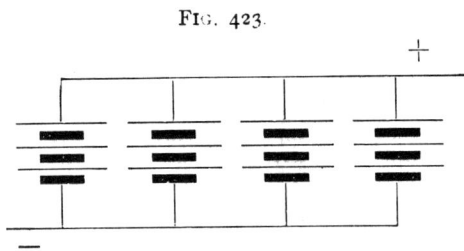

FIG. 423.

total resistance of the circuit would be internal resistance 3.75 + external resistance $3.75 = 7.5$ ohms; and by the formula $C = \frac{E}{R}$ we will have $\frac{3}{7.5} = 0.4$ ampere.

In Fig. 424 the cells are arranged in three parallel series

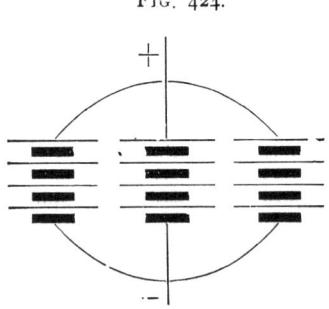

FIG. 424.

of four each. The electro-motive force is 4 volts, the resistance of each series is 20 ohms; this divided by the number of series = 6.66 ohms. Adding the resistance of the external circuit, which should be the same, the total resistance of the circuit would be 13.32 ohms. The electro-motive force, which is 4 volts, divided by this resistance = 0.3 ampere.

Take another example, in which 12 cells are arranged in two series of 6 each. The electro-motive force will be 6 volts, the resistance 15 ohms, and if a similar resistance be added in the external circuit, the total resistance will be 30 ohms, and the current strength will be 0·2 ampere.

If, however, a resistance of 60 ohms be placed in the external circuit, with cells arranged as in Fig. 425, the total resistance of the circuit then being 75 ohms, the current strength would be $\frac{6}{75} = 0·08$ ampere, which is much less than that obtained by the first arrangement, in which all the cells are in series. Or take the first example, in which all of the cells are in series, and make the external resistance 15 ohms, instead of 60. The current strength would be 0·16 ampere, but the extra strength would be attended with an undue loss in the battery.

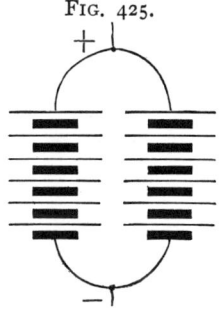

FIG. 425.

It will thus be seen that by connecting cells in series the highest electro-motive force is secured, while cells must be connected parallel for the greatest strength of current.

GALVANOMETERS.

No one can go very deeply into the study of electricity without reaching the subject of electrical measurements; certainly very little can be done in this direction without a galvanometer of some kind. The simple instrument already described answers very well for detecting currents and showing their direction, but it is not sufficiently delicate to be of value in electrical measurements.

Among all the galvanometers yet invented, there is perhaps none possessing so many good qualities as the one shown in Fig. 426. It is very simple. The materials are inexpensive. No great mechanical skill is required in its construction, and its sensitiveness and accuracy are sufficient for most requirements. Besides all this, it is perfectly "dead beat," so that no time need be wasted in waiting for the instrument to come to rest. This galvanometer is the joint invention of MM. Deprez and D'Arsonval, of Paris.

Fig. 426.

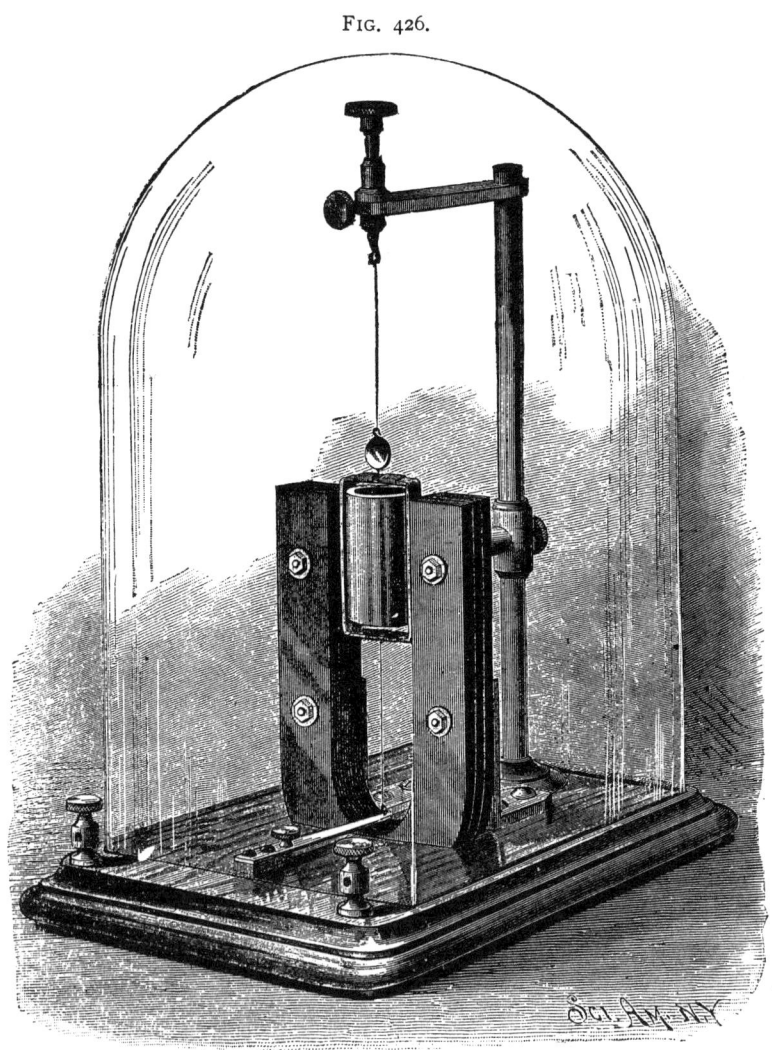

Deprez-D'Arsonval Galvanometer.

It consists essentially of a rectangular coil of fine wire suspended on strained torsional wires in a strong magnetic field. To the base is secured, by means of angle plates, a compound U-magnet, 7 inches high, formed of three steel magnets, one-quarter inch thick, secured together and to the angle plates by bolts. The distance between the inner faces of the poles of the magnet is $1\frac{7}{16}$ inches. Two and three-quarter inches behind the center of the magnet a brass column rises from the base, and is provided near its center with an adjustable brass arm, supporting at its outer end, and exactly in the center of the space between the poles of the magnet, a hollow soft iron cylinder, $2\frac{1}{4}$ inches long, $1\frac{1}{32}$ inches in external diameter, $\frac{3}{16}$ inch in internal diameter. The top of this cylinder is even with the upper ends of the magnet. To the top of the brass column is secured, at right angles, an arm that extends over the hollow iron cylinder, and is provided with a vertical sleeve, in which is clamped a rod having on its lower end a small silver hook, arranged axially in line with the iron cylinder.

To a block attached to the base, opposite the center of the magnet, is secured a tapering spring, $\frac{1}{16}$ inch thick and $3\frac{3}{4}$ inches long, carrying at its free end a small silver hook, which is arranged in line with the axis of the iron cylinder.

A rectangular coil of No. 40 silk-covered copper wire, large enough to swing freely over the iron cylinder, is suspended by a hard-drawn No. 32 (0·008 inch in diameter) silver wire from the hook above, and is connected by a similar wire with the hook on the spring below. The upper wire is $2\frac{1}{4}$ inches long between its connections, the lower one $2\frac{3}{4}$ inches.

The sides of the rectangular coil are flat, being about $\frac{1}{8}$ inch thick and $\frac{5}{16}$ inch wide. The resistance of the coil is 150 ohms. The silver hooks are connected with opposite ends of the coil, in the manner shown at 4 and 5, Fig. 426a. Each hook is provided with a flat head, which is secured between two thick plates of mica, the shank of the hook projecting through a hole in the outer mica plate. Each pair of mica plates is secured in place on the coil by a winding of silk, which is coated with shellac varnish to prevent the plates from slipping. The hooks are arranged exactly in the middle of the ends of the coil, so that when the coil is supported in the position of use by the silver wires, it will oscillate freely between the poles of the magnet and the iron cylinder. The terminals of the coil are soldered to the silver hooks. The upper hook is made a little more than a half inch long, to receive a small concave mirror, as shown at 4, which is secured in place by cement or wax. The mirror has a focus of 1 meter.

The relation of the magnet, A, the coil, C, and the iron cylinder, B, are clearly shown at 3, which is a horizontal section taken through those parts.

A glass shade protects the delicate parts of the instrument. The two binding posts, which are outside of the glass shade, are connected under the base with the brass column and the spring, so that the current passes from one

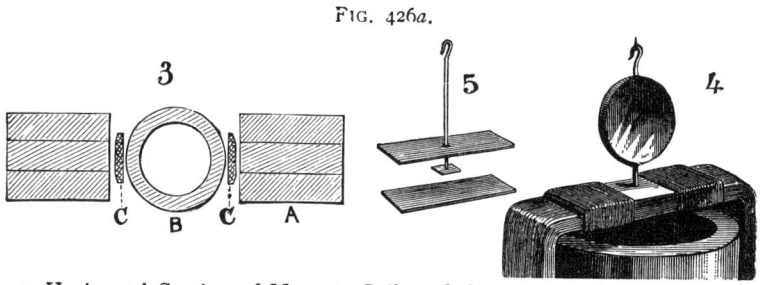

FIG. 426a.

3, Horizontal Section of Magnet, Coil, and Core.   4 and 5, Details of Deprez's Galvanometer.

binding post to the column, thence down the upper silver wire, then through the coil, the lower silver wire, and the spring to the other binding post.

The silver wires are placed under considerable tension, and the coil is adjusted to a central position by turning the hooked rod at the top of the instrument.

When an electric current is sent through the coil, it tends to assume a position at right angles with a line joining the two poles of the magnet, the amount of displacement of the coil from its normal position depending on the strength of the current. As the deflection for a very light current is small, a beam of light reflected from the concave mirror is employed as an index. The scale is arranged as shown in Fig. 427, the light being projected from a lamp,

DYNAMIC ELECTRICITY. 437

supported at the proper height behind the scale, through a slit below the scale and on to the concave mirror. The mirror reflects the beam on to the scale. The mark at the center of the scale is o, and arbitrary numbers, running upward

Fig. 427.

Arrangement of Galvanometer, Lamp, and Scale.

regularly, are arranged on the marks on opposite sides of o. The common paper scale used by draughtsmen answers for this purpose.

When the coil is at rest, the light spot remains at the

center of the scale, but when a current passes through the coil, the beam moves steadily forward and stops without oscillation, the distance through which it moves depending, of course, on the strength of the current. The coil is returned to its normal position by the spring of the silver wires.

By employing shunts, heavy currents may be measured with the aid of this instrument. The sensitiveness of this galvanometer is so great as to indicate a current when the ends of two No. 18 copper wires connected with it are placed on opposite sides of the tongue.

The coil is carefully wound over a form covered with paper, each layer of wire being varnished with shellac varnish as the work of winding progresses. When the coil is complete, the coil, together with the form, is heated in a warm oven until the varnish becomes hard throughout the coil.

The concave mirror may be purchased from the optician, or a very fair mirror may be made by cutting a small disk from a double convex spectacle lens of 20 or 30 inch focus, and silvering it. A simple and quick way of silvering a small surface consists in scraping from the back of a piece of ordinary looking glass all the silvering, except a patch of the size of the mirror to be silvered. A small drop of mercury placed on the patch soon loosens it, so that it may be slid from the glass and transferred to the disk, which must be perfectly clean. After the patch is in position, a piece of tin foil is placed on the back of the disk, pressed down firmly, and allowed to remain long enough to absorb all of the surplus mercury. It is then removed, and the transferred silver will be found adhering strongly to the disk.

The various dimensions above given are taken from an almost exact copy of a Deprez-D'Arsonval galvanometer made by Carpentier, of Paris. The copy operates admirably. It is probable, however, that a considerable deviation from these dimensions might be made without seriously affecting the value of the instrument.

The tangent galvanometer is of great importance in

DYNAMIC ELECTRICITY.

electrical measurements, especially in the class relating to currents. The principle of the instrument is illustrated by Fig. 428. In a narrow coil of wire is suspended a short magnetized needle, whose length does not exceed one-twelfth the diameter of the coil. Two light pointers are connected with the needle at right angles thereto. When a current is sent through this coil, the needle is deflected to the right or left, according to the direction of the current, and the amount of deflection is dependent upon, but not proportional to, the strength of the current. It is, however, proportional to the tangent of the angle of deflection.

A practical tangent galvanometer is shown in Fig. 429. In this instrument the conductor is wound upon a grooved wooden ring 9 inches in diameter, the groove being ¾ inch wide and 1 inch deep. The wooden ring is mounted in a circular base piece, which is pivoted to the lower base to admit of adjustment. The lower base is provided with three leveling screws, which are bored longitudinally to receive pointed wires, which are driven into the table to prevent the instrument from sliding.

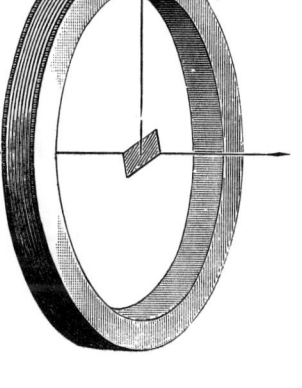

Principle of Tangent Galvanometer.

The lower base is provided with an angled arm, which extends over the upper base piece, and is provided with a screw for clamping the latter when adjusted.

The winding of the ring is divided into five sections having different resistances, so that by means of a plug inserted in the switch on the base the resistance may be made 0, 1, 10, 50, or 150 ohms.

Fig. 430 is a diagram showing the coils and the switch connections stretched out. The first coil, a, is a band of copper ¾ inch wide and $\frac{1}{16}$ inch thick, with practically no resistance. The other coils are of wire. The coils, b and a, together, have a resistance of one ohm. The coils, *c*, *b*, *a*, have a combined resistance of 10 ohms. The coil, *d*, together with the preceding, offers a resistance of 50 ohms, and the combined resistance of all of the coils, *e, d, c, b, a*, is 150 ohms.

The conductors are connected with the binding posts, *f g*, and the current flows through the coils in succession,

Fig. 429.

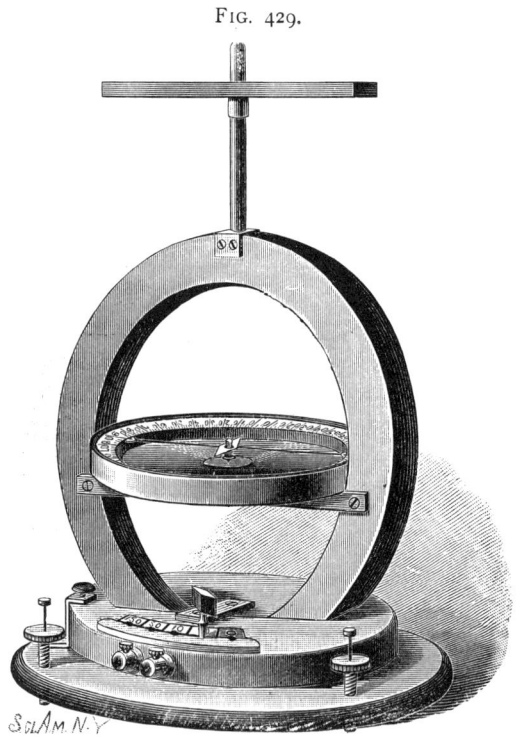

Tangent Galvanometer.

until it reaches one of the smaller switch plates, which is connected with the plate, A, by the plug. In the present case the plug is inserted between the plate marked 1 and the plate, A, causing the current to flow from the binding post, *f*, through the coils, *a, b*, and plate, A, to the binding post, *g*. The resistance of the galvanometer is obviously one ohm.

The magnetic needle, which is $\frac{3}{4}$ inch long, is located

DYNAMIC ELECTRICITY. 441

exactly at the center of the ring, and delicately poised on a fine hard steel point. The needle should be jeweled to reduce the friction and wear to a minimum. To the sides of the needle are attached indexes of aluminum having flat ends, each of which is provided with a fine mark representing the center line of the index. The box containing the scale and the needle is supported by a cross-bar attached to the wooden ring. To the top of the wooden ring is attached a brass standard, which is axially in line with the compass needle.

Upon the standard is mounted a bar magnet, which may be adjusted at any angle or raised or lowered. This

FIG. 430.

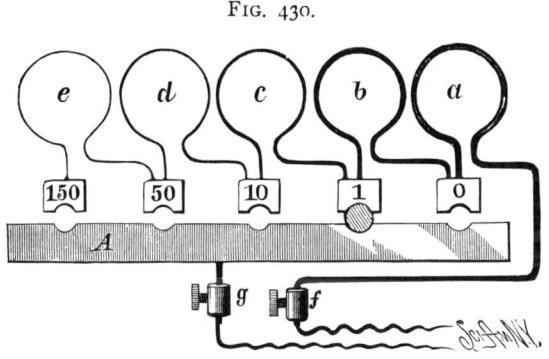

Arrangement of Switch Connections.

magnet serves as an artificial meridian when the galvanometer is used for ordinary work. When it is used as a tangent galvanometer, the magnet is removed.

The Deprez galvanometer is independent of the earth's magnetism, but the tangent galvanometer must be arranged with the coil and the needle in the magnetic meridian, and its adjustment must be such that a current which produces a certain deflection of the needle in one direction will, when reversed, produce a like deflection in the opposite direction. The angle of maximum sensitiveness in the tangent galvanometer is 45°; therefore, when it is possible to do so, the current should be arranged to produce a deflection approximating 45°.

ELECTRICAL MEASUREMENTS.

The resistance of a battery may be ascertained by means of the tangent galvanometer as follows: The battery is connected with the galvanometer, and the deflection of the needle is noted; then a variable resistance is introduced and adjusted until there is a deflection, the tangent of the angle of which is equal to one-half the tangent of the angle of the first deflection. The resistance thus introduced is equal to that of the battery and galvanometer. Take from this quantity the resistance of the galvanometer, and the remainder will be the resistance of the battery.*

For example, when a battery placed in circuit with a tangent galvanometer produces a deflection of 48°, the tangent of that angle being 1·111, half of this quantity would be 0·555, which is very nearly the tangent of the angle of 29°; therefore, resistance is introduced until the needle falls back to 29°. Assuming this resistance to be 15 ohms, and the resistance of the galvanometer to be 10 ohms, the galvanometer resistance deducted from the resistance introduced leaves 5 ohms, which is the resistance of the battery.

To measure the electro-motive force of a battery, a standard cell is necessary. A Daniell or gravity cell, having an E. M. F. of 1·079 volts, is commonly used. This is connected with the tangent galvanometer, and the deflection and total resistance in the circuit, which should be high, is noted. The standard battery is then removed and the one to be measured is inserted in its place, and the resistance of the circuit is adjusted until the deflection of the galvanometer needle is the same as in the first case. It now becomes a matter of simple proportion, which is as follows:

E. M. F. of standard battery.	E. M. F. of battery being measured.	Total resistance in first case.	Total resistance in second case.

Assuming the resistance in the first case to have been 2,500 ohms, and that in the second case 2,000 ohms, the proportion would stand thus:

$$1\cdot079 : \text{Unknown E. M. F.} :: 2{,}500 : 2{,}000$$

* A table of natural tangents is given at the close of this chapter.

or as 5 to 4. The E. M. F. of the battery measured is therefore 0·8632 volt.

A convenient arrangement of the tangent galvanometer scale is to have one side of the scale divided into degrees, the other side being arranged according to the tangent principle, so that the reading will be direct and reference to the table of tangents will be avoided.

The simplest method of measuring resistance is that known as the substitution method, in which the unknown re-

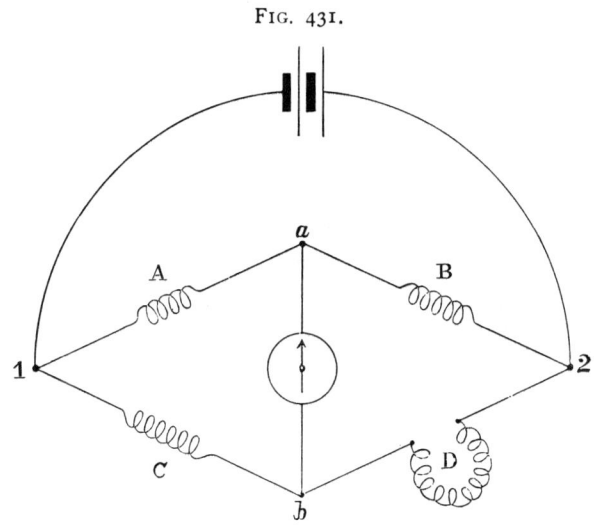

Diagram of Wheatstone's Bridge.

sistance and a galvanometer are placed in the circuit of the battery. The deflection of the galvanometer needle is noted. A variable known resistance is then substituted for the unknown resistance, and adjusted until the deflection is the same as in the first case. The variable known resistance will then equal the unknown resistance. If the current is so great as to cause a deflection of the needle much exceeding 45°, it should be reduced either by removing some of the battery or by the introduction of extra resistance into the circuit. The same conditions must obtain throughout the measurement.

The Wheatstone bridge presents the best known method of quickly and accurately measuring resistances. Any galvanometer may be used in connection with the bridge, that shown in Fig. 428 being the best for most purposes. The bridge method was originally devised by Mr. Christie. The late Sir Charles Wheatstone's name is attached to the invention, in consequence of his having brought it before the public. The principle of this apparatus is illustrated in Fig. 431. A current, in passing from 1 to 2, divides, a part passing over 1, *a*, 2, another part passing over 1, *b*, 2. For every point in 1, *a*, 2 there is a point in 1, *b*, 2 having the same potential. If these two points of equal potential be joined by a conductor, no current will pass through the

FIG. 432.

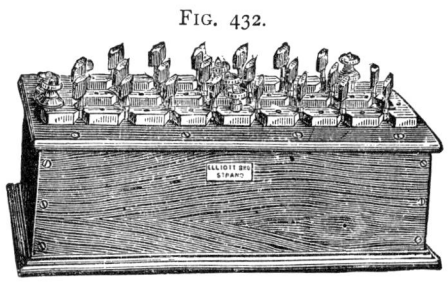

Bridge Resistance Box

conductor. In the diagram the points of equal potential are marked *a*, *b*, and they are connected by a conductor in which is inserted a galvanometer.

A, B, and C are known resistances, and D is the unknown resistance. When A : B :: C : D, the galvanometer needle will stand at 0. The resistance, C, is variable, so that when the unknown resistance, D, is inserted, the resistance, A, is adjusted until the needle falls back to 0.

The commercial form of Wheatstone's bridge is represented in Fig. 432.

In this instrument a number of coils are suspended from the vulcanite cover of the box and connected with brass blocks attached to the cover in the manner shown in Fig. 433, which represents a part of the resistance box.

The terminals of the coils are connected with adjacent

blocks, so that a current entering at A will pass from the first block down through the first coil, thence to the second block. In the present case the second and third blocks are connected electrically by a plug inserted between them, so that the second coil is cut out, the current taking the path of least resistance. The current can pass from the third to

FIG. 433.

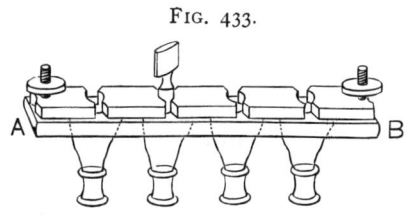

Resistance Box Connections.

the fourth blocks only by going through the third coil, and to pass from the fourth block to the fifth, the current must pass through the fourth coil. Whenever a plug is inserted it cuts out the coil connected with the blocks between which the plug is placed, and when a plug is removed the coil at that point is thrown into the circuit. The coils of the

FIG. 434.

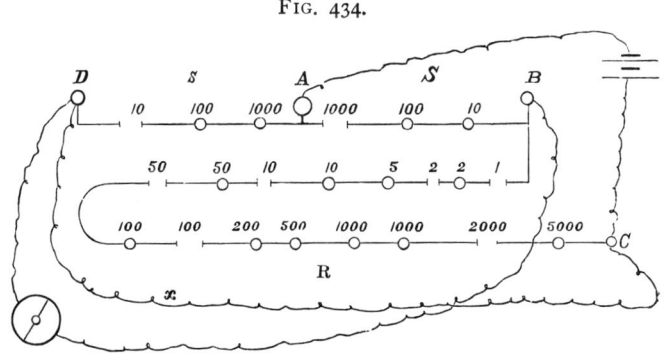

Diagram of Bridge Connection.

resistance box are wound double, so that the current passes into the coil in one direction and out of it in the opposite direction, thus perfectly neutralizing any magnetic effects.

Fig. 434 represents the top of the bridge resistance box, and the circuits diagrammatically. The three branches including the known resistance of the bridge are contained in the resistance box. In this diagram the connections of the battery and galvanometer, as shown in Fig. 431, are transposed for the sake of convenience in calculation, but the results are the same. The resistances, A B, of Fig. 431 are each replaced here by three coils of 10, 100, and 1,000 ohms. These are called the proportional coils. The rest of the resistance box constitutes the adjustable resistance; and x, connected at D and C, is the unknown resistance.

The galvanometer is connected at D B, and the battery at A C. The value of the unknown resistance, x, is determined by simple proportion,

$$x : R : : s : S.$$

As shown in Fig. 434, the variable resistance R = 2163 ohms, s = 10 ohms, and S = 1,000 ohms, therefore x = 21·63 ohms.

The value of the proportional coils may be expressed as follows:

$$\frac{10}{1000} = \frac{1}{100}$$

$$\left.\begin{array}{c}\dfrac{10}{100}\\[4pt]\dfrac{100}{1000}\end{array}\right\} = \frac{1}{10}$$

$$\left.\begin{array}{c}\dfrac{10}{10}\\[4pt]\dfrac{100}{100}\\[4pt]\dfrac{1000}{1000}\end{array}\right\} = 1$$

$$\left.\begin{array}{c}\dfrac{100}{10}\\[4pt]\dfrac{1000}{100}\end{array}\right\} = 10$$

$$\frac{1000}{10} = 100$$

Also

$$\frac{1010}{100}$$

$$\frac{1100}{10}$$

$$\frac{10}{1100}$$

$$\frac{100}{1010}$$

The arrangement of the proportional coils may be 1,000 : 1,000 for large resistances, and 10 : 10 for small resistances. In using the Wheatstone bridge in testing cables or in measuring the resistance of an electro-magnet, or a coil, to avoid delay caused by the deflection of the needle before the current becomes steady, it is best to send a current through the four arms of the bridge (s, S, R, x) before it is allowed to pass through the galvanometer. This is accom-

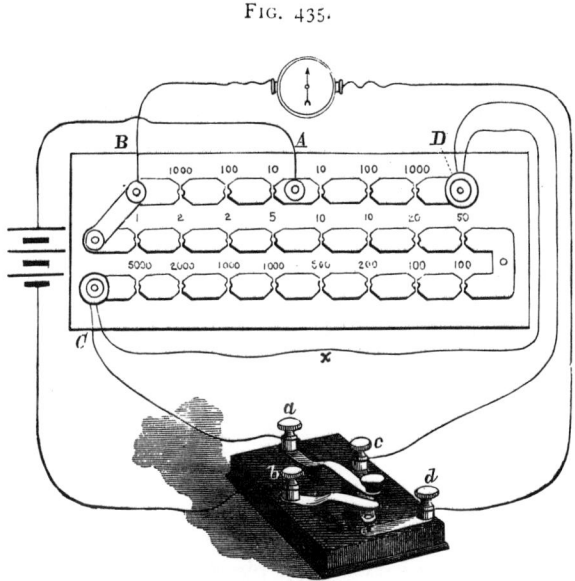

Fig. 435.

Bridge Key and Connections.

plished by means of the bridge key, shown in Fig. 435, together with its connections.

This is in reality nothing more than a double key arranged to control the two parts of the circuit independently, the upper key being arranged so that after it is closed it may be still farther depressed to close the lower one, the two keys being separated by an insulating button.

The binding posts, $a\ b$, of the upper key are inserted in the wire which includes the battery, while the binding posts,

$c\ d$, of the lower key are inserted in the conductor including the galvanometer. When this key is depressed it first sends the current through the arms of the bridge, and then allows it to pass through the galvanometer.*

JOINT RESISTANCE OF BRANCH CIRCUITS.

The resistance of a conductor is directly proportional to its length and inversely proportional to its sectional area, and the conductivity of a wire is the reciprocal of its resist-

FIG 436.

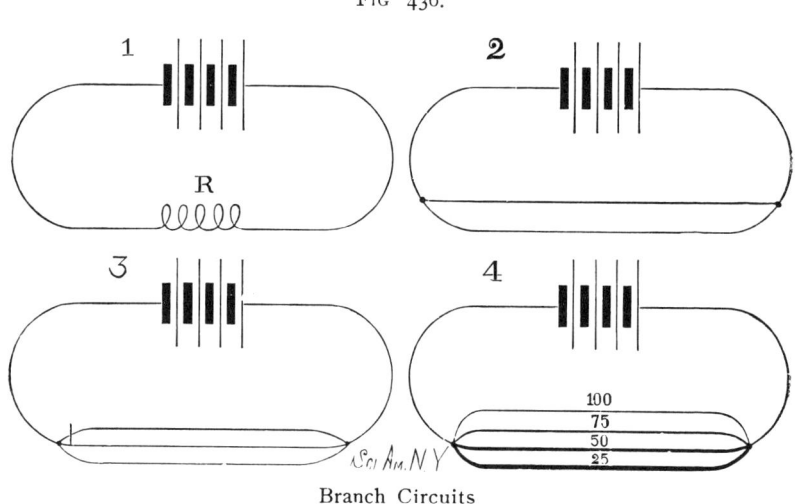

Branch Circuits

ance. The conductivity of a wire having a resistance of 1 ohm is 1; that of a wire having a resistance of 2 ohms is ½; that of a wire having 3 ohms resistance is ⅓, and so on.

The joint resistance of two parallel conductors is, of course, less than that of either taken alone. The joint resistance of a divided circuit is ascertained by finding the conductivities of the different branches. The reciprocal of this result will be the joint resistance.

The method of determining the resistance (R) of a single

---

* "Hand Book of Electrical Testing," by Kempe; "Practical Electricity," by Ayrton; "Elementary Practical Physics," by Stewart and Gee; and "Electrical Measurements and the Galvanometer," by Lockwood, are desirable books on electric measurements.

conductor has already been explained. To find the joint resistance of the divided circuit, 2, Fig. 436, one branch having a resistance of 4 ohms, the other 8 ohms, the reciprocals of these numbers being respectively $\frac{1}{4}$ and $\frac{1}{8}$, these added $= \frac{3}{8}$, which is the joint conductivity. The reciprocal of this is $\frac{8}{3} = 2\cdot 66$ ohms. In a similar manner the joint resistance of three branches (3, Fig. 436) may be ascertained. Assuming the resistances to be 2, 5, and 10 ohms respectively, the reciprocals are $\frac{1}{2}$, $\frac{1}{5}$, and $\frac{1}{10}$, which added $= \frac{8}{10}$, which is the joint conductivity, the reciprocal of this $\frac{10}{8} = 1\cdot 25$ ohms, the joint resistance.

The joint resistance of four or more parallel conductors is found in the same way. In the case of the example shown at 4, Fig. 436, where the resistances are respectively 100, 75, 50, and 25 ohms, the joint resistance is 12 ohms.*

Electrical measurements are made in a commercial way by means of instruments graduated so as to be read directly in ohms, volts, and amperes.

EXPANSION VOLTMETER.†

In the ordinary voltmeter, in which acidulated water is decomposed by electrolysis, and in which the strength of the current is determined by the volume of gas accumulating in a given time, there are several objectionable features which prevent it from coming into general use for the measurement of the strength of electric currents.

In the first place, the electrolytic voltmeter is incapable of indicating the strength of the current at any particular moment, and cannot, therefore, yield anything but a mean result. It offers considerable resistance in the circuit, its indications depend upon the acidity of the water and the size and distance apart of the electrodes; and to secure accurate results, the temperature and barometric pressure must be taken into consideration.

The voltmeter shown in the engraving, Fig. 437, depends

* For simple methods of working out these and analogous problems the reader is referred to "The Arithmetic of Electrical Measurements," by W. R. P. Hobbs.

† Published originally in the *Scientific American* of July 9, 1881.

on the heating effect of the current on a thin wire of platinum or copper, the linear expansion of the wire giving the index more or less motion, according to the strength of the current.

This instrument has one source of error to be compensated for—that is, the increase of the resistance of the wire with the increase of temperature. No account is taken of the environing temperature nor of barometric pressure, and the indication may be read at any moment; and, moreover, the increase of resistance due to increased temperature may be disregarded, since the normal resistance of the wire is almost nothing.

This voltmeter finds its principal application in connection with the stronger currents, such as are employed in electric lighting, in electro-metallurgy, and in telegraphy. It must be adapted within certain limits to the current which is to operate it, but when the instrument is properly proportioned to its duties, its indications may be relied upon.

A vertical plate of vulcanite supports a horizontal stud, upon which are placed two metal sleeves having a glass lining. To one of these sleeves is attached a counterbalanced arm, carrying at its upper end a curved scale, having arbitrary graduations determined upon by actual trial under approximately the same conditions as the instrument will be afterward subjected to in actual use. The other sleeve carries a light counterbalanced metal index, which moves in front of the curved scale. Each sleeve is provided with a curved platinum wire arm, dipping in mercury contained in an iron cup secured to the base. Two platinum or copper wires are stretched along the face of the instrument, and attached at one end to hooks passing through an insulating post, and after passing once around their respective sleeves on the index and scale, are attached to spiral springs, which in turn are connected with wire hooks extending through an insulating post projecting horizontally from the vulcanite plate.

Under each wire there is a horizontal metal bar communicating under the base with one of the binding posts. The two other binding posts are connected separately with the

Fig. 437.

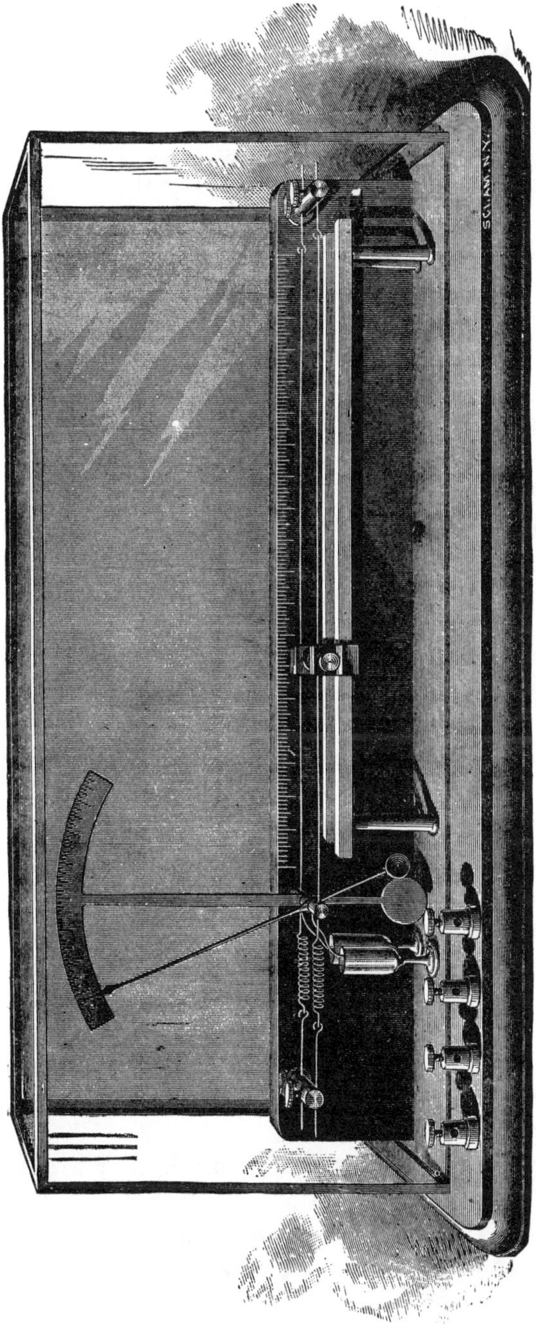

Expansion Voltmeter.

two mercury cups. It will be seen that with this construction the expansion of the rear wire will move the scale, while the expansion of the front wire will move the index. In order to apply the current to any required length of wire, there is upon each of the horizontal bars a clamp, which may be placed anywhere along the bar and screwed up so as to clamp both wire and bar.

Usually the current to be measured will pass from the battery or machine to one of the binding posts, thence to the forward horizontal bar, thence through the expansion wire connected with the index, through the sleeve of the index, and finally through the mercury cup to the other binding post.

It will be observed that both scale and index will be moved in the same direction by the expansion of their respective wires, and that the atmospheric temperature affects both alike. This being true, it is unnecessary to take any account whatever of external temperature. The apparatus is inclosed in a glass case to prevent the cooling action of the draughts of air.

By connecting the index expansion wire with a battery having an electro-motive force of one volt, the deflection is slight, even with a fine wire, but in a stronger current from a battery having an electro-motive force of five volts and upward, slight variations will be readily indicated.

As mentioned before, the instrument must be adapted to the conditions under which it is to be used. For use with a moderate current, a No. 36 platinum wire, about the length of that shown in the engraving, answers a good purpose, but for heavier currents from a dynamo-electric machine, a larger and longer wire of copper will be required. It should be small enough to be heated somewhat by the current, but not so small as to offer any material resistance in the circuit.

When the larger wires are used, they are not wound about the sleeves of the index and scale, but are bent downward before reaching the sleeves, and the mercury cups are placed so as to receive their lower ends. Cords or small chains are attached to the angles of the wires and wrapped

DYNAMIC ELECTRICITY. 453

once around the sleeves and attached to the springs. This instrument, placed directly in the circuit of a dynamo-electric machine, or in a shunt, will indicate the amount of current passing. When it is desired to compare two currents, the expansion wire of the index is placed in one

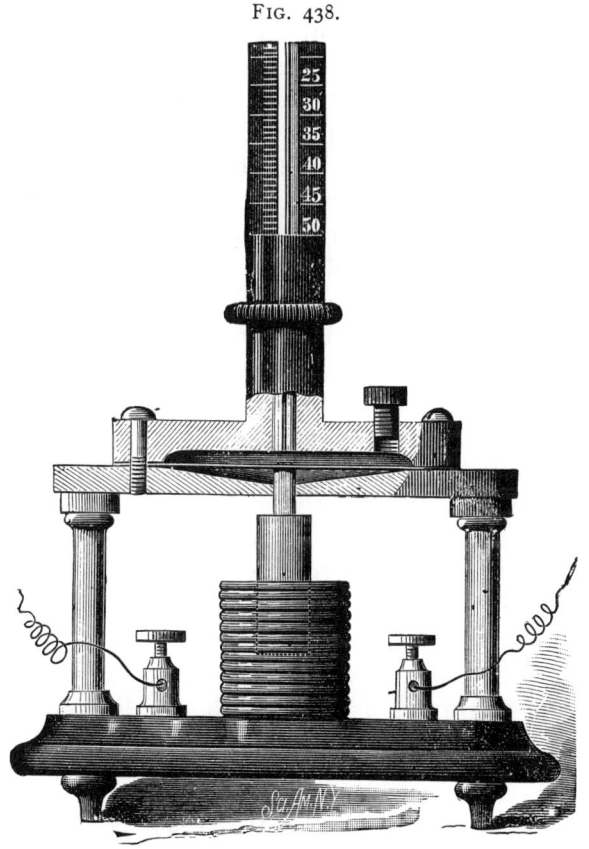

FIG. 438.

Ammeter.

circuit and the expansion wire of the scale is placed in the other circuit. In a delicate instrument of this kind the tension of the expansion wires should be only sufficient to keep the wires taut, as they are readily stretched when considerably heated.

AMMETER.

The instrument shown in Fig. 438 is an ammeter, for indicating the strength of the current when its coil is included in an electrical circuit. The horizontal metallic plate, mounted on the columns, is concaved in the middle and supports a spring steel diaphragm that is held in place by the iron cap secured to the plate by several screws, so as to clamp the diaphragm tightly.

The cap is chambered out to receive mercury, and has a stuffing box for holding a glass tube of small caliber. A vulcanite screw in the cap serves to bring the mercury in the tube to zero before taking a reading, thus avoiding variations by the expansion of the mercury. The graduations on the scale at the side of the tube, which are empirical, represent the amperes of the current passing through the coil. A short rod is attached to the middle of the diaphragm, and projects downward through a hole in the base plate to receive a soft iron cylindrical armature or core which extends into the coil.

The diameter of the diaphragm is 2 inches; the caliber of the glass tube, 0·02 inch; a very slight motion of the diaphragm is indicated by a considerable movement of the mercury in the tube.

This instrument, placed anywhere in the main circuit, will indicate the strength of the current. An increase in the strength of the current results in the drawing of the iron core into the coil, and a consequent deflection of the diaphragm and downward movement of the mercury column. The engraving is five-eighths of the actual size of the instrument. The glass tubes and scale are shown only in part.

RECORDING VOLTMETER.

In making electrical tests it is often desirable to consider the element of time, but, as every electrician knows, to do this with the ordinary appliances is tiresome, and the result is liable to be inaccurate.

The extreme delicacy of the action of the galvanometer renders it difficult to apply to it any device capable of recording the movements of the needle without interfering more or less with its action. In the instrument shown in the en-

DYNAMIC ELECTRICITY. 455

graving a disruptive spark from an induction coil is utilized for making the record. The indicating parts are made and

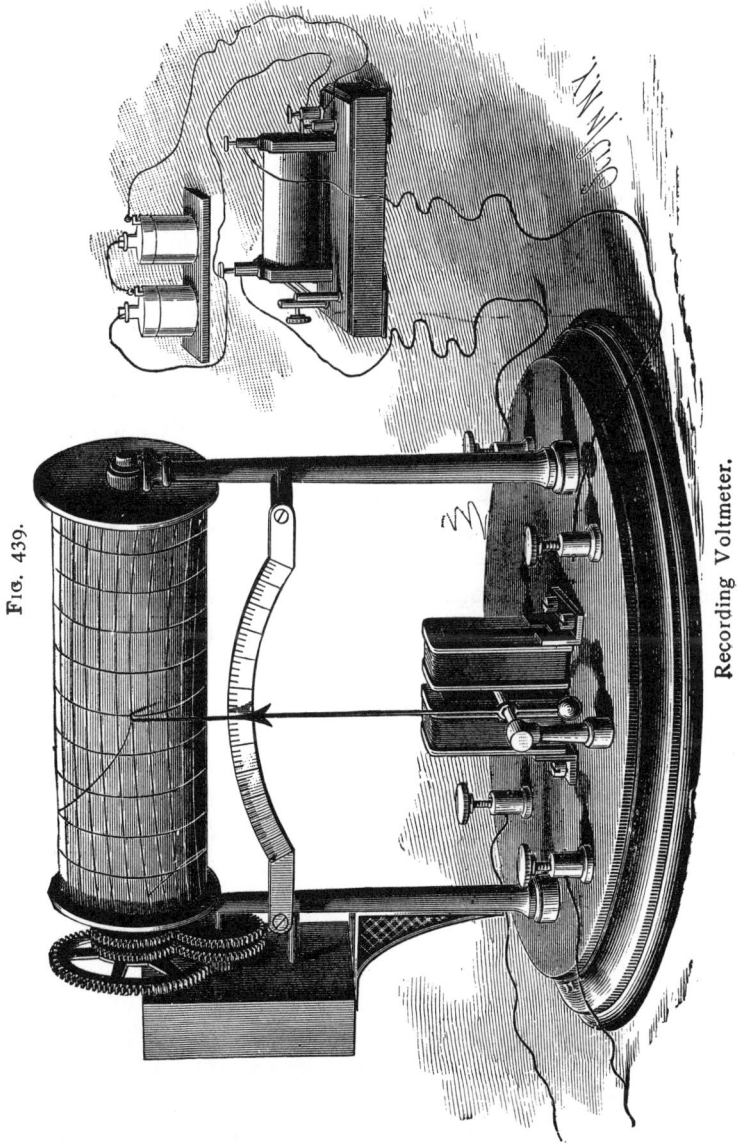

Fig. 439.

Recording Voltmeter.

arranged as in an astatic galvanometer. The helixes are wound with rather coarse wire (No. 22). The needle is astatic, the inner member swinging in the central opening in the helixes in the usual way, the outer member being located behind the helixes. The arbor supporting the needle has very delicate pivots, and carries a long and very light aluminum index, which is counterpoised so that it assumes a vertical position when no current passes through the helixes. The needle is unaffected by terrestrial magnetism.

The upper end of the index swings in front of a graduated scale, and is prolonged so as to reach to the middle of the cylinder, carrying a sheet of paper upon which the movements of the needle are to be recorded. This cylinder is of brass, and its journals are supported by metal columns projecting from the base upon which the other parts of the instrument are mounted. The scale is supported by vulcanite studs projecting from the columns, and to one of the columns is attached a clock movement provided with three sets of spur wheels, by either of which it may be connected with the arbor of the cylinder. One pair of wheels connect the minute hand arbor of the clock with the cylinder, revolving the cylinder once an hour; another pair of wheels connect the hour hand mechanism with the cylinder, so that the latter is revolved once in twelve hours; while a third pair of wheels give the cylinder one revolution in seven days.

This instrument is designed especially for making prolonged tests. It is provided with four binding posts, two of which connect the wires of the batteries under test with the helixes. The other binding posts are connected respectively with the posts supporting the needle and with the journals of the recording cylinder. These posts receive wires from an induction coil capable of yielding a spark from one-eighth to one-quarter inch long.

The induction coil is kept continuously in action by two Bunsen elements, and a stream of sparks constantly pass between the elongated end of the index and the brass cylinder, perforating the intervening paper and making a permanent record of the movement of the needle. To render the line of perforations as thin as possible, the end of the index is made sharp and bent inward toward the cylinder.

The spur wheels are placed loosely on the arbor of the cylinder, and the boss of each is provided with a set screw by means of which it may be fixed to the arbor. This arrangement admits of giving to the cylinder either of the speeds, as may be required.

The paper upon which the record is to be made is divided in one direction to represent volts, and in the other into hours and minutes. The hour and minute lines are curved to coincide with the path of the end of the index.

These records may be duplicated by using the sheet as a stencil and employing the method of printing used in connection with perforating pens. When the tests are of long duration, the action of the induction coil is rendered intermittent by an automatic switch connected with the clock.

ELECTRO-MAGNETS.

A body of iron with an insulated conductor wrapped one or more times around it constitutes an electro-magnet. The power of an electro-magnet depends upon the form, size and quality of its iron core, upon the number of turns the conductor makes around the core, and upon the current passing through the conductor. The number of amperes flowing through the wire of a magnet, multiplied by the number of turns the wire makes around the core of the magnet, gives the number of ampere turns; one ampere flowing ten times around is equal to ten amperes flowing once around. Two amperes flowing five times around is the equivalent of either of the foregoing.

The magnetizing power of the circulating current is proportional to the number of ampere turns. The magnetism produced in the iron core is not always proportional to the ampere turns, as the current produces comparatively little effect when the magnet core approaches saturation.

The battery must be proportioned to the resistance of the magnet to secure the best results; or, if the magnet is arranged with its winding in sections, so that they may be connected up parallel or in series, as will presently be described, the magnet may be adapted to the current.

A large magnet made on this plan is shown in Fig. 440.

It is well adapted for experimental work. With a current from six medium sized bichromate battery cells it is capable of sustaining about one thousand pounds. It is provided with a switch, so that it may readily be adapted to a light or a heavy current by combining the several coils in series or in parallel. It is made separable, to permit of using the coils detached from the core.

For the construction of the magnet 18 pounds of No. 14 double-covered magnet wire are required, also two well annealed cylindrical bars of soft iron, 8 in. long and $1\frac{1}{2}$ in. in diameter for the core, a flat, soft iron bar $2\frac{1}{2}$ in. wide, 8 in. long, and $\frac{3}{4}$ in. thick for the yoke, a bar of the same kind

Fig. 440.

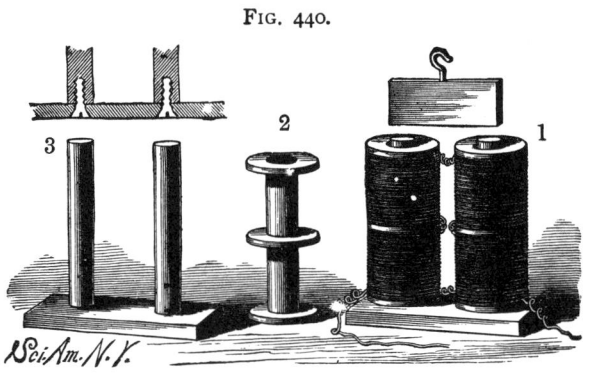

Magnet for Experimentation.

7 inches long for the armature, two double wooden spools 4 in. in diameter and $7\frac{3}{4}$ inches long, with flanges $1\frac{1}{16}$ in. wide and $\frac{7}{16}$ in. thick.

The walls of the spools are $\frac{1}{16}$ in. thick. Each space in each spool is filled with the No. 14 magnet wire. There are two ways of winding the wire. According to one method a hole is drilled obliquely downward in the flange, and one end of the wire is passed from within outward through the hole, and the spool is wound in the same manner as a spool of thread, the wires at the end of the coil being tied together with a stout thread to prevent unwinding. Each section of each spool is filled in the same manner.

DYNAMIC ELECTRICITY. 459

Although this is the quickest way to wind the magnet, it is not the best way, as the inner end of the coil is liable to be broken off, when the entire coil must be rewound to secure a new connection with the inner end. The correct way to wind the wire is to take a sufficient length and wind it from opposite ends on two bobbins. Wind the wire once over the spool from one of the bobbins, then wind from the ends of the coil thus formed toward the middle, first with wire from one bobbin, then from the other bobbin, then wind from the middle back each way toward the ends in the same way, then again toward the center, and so on. By this method both terminals of the coil are made to come out on the outer layer.

FIG. 441.

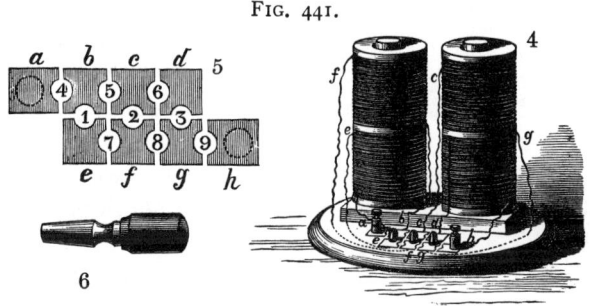

Magnet and Switch.

At 1, Fig. 440, is shown the completed magnet and its armature. 2 is a detail view of the spool. 3 shows the cores and yoke, both in perspective and section, the sectional view exhibiting the method of fastening the cores to the yoke by means of screws. 4 (Fig. 441) shows the magnet mounted on a wooden base provided with a plug switch for connecting the coils in parallel or in series. 5 is an enlarged view of the switch, and 6 shows one of the plugs by which the connections are made.

The switch is formed of brass blocks, *a, b, c, d, e, f, g, h,* arranged in two series, as shown at 5 (Fig. 441). The blocks, *a, h,* are provided with binding posts for receiving the battery wires. The blocks are provided with semicircular notches forming the plug holes, 1, 2, 3, 4, 5, 6, 7, 8, 9.

The block, *a*, is connected with the lower terminal of the lower left hand coil, and the block, *e*, is connected with the upper terminal of the same coil. The block, *b*, is connected with the lower terminal of the upper left hand coil, and the block, *f*, is connected with the upper terminal of the same coil. The block, *h*, is connected with the lower terminal of the lower right hand coil, and the block, *d*, is connected with the upper terminal of the same coil. The block, *g*, is connected with the lower terminal of the upper right hand coil, and the block, *c*, is connected with the upper terminal of the same coil.

When the holes, 1, 2, and 3, are plugged, the current goes in series through all the coils. By plugging the holes, 4, 7, 2, 6, and 9, the current goes through the coils two in parallel and two in series, reducing the resistance to a quarter of the original amount, by halving the length and doubling the sectional area. By plugging the holes, 4, 5, 6, and 7, 8, 9, the current goes through all the coils in parallel, and the resistance is reduced to $\frac{1}{16}$ the original amount, by reducing the length to $\frac{1}{4}$ and increasing the sectional area four times.

The polar extremities of the magnet are drilled axially and tapped to receive screws by which are attached extension pieces for diamagnetic experiments.

To retain the spools on the cores when the magnet is in an inverted position, a thin brass collar is screwed on the end of each core. The armature is provided with a hook for receiving a rope or chain, and the yoke has a threaded hole at the center for receiving the eye for suspending the magnet.

Although this magnet is very complete and desirable, a large proportion of the experiments possible with it may be performed by means of the inexpensive magnet shown in Fig. 442.

The core of this magnet is made of twenty thicknesses of ordinary one inch hoop iron, about $\frac{1}{20}$ inch thick, thus making a rectangular U-shaped core one inch square. The parallel arms of the magnet are five inches long, and the distance between the arms four inches.

The pieces of hoop iron are readily bent and fitted one over the other in succession, the inner one being fitted to and supported by a rectangular wooden block. When the core has reached the required thickness, the layers of which it is formed are fastened together by means of iron rivets passing through holes traversing the entire series of iron strips near the ends of the core. If it is inconvenient to secure the layers in this way, they may be wrapped from the extremities down to the angles with very strong carpet

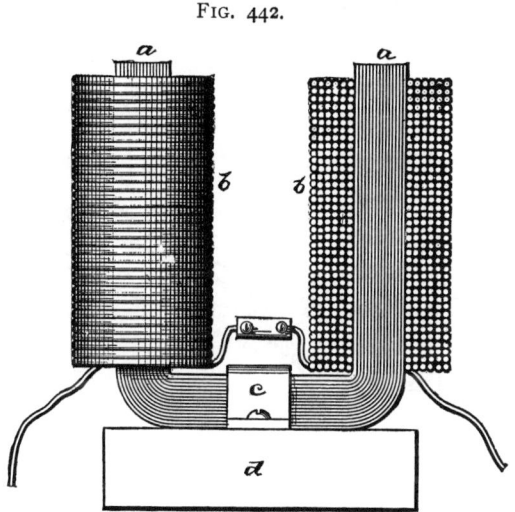

Fig. 442.

Electro-Magnet Partly in Section.

thread or shoe thread and afterward coated with shellac varnish, which holds on the thread and assists in cementing the whole together.

The extremities, *a a*, of the core are filed off squarely. The yoke is clamped to the base, *d*, by the clip, *c*, made of hoop iron or of wood.

To the arms, *a a*, are fitted the coils, *b b*, which are formed by the aid of the device shown in Fig. 443. This consists of two wedge-shaped wooden bars, A B, which together form a bar a little larger than the core of the magnet, and two mortised heads, C D, fitted to the bar with a space of 4¾ inches between them. The head, D, is provided with a screw for clamping the wedge bars, A B, and with an aperture, *a*, for the inner end of the wire. The heads are lined with thick paper, and the bar between the heads is covered with a single thickness, E, of heavy paper.

The winding is begun by passing the end of the wire (No. 16 copper cotton-covered magnet wire) through the aperture, *a*, allowing it to project about three inches, then winding the wire evenly over the bar from one end toward the other until the head, C, is reached. Before the second layer of wire is wound, the first one is brushed over with

FIG. 443.

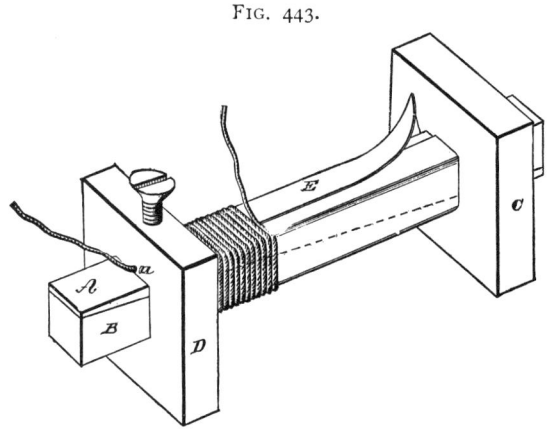

Form for Coils.

thin glue. The second layer is then wound, starting from the head, C, and winding in the same direction toward the head, D, and when the second layer is complete it is brushed over with the glue, after which the third layer is wound and glued, and so on, laying the wire on like thread on a spool until six or eight layers have been applied.

To prevent the destruction of the coil by the loosening of the ends of the wire, a loop of tape should be placed on the beginning of the first convolution and laid over the first layer of wire, so that it may be clamped by the second layer, and in a similar manner some stout threads should be placed between the outer layer and the adjacent layer, so that they

DYNAMIC ELECTRICITY. 463

may be tied over the last convolution of the last layer. After the glue has become thoroughly dry and hard, the heads, C D, are removed from the bars, A B, and the tapering bars are knocked out of the coil in opposite directions, their wedge shape facilitating this removal. Two coils precisely alike are required. When they are placed on the core, the outer end of one coil is connected with the outer

FIG. 444.

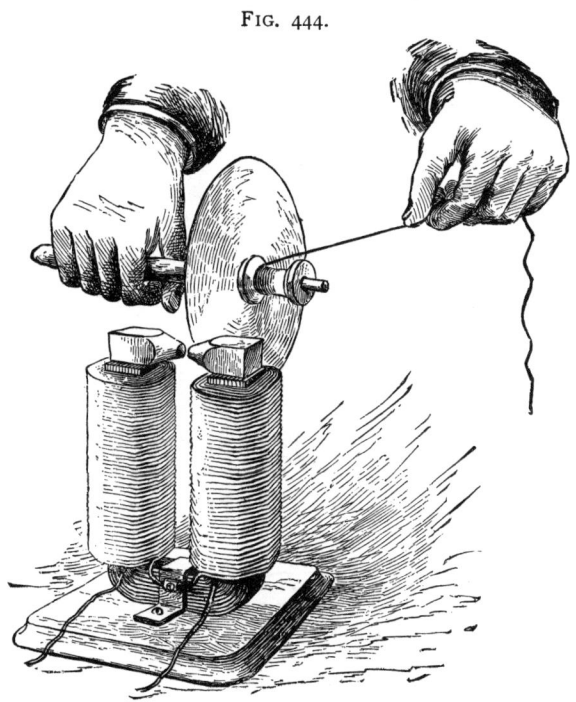

Foucault's Experiment.

end of the other, and the remaining ends are connected with a battery.

To give the coils a finished appearance, they may be coated with shellac varnish colored with a pigment of suitable color, vermilion for example.

Almost any battery may be used in connection with this magnet. The simple plunge battery shown in Fig. 393 will answer admirably.

EXPERIMENTS WITH THE ELECTRO-MAGNET.

To the poles of the magnet should be fitted two short iron bars having conical ends. These bars will need no special fastening, as the attraction of the magnet will hold them in place.

In Fig. 444 is shown a simple way of reproducing Foucault's experiment. A centrally apertured copper disk, 6 inches in diameter, is attached by means of small nails to the end of a common spool, and the spool is mounted so as to turn on a screw inserted in a handle. The short iron

FIGS. 445 AND 446.

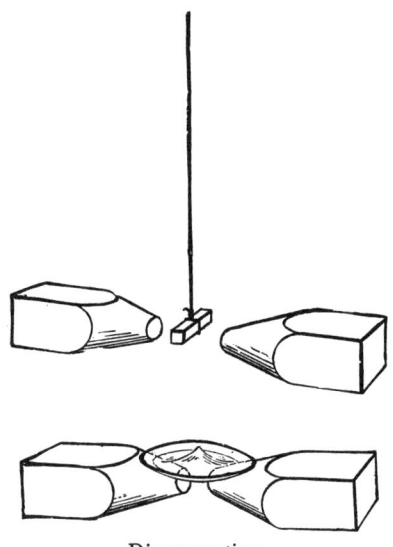

Diamagnetism.

bars are arranged on the poles of the magnet, as shown in the engraving, with the conical ends about one-fourth inch apart. A strong current is sent through the magnet, and the copper disk is whirled rapidly by quickly unwinding a string from the spool, after the manner of top spinning. The edge of the disk is then inserted between the conical pole pieces, but without touching them. The rotation of the disk is almost instantly stopped. A sheet of copper moved back and forth between the pole pieces offers a sensible resistance.

DYNAMIC ELECTRICITY. 465

Most experiments in diamagnetism may be performed with this magnet. Short bars of various metals may be suspended, by means of a silk fiber, between the poles. Iron, nickel, cobalt, manganese, etc., will arrange themselves in line with the poles, while bismuth, antimony, and several other metals will arrange themselves across the line of the poles. The former are known as paramagnetic bodies, the latter as diamagnetic.

Liquids placed in a watch glass, as shown in Fig. 446, exhibit paramagnetic or diamagnetic properties: by piling up at the center of the glass, as shown in the engraving, if paramagnetic, or by piling up on opposite sides of the center, if diamagnetic.

The coils of this magnet, being removable, may be used in magnetizing steel bars, and for other purposes requiring the coils only.

There are about three pounds of wire in each coil of the magnet.

EXPERIMENTS ILLUSTRATING THE PRINCIPLE OF THE DYNAMO.

The great development of electricity in recent years, especially in the line of electric illumination, has served to add luster to the name of the immortal Faraday, and to show with what wonderful completeness he exhausted the subject of magneto-electric induction.

Since the close of his investigations no new principles have been discovered. Physicists and electrical inventors have merely amplified his discoveries and inventions, and applied them to practical uses.

The number of those who are familiar with the discoveries of Faraday and their bearing on modern electrical science is not only large, but rapidly increasing, but there are those who are still learners, to whom new things, or old things placed in a new light, are ever welcome. To such the simple experiments here given may be an aid to the understanding of induction as developed in dynamos and motors.

Any one at all acquainted with electrical phenomena

knows that a hardened steel bar surrounded by a coil of wire which is traversed by an electric current becomes permanently magnetic. It is perhaps unnecessary to reiterate

FIGS. 447 AND 448.

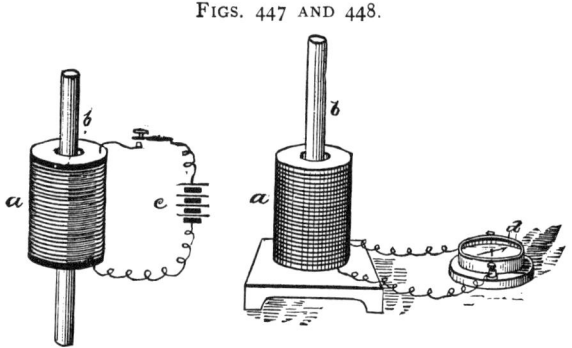

Magnetization of Steel Bar.   Magneto-Electric Induction.

the accepted theories of this action, as they are well established and appear in almost every text book of physics.

The fundamental magneto-electrical experiment of Faraday was exactly the reverse of the operation of producing a magnet by means of an electrical current. That is, it was

FIG. 449.

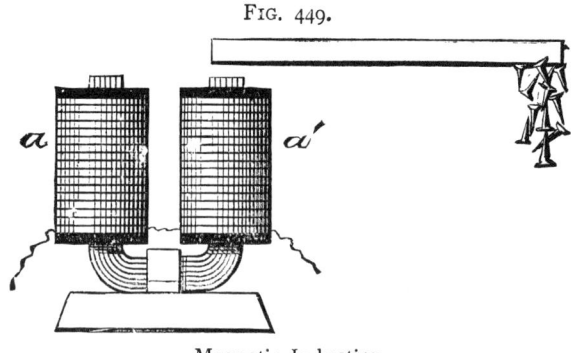

Magnetic Induction.

the production of an electrical current by means of a magnet and coil. In the first instance the magnetizing power of the electric current is employed to bring about the molecular change in the steel bar, which manifests itself in polar-

DYNAMIC ELECTRICITY. 467

ity. In the second instance the magnetized steel bar is made to generate an electric current in the wire of the coil. In the first instance the current moving in the wire of the coil induced magnetism in the steel. In the second instance the movement of the magnetized steel within the coil induced a current in the wire.

The method of magnetizing a bar of steel is clearly shown in Fig. 447, in which *a* is a helix of six or eight ohms resistance, *b* the bar of hardened steel, and *c* a battery of four or five elements. A key is placed in the circuit, but the ends of the wires may be made to serve the same purpose.

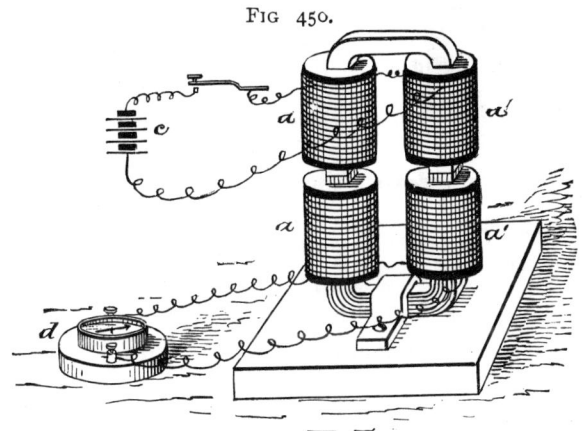

Induced Current from Induced Magnetism.

By closing and opening the circuit while the steel bar is within the coil, as shown, the bar becomes instantly magnetic. When the coil is disconnected from the battery and connected with a galvanometer, *d*, as shown in Fig. 448, and the magnet, *b*, is suddenly inserted in the coil, the needle of the galvanometer will be deflected; but the action is only momentary. The needle returns immediately to the point of starting. When the magnet is quickly withdrawn from the coil the needle is deflected for an instant, but in the opposite direction, and as before it immediately returns to the point of starting. It is obvious that if these electric pulsations can be made with sufficient rapidity to render them practically continuous, and if they can be corrected so that pulsations of the same name will always flow in the same direction, the current thus produced may be utilized.

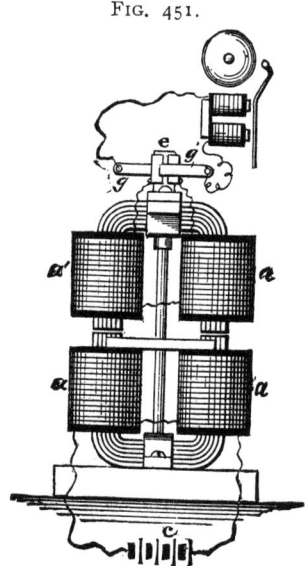

Fig. 451.

Simple Current Generator.

Before proceeding further with the consideration of magneto-electric induction, it will be necessary briefly to examine the subject of magnetic induction, as it is intimately connected with the action of the dynamo. In Fig. 449 is illustrated the usual experiment exhibiting this phenomenon. The electro-magnet is connected with a suitable battery and a bar of soft iron is held near but not in contact with one of the poles of the magnet. It becomes magnetic by induction, the end nearest the magnet being of a name different from that of the pole by which the induction is effected. The end of the bar remote from the magnet exhibits magnetism like that of the pole of the magnet. The relation of magnetic induction to magneto-electric induction is clearly shown by the experiment illustrated in Fig. 450. In this case two electro-magnets are arranged with their poles near each other or in contact. One of them is connected with a galvanometer, and the other with a battery. When the circuit of the upper magnet is closed, the core of that magnet becomes magnetic, the core of the lower magnet becomes magnetic by induction, and the galvanometer needle is deflected. When the circuit of the upper magnet is broken, the galvanometer needle is deflected in the opposite direction, showing that the results are precisely the same

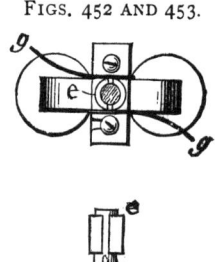

FIGS. 452 AND 453.

Details of Generator

DYNAMIC ELECTRICITY. 469

as in the experiment illustrated by Fig. 448. In this case no mechanical movement is necessary, as the magnetism is introduced into the coils of the lower magnet by induction. It is thus shown that it is not necessary to move any matter in the neighborhood of a magnet to secure magneto-electric induction. In Fig. 451 is shown an arrangement of electro-magnets in which one is fixed, while the other can be revolved. It is a device intended simply for showing how two ordinary electro-magnets may be utilized to advantage in experiments in induction.

To the polar extremities of the fixed magnet is fitted a wooden cross bar, having in its center an aperture for receiving the vertical spindle, the lower end of which is journaled in the clamp that holds the fixed magnet to the base. The upper end of the spindle is provided with a yoke for holding the movable magnet. The cross bar which clamps the magnet in the yoke is held in place by two screws, as shown in Fig. 453, and to the center of the cross bar is attached a wooden cylinder, e, axially in line with the spindle. To the wooden cylinder are secured two curved brass plates which are connected electrically with the terminals of the coils, $a\ a'$, of the movable magnet, one plate to each coil. Two strips of copper, $g\ g,'$ held upon opposite sides of the cylinder, complete the commutator. The copper strips are connected with any device capable of indicating a current—in the present case an electric bell—and the coils, $a\ a'$, of the fixed magnet are connected with the battery, c.

By turning the upper magnet, the following phenomena will be observed: 1. When, by turning the movable magnet, its poles are pulled away from the fixed magnet, the departure of the induced magnetism from the core of the movable magnet produces an electric pulsation in the coil which operates the electric bell. When, by a continued movement of the magnet in the same direction, the poles exchange position, another electrical impulse will be induced by the remagnetization of the movable core, and the bell will be again operated. 2. By examining these impulses by means of a galvanometer introduced into the circuit, it will be found that they are of the same name. 3. When the magnet is turned a little faster, these two impulses will blend into one, so that for each half of the revolution of the magnet the bell yields but one stroke. 4. By whirling the magnet quite rapidly, the current through the bell magnet is made practically continuous, so that the bell armature is drawn forward toward the magnet and held there.

From what has been said, it will be seen that all of the positive electrical impulses are generated upon one side of the poles of the fixed magnet, and all the negative impulses are generated upon the other side of the fixed magnet, and that the curved plates of the commutator conduct all of the positive electrical impulses to one of the strips, $g\ g'$, and all of the negative impulses to the other strip.

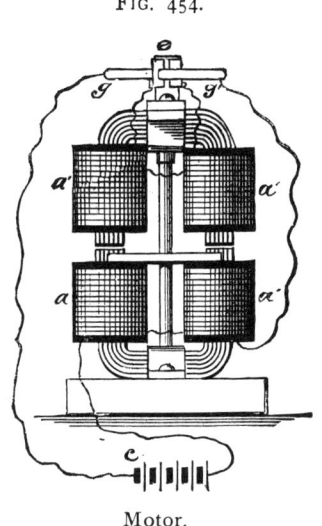

FIG. 454.

Motor.

In Fig. 454 is shown an arrangement of connections to convert the device into a motor.

In this case the battery current passes through both the stationary and movable magnet, but it is commuted in the upper magnet so that it changes direction twice during each revolution, the change occurring when the poles are in opposition, or a little in advance of reaching this position, to compensate for the time required to magnetize and demagnetize the core of the movable magnet or armature. The direction of the current in the movable magnet or armature is such that as its poles are approaching those of the fixed magnet, its magnetic charge will be of opposite name to that of the fixed magnet, so that attraction will result, but as soon as the poles are in opposition, the current in the movable magnet being reversed, there is repulsion. In a motor of this kind these operations succeed each other with great rapidity.

DYNAMIC ELECTRICITY. 471

The fifty cent toy electro-motor is shown in the annexed engraving. It embodies the main features of the apparatus shown in Fig. 454, differing only in having a permanent fixed magnet instead of an electro-magnet.

The vertical spindle which carries the armature is journaled at the lower end in the middle of a U-magnet and at the upper end in a brass cross piece attached to the poles of the magnet. The armature consists of a cross arm of soft iron wound with four or five layers of fine wire. The terminals of the winding of the armature are connected with a two-part commutator carried by the spindle, and touched by two commutator springs supported by wires driven into the base. A metal stud, rising from the base, is connected with

FIG. 455.

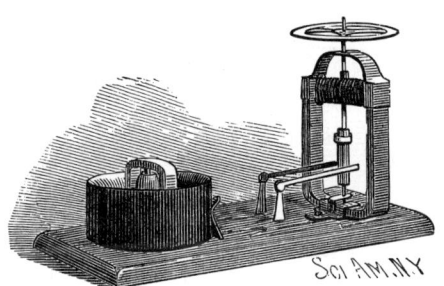

Fifty Cent Electric Motor.

one of the commutator springs, and is provided with an insulating covering on its sides, while its upper end is bare. Upon the stud is placed an annular cell of carbon, which is touched on its outer surface by a spring connected with the remaining commutator spring. The cell forms one of the elements of the battery. The other element consists of a bar of zinc provided with a central aperture for receiving the upper end of the stud, and having its ends bent downward. The cell is filled with a solution of bisulphate of mercury in water. As the salt is reduced by chemical action, a current is produced which will run the motor at a high rate of speed. The motor is fitted with a wheel or plate for carrying color disks, similar to those accompanying the chameleon top.

EXPERIMENTS ILLUSTRATING THE PRINCIPLE OF THE DYNAMO.

After noticing the effect of plunging a magnet into a coil of wire, it is not very difficult, in the light of present electrical knowledge, to understand how the process of induction is carried on in a continuous way in the armature of a dynamo.

The simplest form of armature for illustrating this point is undoubtedly that known as the Gramme ring armature. It is perhaps unnecessary to go into the details of the construction of the Gramme ring, as commonly used in

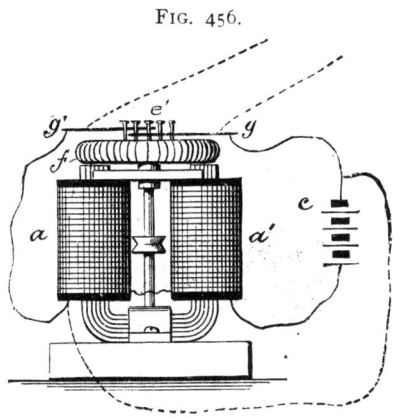

FIG. 456.

Gramme Machine for Illustration.

dynamos. A very crude ring answers the present purpose. Its core is formed of a compact circular coil of soft iron wire, which, in cross section, may be circular or of any other form. The core is wrapped with tape and varnished to insure insulation.

Around this iron ring or core is wound an insulated copper wire, arranged in a spiral coil, f, like the winding of an ordinary electro-magnet. The ends of the copper winding are joined by soldering, thus forming a closed coil. The ring is mounted upon a circular wooden support attached to a spindle, so that the armature may be revolved in front of the poles of a magnet, $a\ a'$, as shown in Fig. 456. In the

DYNAMIC ELECTRICITY. 473

wooden support, in a circle concentric with and near the spindle, are inserted six or eight wire nails, e', arranged at equidistant points. The copper winding of the ring is spaced off into as many sections as there are nails in the circular row, and at the end of each section the insulation of the copper wire is removed a short distance, and a wire, i, is attached by soldering. These attached wires are each connected with one of the wire nails. Now, all that remains to complete the Gramme dynamo or motor is the application of two conductors, $g\ g'$, to the circular row of wire nails, as shown in Figs. 457 and 458.

FIGS. 457 AND 458.

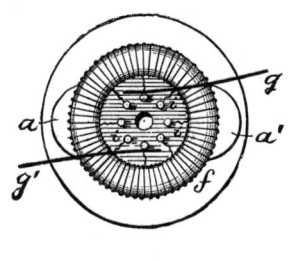

Details of Armature.

This dynamo has all of the essential features of the regular machine—the field magnet, the iron armature core, the conductor wound upon the core, the commutator cylinder formed of the wire nails, and the brushes consisting of wires held on opposite sides of the commutator cylinder.

This dynamo is constructed for illustration only, and not for practical use. It will generate a current, and may be driven as a motor by a current, but of course not with the same advantage as a more complete machine.

In investigating the phenomena of the armature, it is well to begin with the simplest case of magnetic induction. When a bar of soft iron is held before the poles of a magnet, as shown in Fig. 459, it becomes itself a magnet. The magnetism developed in the bar by the action of the magnet is opposite that of the magnet. That is, the magnetism developed in the end of the bar opposite the N pole of the magnet is S, and, similarly, the magnetism developed in the end of the bar opposite the S pole is N. The center of the iron bar is neutral.

FIG. 459.

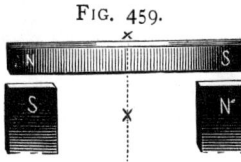

Magnetic Induction.

By substituting an iron ring for the straight bar, as shown in Fig. 460, the effect will be the same. The portions of the ring opposite the poles of the magnet acquire polarity by induction, as in the first instance, and the magnetism extends in the ring from the vicinity of the poles toward the neutral line, X X, which forms a right angle with a line joining the poles of the magnet. In the figure of the ring the location of the magnetism in the ring is indicated by the shading.

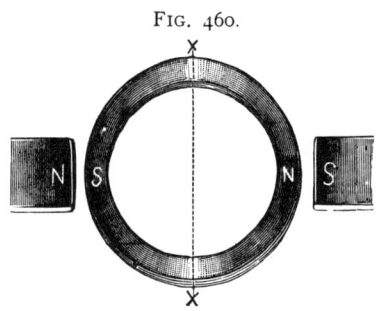

FIG. 460.

Induction in an Iron Ring.

By turning the ring upon its axis, the material of the ring moves, but the polarity of the ring maintains a fixed position relative to the poles of the magnet.

When the ring carries a coil, as shown in Fig. 461, the magnetic poles of the ring remaining stationary while the material of the ring and coil are revolved, there is a continual passing of the sections of the coil through the magnetic field surrounding the polarized portions of the armature core and the poles of the magnet, which is somewhat the same in effect as the passing of a magnetic bar through the coil of the armature.

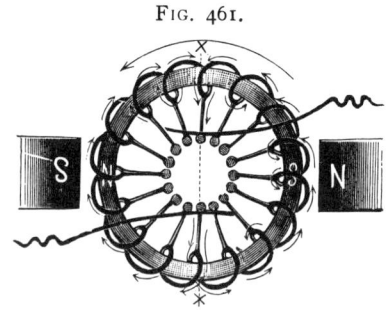

FIG. 461.

Armature in Magnetic Field.

The principal inductive effect is produced by the passing of the conductor through the magnetic field of the inducing magnet. Each half of the armature between the neutral points is practically a single coil of wire, terminating at two of the commutator bars—which in the present case are the two nails—at diametrically opposite sides of the commutator cylinder; all of the remaining commutator bars and their connections being idle. In Fig. 456 two circuits are shown in connection with the machine—one in full lines, the other partly in dotted lines, both connected with the battery, *c*. When the circuit represented in full lines only is employed, the machine runs as a motor. When the wires shown by full lines are disconnected from the brushes, $g\,g'$, the rotation of the armature in the field of the magnet, $a\,a'$, produces a current in the manner already indicated, and this current is taken from the armature by the way of the wires, *i*, the nails, e', and the brushes, $g\,g'$.

This machine when used as a generator is strictly a magneto-electric machine, although an electro-magnet is em-

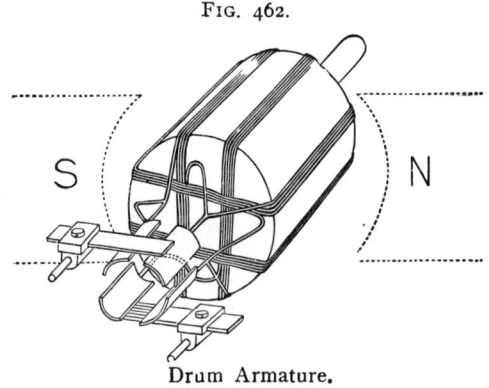

FIG. 462.

Drum Armature.

ployed as a field magnet. A permanent magnet might be substituted for the electro-magnet.

In the Siemens, Edison, Weston, and many other dynamos, the drum armature is used. This is shown diagrammatically in Fig. 462.

In this case the beginning of one coil and the end of the next preceding coil is connected with one commutator bar, and this order is maintained throughout the entire winding. This arrangement causes the current to flow always in the same direction, as the armature is revolved in the magnetic field.*

---

* *Scientific American Supplement* 600 contains full information on the construction of an eight-light dynamo having a drum armature. Thompson's "Dynamo-Electric Machinery" and Hering's "Dynamo-Electric Machines" may be consulted for information on dynamos.

In the diagram only four coils and four commutator bars are shown. In the actual machine the armature is divided up into a large number of sections.

MAGNETO-ELECTRIC MACHINES.

It has been already shown that it makes no material difference in the result whether a magnetized steel bar is introduced into the coil, as in Fig. 448, or whether the coil is provided with a soft iron core capable of being magnetized by induction, by contact with, or proximity to, a permanent magnet.

Fig. 463 illustrates an experiment of this kind, in which ths coil, A', of very fine wire, is provided with a permanent

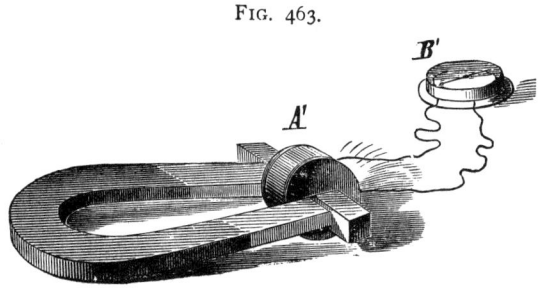

FIG. 463.

soft iron core, and is connected with the galvanometer, B'. By placing the poles of a permanent horseshoe magnet in contact with the projecting ends of the soft iron core of the coil, the core instantly becomes a magnet by induction, and a current is set up in the coil in the same manner as in the former experiment. When the magnet is removed, the magnetism of the core departs. This is equivalent to the removal of the magnet from the coil in the experiment illustrated in Fig. 448, and the result is a momentary current in a direction opposite to that of the first.

The inductive effect of the magnet is much the same if the bobbin of fine wire be placed around a permanent magnet and the magnetic tension be disturbed by the application and removal of an armature. The Bell telephone (the essential parts of which are shown in Fig. 464) is a familiar example of this species of generator of induced currents. When the diaphragm, acting as an armature, approaches the magnet, a momentary current is set up in the bobbin, A″, in one direction, as indicated by the galvanometer, B″, and when the diaphragm recedes from the magnet the current set up in the bobbin is in the opposite direction. In the telephone these currents have sufficient power to operate a second instrument of the same sort; but owing to the fact that the armature is very light, and never touches the magnet nor recedes very far from it, and the further disadvantage arising from the use of a bar magnet, the apparatus cannot rank high as a generator of electric currents, however well it may serve the purpose of a telephone.

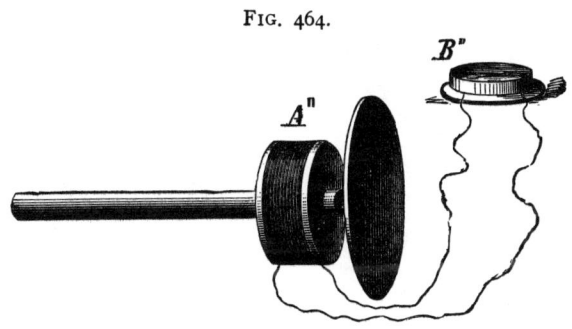

FIG. 464.

Another form of apparatus (Fig. 465), operating on the same principle, generates currents sufficiently powerful to work a polarized bell or annunciator over a line several miles long. This magneto key is made by clamping two six-inch horseshoe magnets upon opposite sides of two soft iron pole extension pieces, a, one-half inch in diameter, one and a half inches long, and projecting one inch beyond the poles of the magnets. Each extension piece is provided with a bobbin, D, one inch long and one and a quarter inches in diameter, filled with No. 36 silk-covered wire. These bobbins are wound and connected like the spools of an electro-magnet, and have a combined resistance of 200 ohms.

In front of the poles of the magnet an armature, E, one-quarter inch thick, a little longer than the width of the extremities of the magnet, and about one inch wide, is pivoted at its lower edge, and provided with a key lever by which it may be drawn from the poles of the magnet. A spring under the key lever throws the armature back into contact with the magnet. This is a simplified form of Breguet's exploder, used in firing blasts in mines, and although much smaller than the apparatus referred to, it is capable of ringing a polarized bell over fifteen or twenty miles of wire, and will give a powerful shock.

It is a convenient and inexpensive apparatus for signaling, and is particularly adapted to the telephone when used in connection with the polarized annunciator or polarized

FIG. 465.

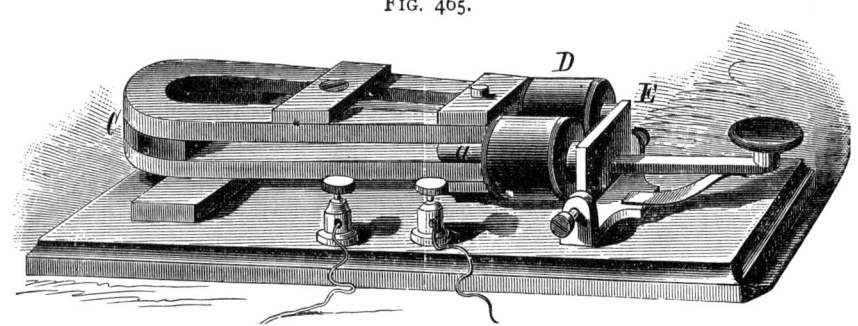

Magnetic Key.

bell, presently to be described. In this apparatus like poles of the magnets must oppose each other, and the clamping pieces and screws should be of non-magnetic material. If two magnets do not produce a current of sufficient strength for the intended use, two more may be added.

In this form of magneto-induction apparatus the action of the magnet and coil is identical with that of the Bell telephone. This action is similar to that of two permanent horseshoe magnets having their unlike poles in contact. In this case the opposing poles neutralize each other to such an extent as to almost destroy all magnetic effects. On separating the poles of the two magnets they regain their normal magnetism. The case is much the same with the magnetic key. The armature, E, when applied to the pole ex-

DYNAMIC ELECTRICITY. 479

tensions, becomes a magnet by induction, and by its reaction upon the magnet neutralizes the power of the magnet and produces nearly the same result as withdrawing the magnet from the bobbin. When the armature is withdrawn suddenly from the magnet, the effect upon the wires of the bob-

FIG. 466.

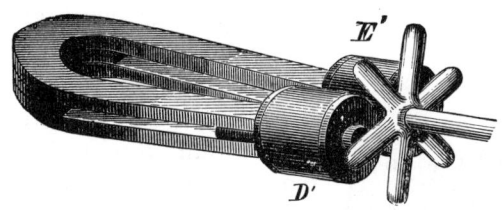

bins is the same as would be produced by introducing into them the poles of the magnet.

To render the electrical pulsations of this class of machines very frequent, the armature may be rotated, as shown in Fig. 466, which represents a modification of an old

FIG. 467.

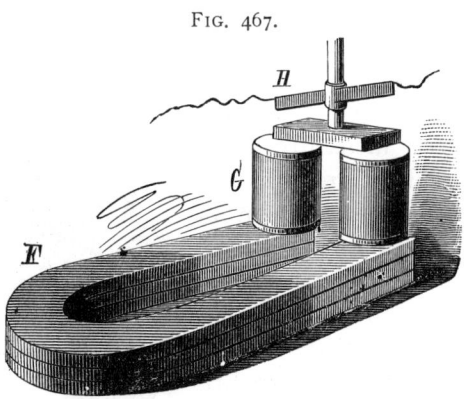

and well known magneto-induction machine, in which the bobbins, D', are placed on pole extensions of the magnets, C', and the variations in magnetic force are produced by the wheel armature, E'.

Another method of generating currents by a rotary

movement of the armature is to make the armature in the form of an electro-magnet, and mount it upon a rotating spindle so that it may revolve in close proximity to the poles of a strong permanent horseshoe magnet. This form of machine, which is the invention of Clarke, is shown in Fig. 467. It has long been used for medical purposes, and before the invention of the more recent machines was employed for electro-metallurgy and for other purposes.

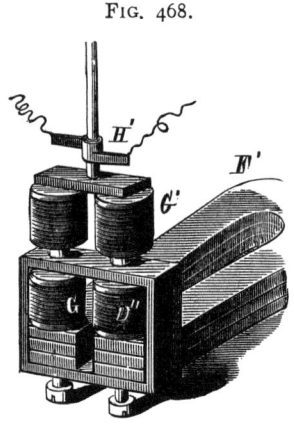

FIG. 468.

The electro-magnetic armature, G, is mounted on a shaft, so that it may revolve very near but not in contact with the poles of the compound magnet, F. One of the terminals of the bobbins is in electrical connection with the shaft, the other is connected with an insulated ferrule on the shaft. The alternating current is taken off by two springs, one touching the insulated ferrule, the other bearing against the shaft. When the current is required to flow in one direction, the insulated ferrule is split longitudinally into two equal separate halves, each of which is connected with one terminal of the armature wire. This split ferrule, together with springs, H, which press upon its diametrically opposite sides, forms a commutator which sends the momentary currents of like name all in one direction.

FIG. 469.

The slots of the ferrule are arranged relative to the springs, H, and armature, so when the polar faces of the armature cross a line joining the poles of the permanent magnet, the springs will leave one-half of the ferrule and touch the other half.

Fig. 468 shows a modification of Clarke's machine, in

DYNAMIC ELECTRICITY. 481

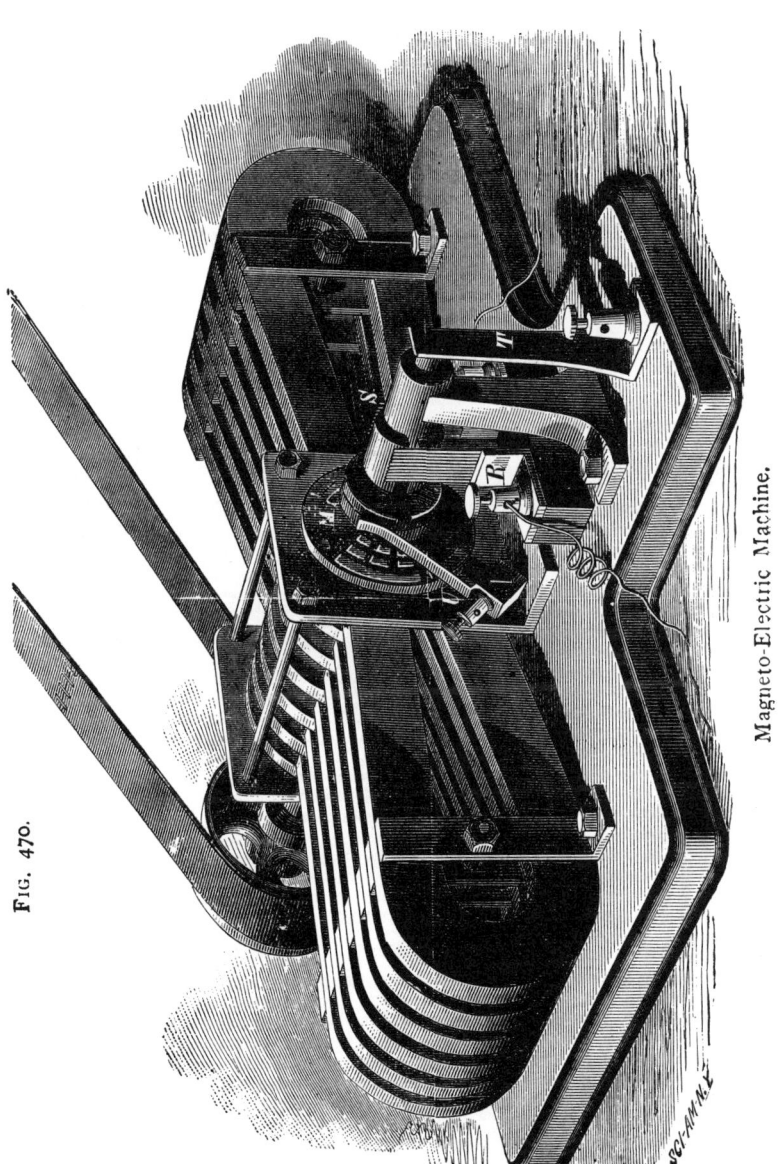

Fig. 470.

Magneto-Electric Machine.

which the permanent magnet, F', is provided with pole extensions of soft iron surrounded by fine wire bobbins, D''. These bobbins are connected like an electro-magnet, and when the armature, G', is turned so as to send a direct current through the springs, H', an alternating current may be taken from the bobbins, D''.

Fig. 469 shows a kind of commutator designed for short-circuiting the machine through a part of the revolution, so that when the short circuit is broken a direct extra current

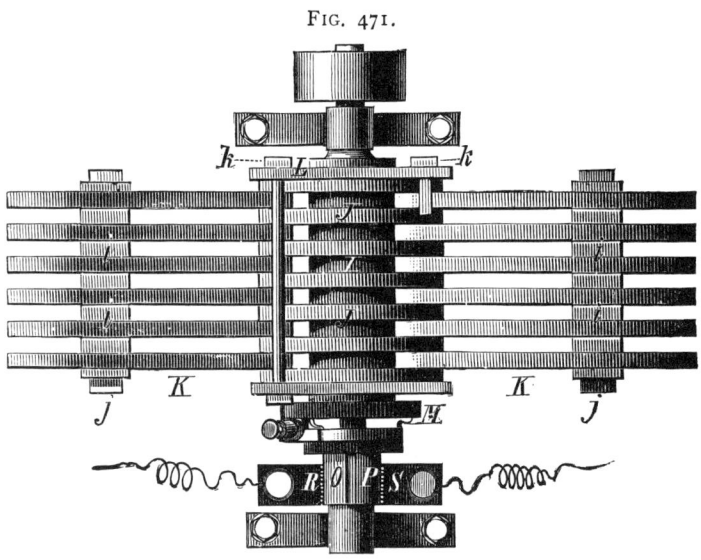

FIG. 471.

Plan View of Magneto-Electric Machine.

capable of giving powerful shocks will pass over the conductors leading from the machine. Each half, d, of the commutator ferrule is provided with an arm, e, terminating in a curved piece, g, attached to opposite sides of the insulating cylinder, c. The curved pieces, g, are pressed by springs which are electrically connected with the commutator springs on their respective sides of the cylinder, so that when the piece, g, is touched by its spring and the ferrule, d, is touched by its spring—the two springs being in electrical communication with each other—the machine is for the moment short-circuited, but when contact with g is broken, the extra current passes by the usual channels from the machine.

A magneto-electric machine equal in power to three or four Bunsen elements is shown in Figs. 470, 471, and 472. The compound field magnet is composed of twelve six-inch horseshoe permanent magnets, K, arranged in two groups of six, with their like extremities clamped between curved soft iron bars, J, as shown in the vertical longitudinal section, Fig. 472. These bars consist of sections cut from common wrought iron washers, 3 inches external diameter, ¼ inch thick, and having a 1⅝ inch hole through them. The washers are all drilled to receive the bolts, $h\ h$, before they are cut in two. The washers, J, and magnets, K, are

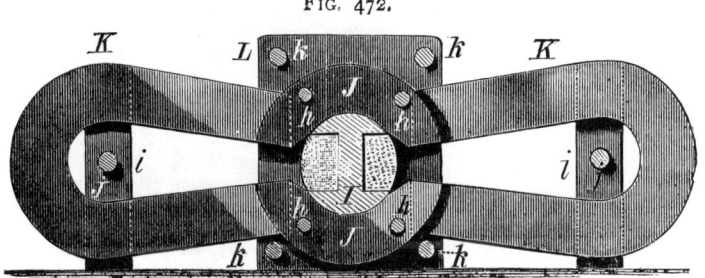

FIG. 472.

Transverse Section of Magneto-Electric Machine.

placed in alternation and clamped between brass angled plates, L, by which the middle portion of the field magnet is fastened to its base. The magnets are further secured to the base by standards, j, which clamp the sides of each group of magnets, the magnets being kept the proper distance apart by interposed strips, i.

The bars, J, are cut away on the inner edges, forming an approximately elliptical opening for receiving the armature, I, which is a very little less than 1⅝ inches in diameter, and is 3½ inches long. It is of the Siemens **H** type, and is wound with four parallel silk-covered No. 32 wires, which terminate in eight insulated metallic blocks on the switch, M, one block to each end of each wire.

The switch is shown in detail in Fig. 473—1, 2, 3, 4, 5, 6,

7, 8 being the terminals of the wires of the bobbin. The blocks, 1 and 5, represent the ends of the first wire, 2 and 6 representing the ends of the second wire, 3 and 7 the third, and 4 and 8 the fourth; 15 and 16 are curved brass pieces capable of being plugged into connection with the blocks just mentioned, by means of screw plugs, shown in place in the engraving. The pieces, 15 and 16, are connected respectively with the two halves, O P, of the commutator cylinder.

At the ends of the curved pieces, 15 and 16, there are metallic blocks, 17, 18—the block, 17, being connected by a

FIG. 473.

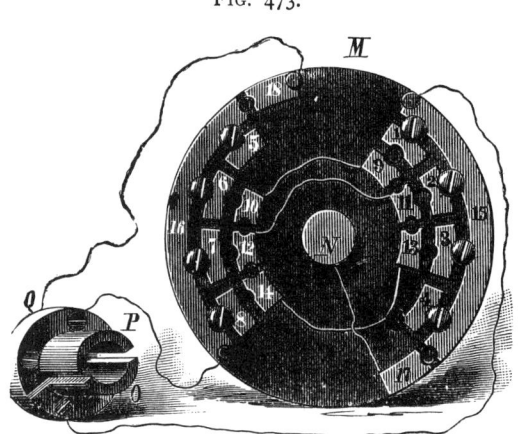

Switch of the Magneto-Electric Machine.

wire with the metallic boss of the rubber wheel upon which the switch is mounted; the block, 18, being connected by a wire with a brass ring, Q, on the rubber support of the commutator.

Inside the blocks, 1 to 8, there are six metallic blocks, 9, 10, 11, 12, 13, 14, connected together by wires as shown. The opposite sides of the commutator cylinder are pressed by springs or brushes, R, which are sustained by an insulating support and are provided with binding posts for receiving the wires for conducting away the direct current. A spring, T, touches the end of the armature shaft, and has a binding post for receiving a wire conductor, and a spring, U, sus-

DYNAMIC ELECTRICITY. 485

tained by an insulator attached to the angle plate, L, has a binding post for receiving a conductor.

This machine will yield currents of three different intensities, and will deliver them either direct or alternating, and it answers admirably as a motor.

To obtain a quantity current the screw plugs are inserted as shown in Fig. 473, so as to connect 1, 2, 3, 4 with 15, and 5, 6, 7, 8 with 16. In this condition it may be used as a motor.

The success of the machine as a motor depends in a great measure on the adjustment of the commutator. Its

FIG. 474.

Polarized Bell.

slit should be nearly opposite the center of the open space or groove in the armature.

To secure a current of higher voltage, connect 5 and 6 with 16, connect 1 to 2 and 2 to 11, connect 12 to 7 and 7 to 8, and finally connect 3 and 4 with 15. To get the highest voltage, connect 5 to 16, 1 to 9, 10 to 6, 2 to 11, 12 to 7, 3 to 13, 14 to 8, and 4 to 15. Direct currents are taken from the springs, R, alternating currents are taken from the springs, T, U, after connecting 15 to 17 and 16 to 18. The quantity current is obtained from four parallel wires, which are equivalent to one wire having four times the sectional area of the single wire and one-fourth the length. When the medium current is secured the wire is doubled, so that it is equivalent to a wire having twice the sectional area of the single wire and one half the length. For the high voltage current the full length of wire is used single.

Fig. 474 represents a Siemens polarized bell, in which an iron yoke, *m*, is supported from the elongated ends of the yoke of the magnet, *l*, by two brass studs. The yoke, *m*, supports the pivots of the bell armature, *n*, also the studs upon which the bells are placed, and to it is secured the magnet, *p*, which is bent under the yoke of the magnet, *l*, without touching it.

Fig. 475 shows a similar but simpler device, in which

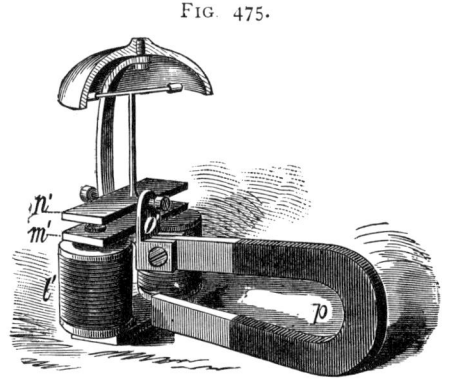

FIG. 475.

Simple Polarized Bell.

the poles of the magnet, *l'*, are fitted with a brass yoke, *m'*, which supports an iron frame in which is pivoted the armature, *n'*, and to which the bell is attached. This frame has a socket, *o'*, for receiving one of the poles of a horseshoe magnet, *p*, the other pole of which touches the yoke of the magnet, *l'*.

The polarized annunciator shown in Fig. 476 has two soft iron cores, *r*, carrying two bobbins of fine wire connected like the spools of an electro-magnet. In front of these soft iron cores there is a light delicately pivoted plate, *s*, of iron, which is held in contact with the cores, *r*, by magnetism induced in them by a magnet, *t*, clamped in the middle and capable of being adjusted by a spring and screw

DYNAMIC ELECTRICITY. 487

at the bottom. The iron annunciator plate, *s*, has sufficient inclination to cause it to drop if released from the cores, *r*. The magnet is placed so near the cores, *r*, as to impart to them just enough attractive force to hold the plate, *s*, and no more.

The polarized bells and annunciator may be worked by either of the instruments shown in Figs. 465, 466, and 467, and will be found for many uses preferable to electric bells and annunciators operated by battery currents.

HAND POWER DYNAMO.

Fig. 477 is a perspective view of a small hand dynamo, which is shown half size in detail in Figs. 478, 479, 480, and 481. This is a Siemens **H**-armature machine, which is as efficient as any small dynamo, while it has the advantage of being readily understood and easily constructed. The field magnet is, for the sake of convenience, composed of two pieces, A B, which are exactly alike excepting that the connecting piece, C, is cast with the piece, A. The parts, A B, are planed at their juncture at the top, and secured together by two bolts which pass through the part, C. The lower ends of these parts are also planed to receive the brass

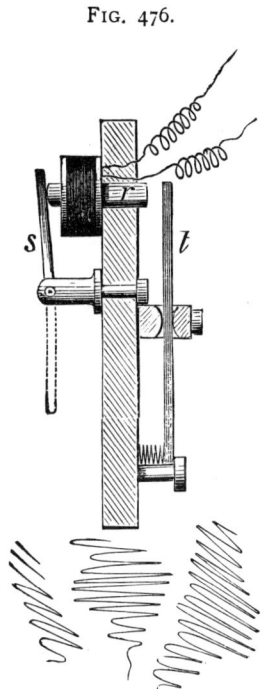

FIG. 476.

Annunciator.

plate, E, which is secured in place by dowels and screws, two of each entering each part. The cylindrical cavity which receives the armature, G, is bored out truly and smoothly of a uniform caliber from end to end. The edges of that portion of the field magnet around which the wire, D, is wound are rounded and a piece of cotton cloth is wrapped around each core, and secured by means of shellac varnish. Upon this is wound seven layers of No. 16 cotton-covered copper wire. The limbs of the magnets are wound in the same direction, or in such a way that when the two portions, A B, are placed end to end, one coil would be simply a continuation of the other. The inner ends of the coils are connected together, while their outer ends are of sufficient length to run downward through the base, and bend outward at *m o*, and are connected with the binding posts, *n p*.

The armature, G, consists of a cylindrical piece of soft

FIG. 477.

Hand Power Dynamo.

cast iron grooved longitudinally and across the ends, and wound with No. 18 cotton or silk covered copper wire. It is, in fact, a very short and wide bar electro-magnet, having enlarged and elongated ends of the form of a segment of a cylinder. In diameter the armature is only a very little less than that of the cylindrical space between the parts, A B, of the field magnet, and its length is little less than the width of the field magnet. In Figs. 478 and 480, G' is the core of the armature around which is wound the wire, H.

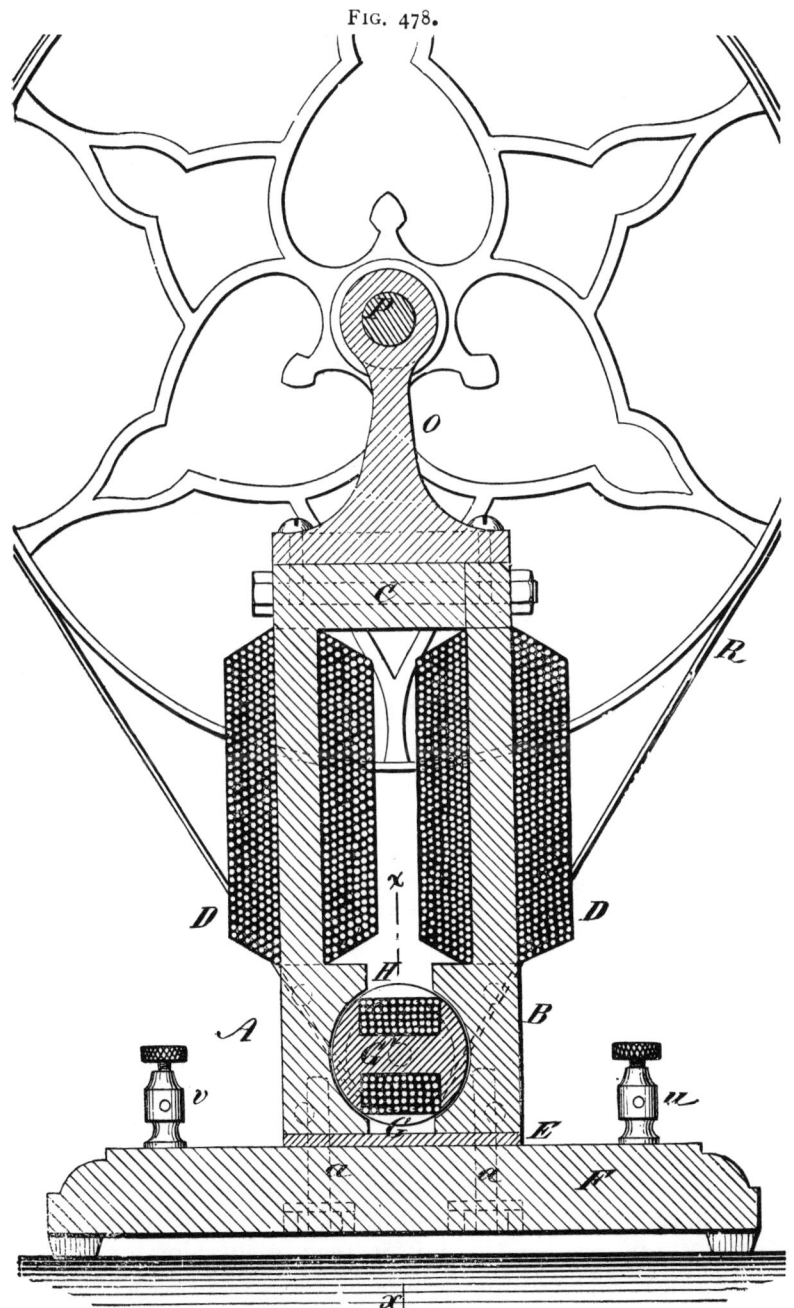

Hand Power Dynamo—Vertical Section—Half Size.

To opposite ends of the armature are fitted the brass heads, I J, into which are screwed the shafts, $b\ c$. The core, G', of the armature is filed to remove roughnesses and hard scale, and the heads and shafts are fitted to the ends of the armature before it is turned and fitted to the cylindrical space in the field magnet. The shaft, b, is journaled in a brass support, L, which is attached by screws to the edges of the parts, A B, of the field magnet. The shaft, c, is journaled in a similar support, M, which is secured to the opposite side of the electro-magnet. Outside of the bearing, L, upon the shaft, b, is secured the pulley, d, and between the support, M, and the head, J, the commutator is placed upon the shaft, c. The commutator consists of a vulcanite cylinder, e, having upon its periphery a copper or brass ferrule, which is slit longitudinally at diametrically opposite points, forming the insulated segments, $f\ g$. These are secured to the vulcanite cylinder by small brass screws, and the slits are placed exactly opposite the center of the longitudinal grooves in the armature. The commutator is prevented from turning on the shaft by a set screw, and with the segments, $f\ g$, are connected the terminals of the armature coil, H. These terminals pass through holes in the head, J, which are lined with an insulating material.

To opposite sides of the support, M, are secured the copper commutator springs, $h\ i$, each consisting of five or six thicknesses of thin, hard-rolled copper. They are both secured by screws, and insulated from the support by vulcanite buttons, j. The spring, h, is bent forward over the commutator and bears upon it with a slight pressure. The spring, i, is bent so that it touches the commutator at a point diametrically opposite the contact point of the spring, h. To the spring, h, a wire (No. 14) is soldered, and extends downward through the wooden base of the machine; a similar wire runs from the spring, i. As the design of this machine is such that the field magnet may be connected with a battery, so that all of the current from the armature may be utilized in the external circuit, instead of allowing a portion of it to pass through the helices of the magnet, two extra binding posts, $u\ v$, and a switch, N, are added.

DYNAMIC ELECTRICITY. 491

The connections under the base are as follows:

The terminals, *m o*, of the field magnet are connected with the binding posts, *n p*. The commutator spring, *h*, is

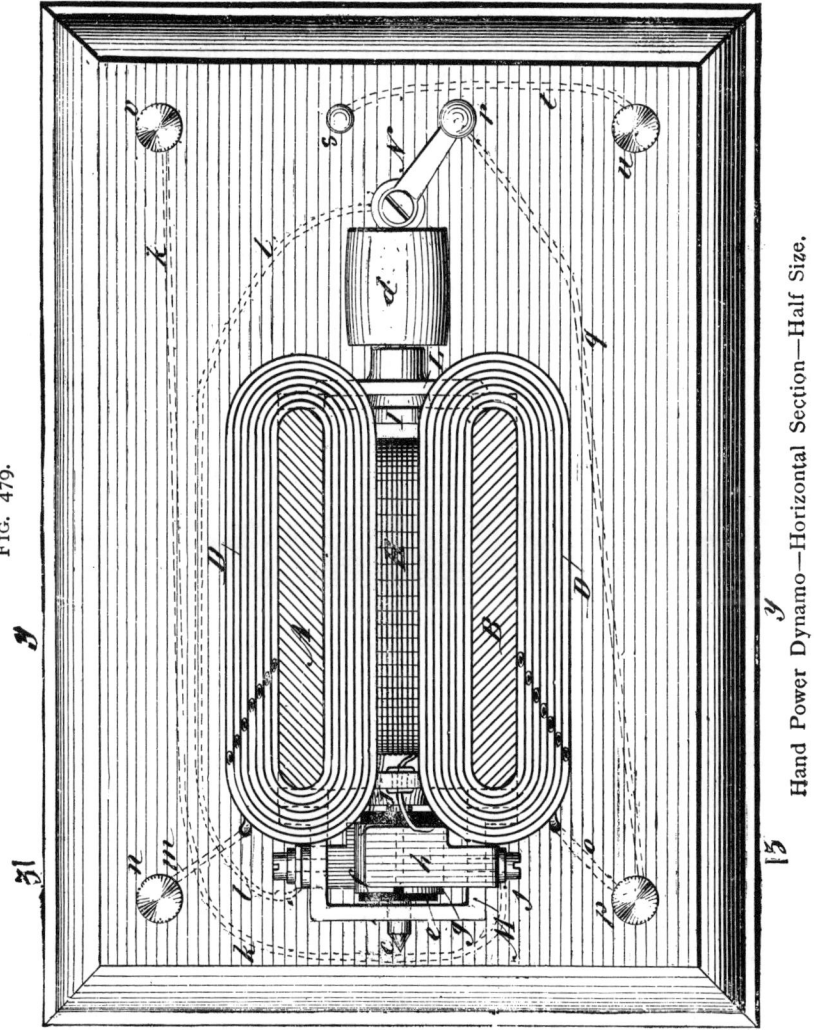

Fig. 479. Hand Power Dynamo—Horizontal Section—Half Size.

connected by the wire, *k*, with the binding post, *v*; the commutator spring, *i*, is connected with the switch, N, by the wire, *l*. The switch button, *r*, is connected with the binding post, *p*, by a suitable wire, and the switch button, *s*, is connected with the binding post, *u*, by the wire, *t*. All of these connections should be made with No. 14 wire. A support, O, for the shaft, P, is secured to the top of the electro-magnet. The shaft, P, has at one end the driving wheel, Q, and at the other end a crank for operating the machine. A one inch belt, R, runs around the pulley, *d*, and the wheel, Q.

When the machine is driven by power the pulley, *d*, may with advantage be larger. The size of the wire on the magnet and armature may be varied for some special purpose, but for general use the sizes here given are recommended. The slit in the commutator should be made slightly diagonal, so that one section of the copper ferrule will touch the spring before the other section leaves it. The armature should fit in the magnet as closely as possible without rub-

FIG. 480.

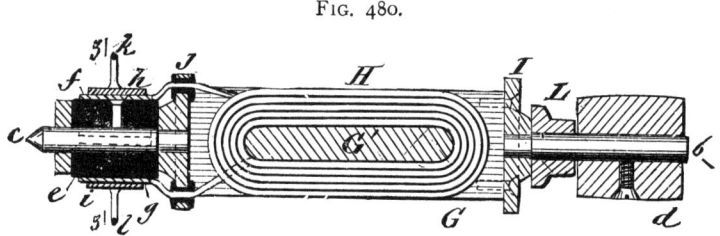

Armature and Commutator—Longitudinal Section.

bing. The parts indicated as brass or copper should be made of these metals, as a magnetic insulation is required wherever they are used.

When the switch, N, is in the position shown in the drawing, the binding posts, *n v*, being connected by a wire, the current passes from the post, *v*, through the commutator and the armature, thence by the wire, *l*, to the switch, thence through the button, *r*, and by the wire to the post, *p*, thence through the field magnet to the post, *n*, through the terminal, *m*. When the machine is arranged in this manner, the wires leading from the machine are taken from the posts, *v n*. The full power of the machine is developed an instant after the connection of the posts, *v n*.

By moving the switch, N, into contact with the button, *s*,

DYNAMIC ELECTRICITY. 493

and connecting a battery of six or eight Bunsen cells with the posts, *n p,* the magnets are excited without detracting from the power of the armature, and the current from the latter is taken through the wire, *k,* as before, to the post, *v,* but the wire, *l,* is now in electrical connection with the binding post, *u,* through the switch, N, button, *s,* and wire, *t*; therefore the current is taken away from the machine by inserting wires in the posts, *u v.*

When not connected with a battery, this machine will heat from four to six inches of No. 36 platinum wire. It will rapidly decompose water when the ends of the wires are dipped in water slightly acidulated. It will run an

FIG. 481.

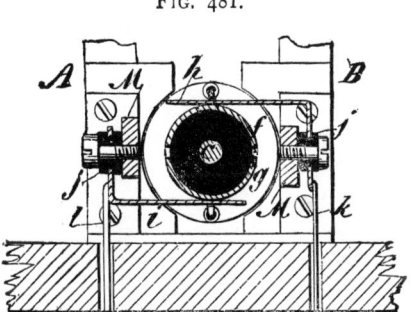

Transverse Section of Commutator.

induction coil. The extra current from this machine is sufficient to give strong shocks, ignite powder, etc. By connecting it with a helix or electro-magnet, small permanent magnets may be charged. For many purposes this machine will be found equal to four or six Bunsen cells.

When a battery is employed to excite the field magnet, the current is very much increased. For example, it will then heat twelve inches of platinum wire instead of four or six inches, and it will afford a current sufficient for a strong electric light. The speed has much to do with the amount of current produced by the machine. The speed should be from 1,200 to 1,500 turns of the armature per minute. The drive wheel in the example given may with advantage be made much larger, say two feet in diameter.

ELECTRO-PLATING DYNAMO.

The electro-plating dynamo differs from the one already described chiefly in its winding. For metallurgical work a large current of low voltage is required. For electro-typing, an electro-motive force of three to four volts is suffi-

FIG. 482.

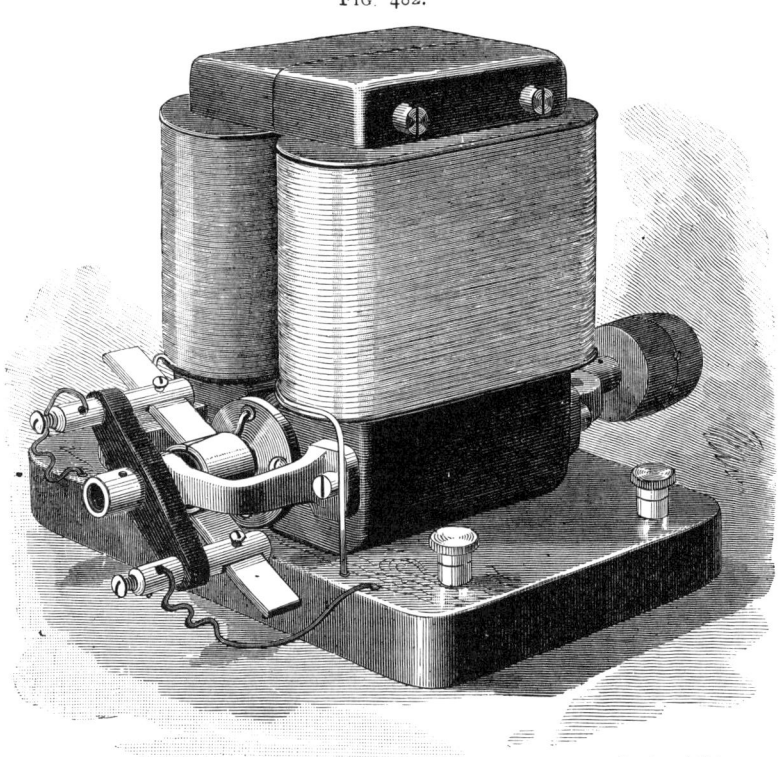

Electro-Plating Dynamo.

cient, while for nickel plating it should run up to about six volts, and for silver plating to about five.

In a small dynamo, like the one illustrated in Fig. 482, it is impossible to secure as wide a range of electro-motive force or of current as can be realized in a larger machine, but by varying the speed and by introducing more or less resistance in the external or internal circuit, the current can

DYNAMIC ELECTRICITY.

be adapted to most uses of the amateur. In the construction of this dynamo all of the dimensions of the cores and polar extremities of the field magnet and of the armature core, as given in the description of the hand power dynamo, are followed except in regard to the thickness of the waists of the field magnets and their polar extremities. These dimensions are here increased by adding $\frac{1}{8}$ inch to the thickness of the waists and $\frac{1}{4}$ inch to the thickness of the polar extremities, thus increasing the amount of iron in the field magnet.

The armature is wound with five layers of No. 12 cotton-covered magnet wire, and the terminals of the coil are connected with the halves of the commutator cylinder as shown in Fig. 483.

The commutator cylinder is formed of two sections cut

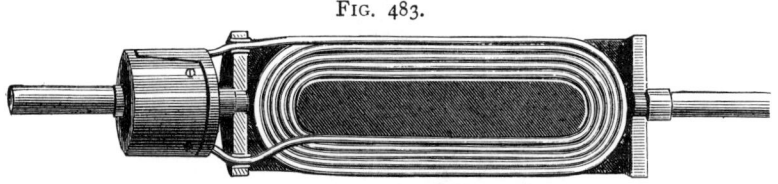

FIG. 483.

Armature of Electro-Plating Dynamo—Half Size.

from a copper tube and mounted upon a hub of vulcanite or vulcanized fiber, the tube sections being separated from each other so as to form diagonal slits in diametrically opposite sides of the cylinder, as shown.

The brushes are supported by mortised studs inserted in the ends of a cross bar of vulcanized fiber mounted on the journal box of the armature shaft. The threaded ends of the mortised studs project through the cross bar to receive binding posts which are screwed down tightly on the bar. In the mortises of the studs are placed the brushes, which press lightly upon the commutator cylinder. The brushes are formed of several thicknesses of thin hard-rolled copper. The field magnet is wound with twelve layers of No. 18 magnet wire, and is connected as a shunt to the armature. That is to say, the terminals of the field magnet wires are connected with the same binding posts that receive the wires from the commutator brushes, as shown in Fig. 484.

The conductors of the external circuit are also connected with these binding posts. When the connections are arranged in this way, the current divides at the binding posts referred to, a part going through the wire of field magnet, another part going through the external circuit, which in the present case includes a plating solution.

To the negative conductor is attached the cathode or the plate or object which is to receive the deposit, and upon

FIG. 484.

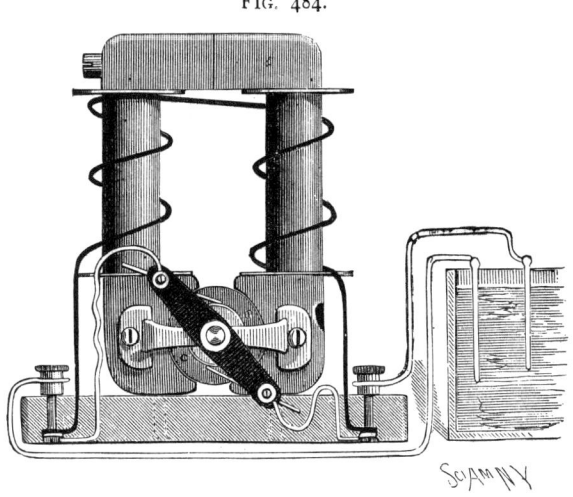

Connections of Plating Dynamo.

the positive conductor is suspended the anode or plate from which the metal for the deposit is supplied to the solution.

Unless the dynamo is at first started with a battery in circuit, it will be impossible to tell, without a test of some sort, which is the positive and which the negative binding post. This can be determined in a moment by trial in the plating solution.

If on starting the machine a deposit is made on the cathode, the connections are correct. If, however, no deposit appears, the conductors should be transposed either at the dynamo or at the plating bath.

Large wire should be used for carrying the current. Within certain limits the electro-motive force of the current may be varied by changing the speed of the machine, and the current may be controlled by inserting resistance into the external circuit or into the shunt.

The hand-power dynamo may be converted into a shunt machine by arranging the connections according to Fig. 484, but it will be necessary to introduce resistance into the shunt or field magnet circuit to prevent too much current from going through the field magnet.

The electro-plating dynamo may be used successfully in copper, nickel, and silver plating on a small scale, also for electro-typing.* This dynamo acts well when used as a motor in connection with the plunge battery shown in Fig. 394.

The length of wire on the armature is 40 feet and on the field magnet about 500 feet.

SIMPLE ELECTRIC MOTOR.

It is generally understood that an efficient electric motor cannot be made without the use of machinery and fine tools. It is also believed that the expense of patterns, castings, and materials of various kinds required in the construction of a good electric motor is considerable.

The little motor shown in the engravings was devised and constructed with a view to assisting amateurs and beginners in electricity to make a motor which might be driven to advantage by a current derived from a battery, and which would have sufficient power to operate an ordinary foot lathe or any light machinery requiring not over one man power.

The only machine work required in the construction of the motor illustrated is the turning of the wooden support for the armature ring. The materials cost less than four

---

* "Electro-plating" and "Electro-typing," by Urquhart, and "Electro-Deposition," by Watt, are excellent works on their respective subjects. The *Scientific American Supplement* also contains valuable information on these subjects and on the construction of dynamos for these uses.

498   EXPERIMENTAL SCIENCE.

Fig. 485.

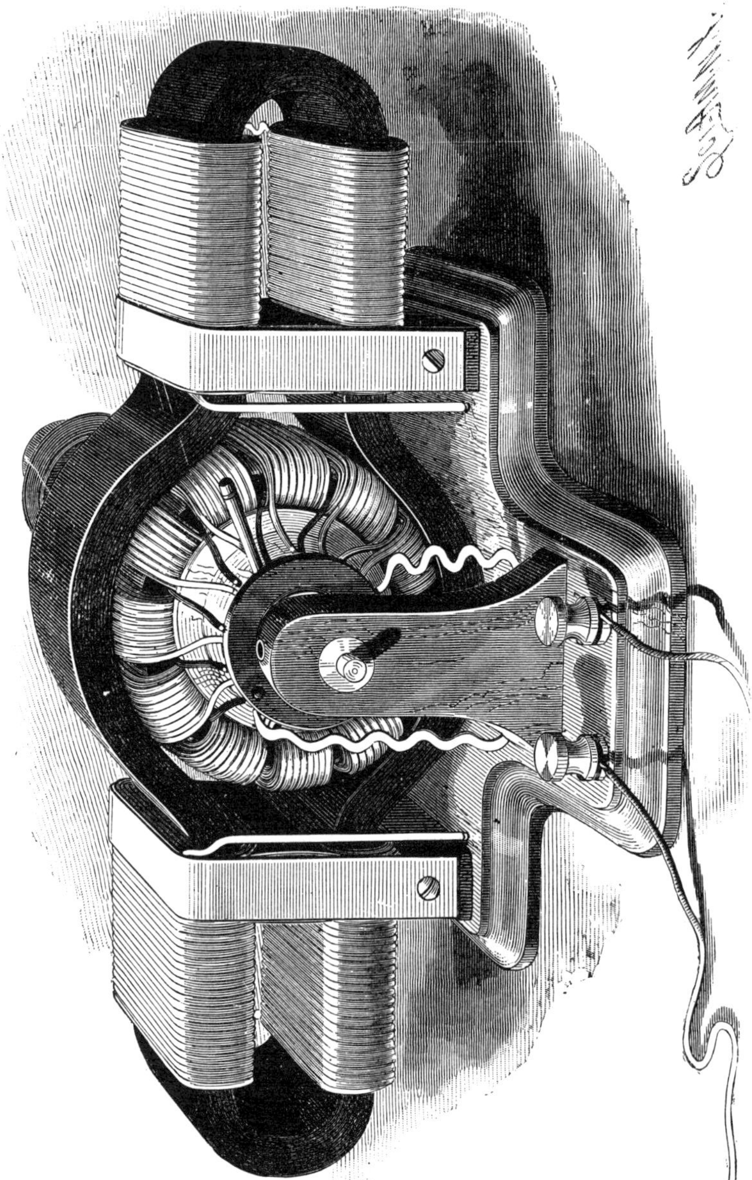

Simple Electric Motor.—About Half Size.

dollars, and the labor is not great, although some of the operations, such as winding the armature and field magnet, require some time and considerable patience. On the whole, however, it is a very easy machine to make, and, if

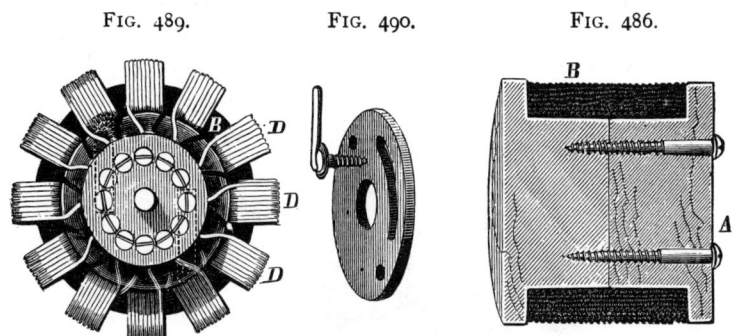

FIG. 486.—Armature Core. FIG. 489.—End View of Armature, showing Commutator. FIG. 490.—Brush-holding Disk.

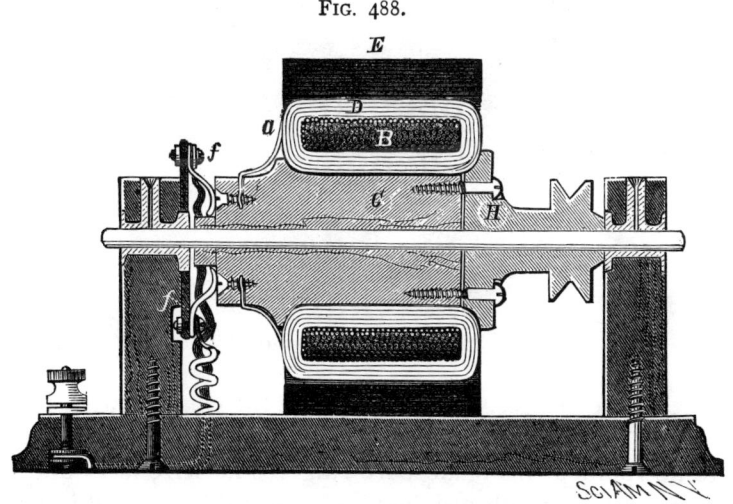

Transverse Section.

carefully constructed, will certainly give satisfaction. Only such materials as may be procured anywhere are required. No patterns or castings are needed.

Beginning with the armature, a wooden spool, A (Fig. 486), should be made of sufficient size to receive the soft iron wire of which the core of the armature is formed. The wire, before winding, should be varnished with shellac and allowed to dry, and the surface of the spool on which the wire is wound should be covered with paper to prevent the sticking of the varnish when the wire is heated, as will presently be described. The size of the iron wire of the core is No. 18 American wire gauge. The spool is $2\frac{3}{16}$ inches in diameter in the smaller part, and 2 inches in length

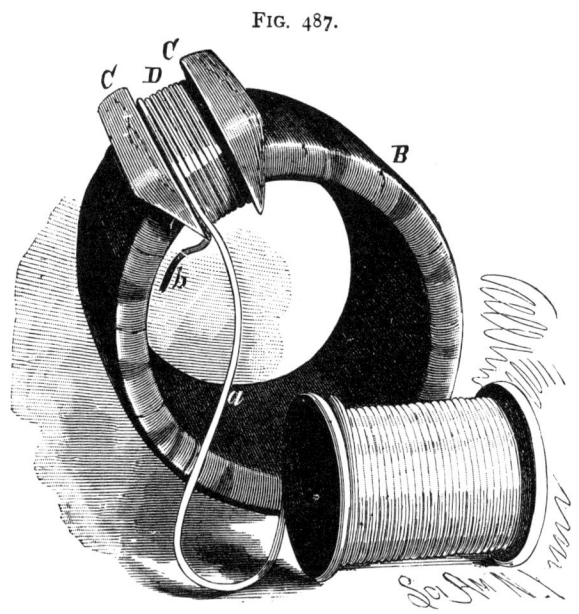

Fig. 487.

Winding the Armature.

between the flanges. It is divided at the center and fastened together by screws. Each part is tapered slightly to facilitate its removal from the wire ring. The wire is wound on the spool to a depth of $\frac{3}{8}$ inch. It should be wound in even layers, and when the winding is complete, the spool and its contents should be placed in a hot oven and allowed to remain until the shellac melts and the convolutions of wire are cemented together.

After cooling, the iron wire ring, B, is withdrawn from the spool and covered with a single thickness of adhesive

DYNAMIC ELECTRICITY. 501

tape, to insure insulation. If adhesive tape is not at hand, very thin cotton tape or strips of cotton cloth may be substituted. A single coat of shellac varnish will hold the coving in place.

The ring is now spaced off into twelve equal divisions, and lines are drawn around the ring transversely, dividing it into twelve equal segments, as shown in Fig. 487.

Two wedge-shaped pieces, C, of hard wood are notched and fitted to the ring so as to inclose a space in which to wind the coil. These blocks may be clamped in any convenient way. The coil, D, consists of No. 18 cotton-covered copper magnet wire, four layers deep, each layer having eight convolutions. The end, a, and the beginning, b, of the winding terminate on the same side of the coil. The last layer of wire should be wound over two or three strands of shoe thread, which should be tied after the coil is complete, thus binding the wires together.

When the first section of the winding is finished, the wire is cut off and the ends (about two inches in length) are twisted together to cause the coil to retain its shape. After the completion of the first section, one of the pieces, C, is moved to a new position and the second section is proceeded with, and so on until the twelve sections are wound. The coils of the ring are then varnished with thin shellac varnish, the varnish being allowed to soak into the interior of the coils. Finally, the ring is allowed to remain in a warm place until the varnish is thoroughly dry and hard.

Care should be taken to wind all of the coils in the same direction and to have the same number of convolutions in each coil. A convenient way of carrying the wire through and around the ring is to wind upon a small ordinary spool enough wire for a single section, using the spool as a shuttle.

The ring is mounted upon a wooden hub, G, Fig. 488, and is held in place by the wooden collar, H, both hub and collar being provided with a concave flange for receiving the inner edges of the ring. The collar, H, is fastened to the end of the hub, G, by ordinary brass wood-screws. Both hub and collar are mounted on a $\frac{9}{32}$ steel shaft formed of Stubs' wire, which needs no turning. A pulley is formed integrally with the collar, H. The end of the hub, G, which is provided with a flange, is prolonged to form the commutator, and the terminals, a b, of the ring coils are arranged along the surface of the hub and inserted in radial holes drilled in the hub in pairs. The wires are arranged so that one hole of each pair receives the outer end of one coil and the other hole receives the inner end of the next coil, the extremities of the wire being scraped before insertion in the holes. The distance between the holes of each pair is sufficient to allow a brass wood-screw to enter the end of the hub, G, and form an electrical contact with both wires of the pair, as shown in Fig. 488.

There being twelve armature sections and twelve pairs of terminals, there will of course be required a corresponding number of brass screws. These screws are inserted in the end of the hub, G, so as to come exactly even with the end of the hub without touching each other. This completes the armature and the commutator.

Before proceeding to mount the armature shaft in journal boxes, it will be necessary to construct the field magnet, as the machine must, to some extent at least, be made by "rule of thumb."

The body, E, of the field magnet consists of strips of Russia iron, such as is used in the manufacture of stoves and stove pipe. The strips are $2\frac{1}{2}$ inches wide, and of any convenient length, their combined length being sufficient to build up a magnet core seven-sixteenths inch thick, of the form shown in Fig. 485. The ends of the strips are simply abutted. The motor illustrated has fifteen layers of iron in the magnet, each requiring about 26 inches of iron, approximately 33 feet altogether.

The wooden block, F, on which the magnet is formed is secured to a base board, G, as shown in Fig. 491, and grooves are made in the edges of the block, and corresponding holes are formed in the base to receive wires for temporarily binding the iron strips together. Opposite each angle of the block, F, mortises are made in the base board, G, to receive the keys, d, and wedges, c. Each key, d, is retained in its mortise by a dovetail, as shown in Fig. 492. By this arrangement each layer of the strip of iron may be held in position, as the formation of the magnet proceeds, the several keys, *d*, and wedges, *c*, being removed and replaced in succession as the iron strip is carried around the block, F. When the magnet has reached the required thickness, the wedges, *c*, are forced down so as to hold the iron firmly, then the layers of iron are closely bound together by

FIGS. 491 AND 492.

Forming the Field Magnet.

iron binding wire wound around the magnet through the grooves, *e*, and holes in the base board, G.

The next step in the construction of the machine is the winding of the field magnet. To insure the insulation of the magnet wire from the iron core of the magnet, the latter is covered upon the parts to be wound by adhesive tape or by cotton cloth attached by means of shellac varnish.

The direction of winding is clearly shown in Fig. 493. Five layers of No. 16 magnet wire are wound upon each section of the magnet. The winding begins at the outer end of the magnet, and ends at the inner end of the section.

When the winding is completed, the temporary binding is removed. The outer ends of coils 1 and 2 are connected together, and the outer ends of 3 and 4 are connected.

The inner ends of 2 and 4 are connected. The inner end of 3 is to be connected with the commutator brush, f. The inner end of 1 is to be connected with the binding post, g, and the binding post, g', is to be connected with the commutator brush, f'.

The field magnet is now placed upon a base having blocks of suitable height to support it in a horizontal position. Blocks are placed between the coils, to prevent the top of the magnet from drawing down upon the armature, and the magnet is secured in place by brass straps, as shown in Fig. 485.

The armature is wrapped with three or four thicknesses of heavy paper, and inserted in the wider part of the field magnet, the paper serving to center the armature in the magnet. The armature shaft is leveled and arranged at right angles with the field magnet. The posts in which the armature shaft is journaled are bored transversely larger than the shaft, and a hole is bored from the top downward, so as to communicate with the transverse hole. To prevent the binding of the journal boxes, the exposed ends of the armature shaft are covered with a thin wash of pure clay and allowed to dry.

The posts are secured to the base, with the ends of the armature shaft projecting into the transverse holes. Washers of pasteboard are placed upon the shaft on opposite sides of the posts, to confine the melted metal which is to form the journal boxes. Babbitt metal, or, in its absence, type metal, is melted and poured into the space around the shaft through the vertical hole in the post. The journal boxes thus formed are each provided with an oil hole, extending from the top of the post downward. If, after cleaning and oiling the boxes, the shaft does not turn freely, the boxes should be reamed or scraped until the desired freedom is secured.

All that is now required to complete the motor is the

DYNAMIC ELECTRICITY. 505

commutator brushes, $f f'$. They each consist of three or four strips of thin hard-rolled copper, curved, as shown in Fig. 488, to cause them to bear upon the screws in the end of the hub, G. The brushes are secured by small bolts to a disk of vulcanized fiber or vulcanite at diametrically opposite points, as shown in dotted lines in Fig. 489, and the brushes are arranged in the direction of the rotation of the armature.

In the brush-carrying disk is formed a curved slot for receiving a screw, shown in Fig. 492, which passes through the slot into the post and serves to bind the disk in any

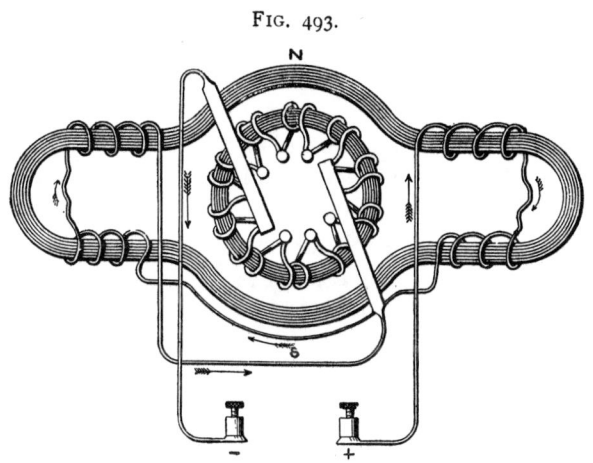

FIG. 493.

Circuit of Simple Electric Motor.

position. The disk is mounted on a boss projecting from the inner side of the post concentric with the armature shaft. The brushes are connected up by means of flexible cord, or by a wire spiral, as shown in Figs. 485 and 493. The most favorable position for the brushes may soon be found after applying the current to the motor. The ends of both brushes will lie approximately in the same horizontal plane.

When the motor is in operation, the direction of the current in the conductor of the field magnet is such as to produce consequent poles above and below the armature, as indicated in Fig. 493.

The dimensions of the parts of the motor are tabulated below:

Length of field magnet (inside)	10 inches.
Internal diameter of polar section of magnet	$3\frac{9}{16}$ "
Width of magnet core	$2\frac{1}{2}$ "
Number of layers of wire to each coil of magnet	5
Number of convolutions in each layer	34
Length of wire in each coil (approximate)	95 feet.
Size of wire, Am. W. G	No. 16
Outside diameter of armature	$3\frac{1}{2}$ inches.
Inside diameter of armature core	$2\frac{3}{16}$ "
Thickness " " "	$\frac{3}{8}$ "
Width " " "	2 "
" " " wound	$2\frac{1}{2}$ "
Number of coils on armature	12
Number of layers in each coil	4
Number of convolutions in each layer	8
Length of wire in each armature coil (approximate)	15 feet.
Size of wire on armature, Am. W. G	No. 18
Length of armature shaft	$7\frac{1}{4}$ inches.
Diameter of armature shaft	$\frac{9}{32}$ "
" " wooden hub	$1\frac{11}{16}$ "
Distance between standards	$5\frac{1}{2}$ "
Total weight of wire in armature and field magnet	6 lb.

This motor is designed for use in connection with a battery of low resistance, preferably one of the plunging type (Fig. 394), as such a battery permits of readily regulating the speed and power of the motor by simply plunging the plates more or less.

This form of battery has the additional advantages of being more powerful for its size than any other and of being very easily cleaned and kept in order. It has, however, the disadvantage of becoming exhausted in three or four hours, but this is partly compensated for by the ease with which it may be renewed.

DYNAMIC ELECTRICITY. 507

Eight cells of plunging bichromate battery like that shown in Fig. 394 will develop sufficient power in the

Fig. 494.

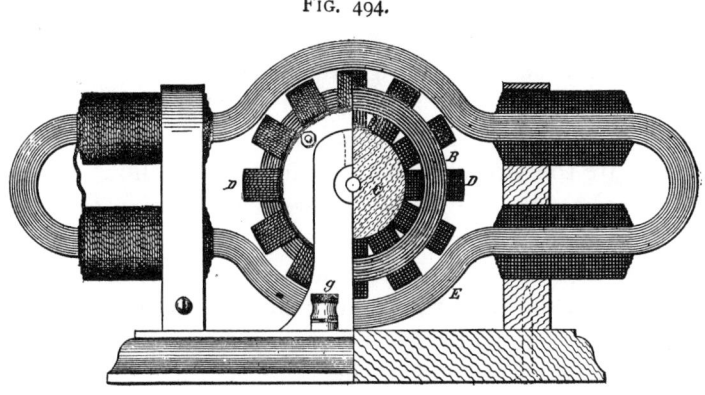

Side Elevation, Partly in Section, of Simple Electric Motor—One-third Size.

motor to run an ordinary foot lathe or two or three sewing machines. If it is desirable to adapt the motor to a battery of higher resistance, the armature and field magnet may be

Fig. 495.

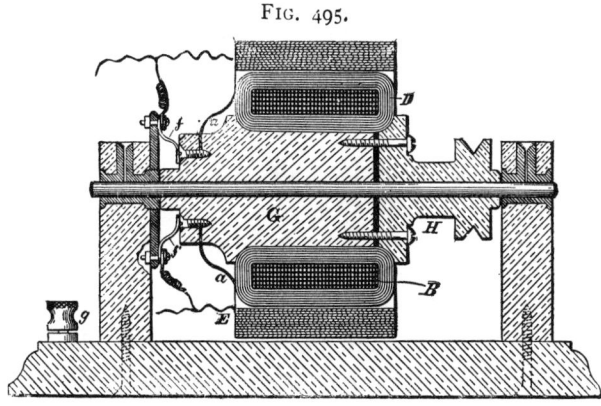

Vertical Transverse Section of Motor, taken through the Center of the Armature—One third Size, showing the Field Magnet in a Shunt.

wound with finer wire. For a dynamo circuit the field magnet of the motor should be placed in a shunt. (See diagram of Plating Dynamo.) If the motor is wound with wire of any size between Nos. 16 and 20, a battery may be adapted to it. When the field magnet is wound with finer wire and connected as a shunt around the armature, the motor becomes self-regulating.

The foregoing description of the small motor was written for the purpose of assisting amateurs who have few tools and no machinery. If all necessary tools are available, the motor may undoubtedly be modified in several particulars, to facilitate the work of construction, but without securing better final results. Fig. 496 shows a magnet made of cast iron. Instead of being formed of a single casting, it consists of two like halves, both made from the same pattern. The ends, which are square, are fitted together accurately either by planing or filing, and fastened together by screws or bolts, two at each end. The body of the cast iron field

FIG. 496.

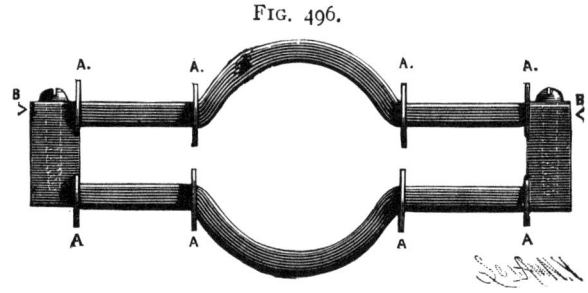

Cast Iron Field Magnet.

magnet should be fully one-half inch thick, and the ends one inch thick.

The flanges, A, which confine the wire as well as the portions of the magnet on which wire is wound, should be covered with thin cloth and shellacked before winding. The halves of the magnet are wound separately in a lathe, the ends being supported by the centers, B B, as shown.

When the cast iron field magnet is adopted, the motor may be used as a dynamo. In this case, however, it would be advisable to use smaller wire, say No. 20 or 22 on the armature and No. 18 on the field magnet. It would also be well to double the number of coils on the armature, at the same time doubling the number of convolutions and layers, so as to greatly increase the length of the wire in each section. Where the exact dimensions of the machine are not known the armature should be made first, the field magnet being adapted to the armature.

CHAPTER XIX.

A QUARTER-HORSE POWER ELECTRIC MOTOR.

The electric motor described on page 497 and the following pages was designed to be made from ordinary materials without the necessity of employing fine tools or machinery in its construction. It is operated by a current from the battery, everything connected with it being within the reach of any amateur. In this chapter an entirely different

FIG. 497

Quarter-Horse Power Electric Motor.

motor (Fig. 497) is shown and described, in which the best obtainable materials are used, and in the construction of which considerable mechanical skill is required and good tools are a necessity.

This motor was designed and built especially for the writer by Mr. W. S. Bishop for this edition of "Experimental Science," with the intention of giving the reader full particulars regarding the construction of a complete modern quarter-horse power self-regulating electric motor for use on a direct-current 110-volt circuit.

A QUARTER-HORSE POWER ELECTRIC MOTOR. 511

This motor is well proportioned, so that by enlarging or reducing it in proportion to sectional areas, a motor smaller or larger than the one illustrated may be constructed.

The motor is of the inclosed type, the field magnet forming a drum or cylindrical casing, with inwardly projecting pole pieces on which are placed the coils of the field magnet (Fig. 497a). The heads support self-oiling journals, a a^1 (Fig. 497) which receive the shaft of the armature, which latter revolves between the poles. The heads (Fig. 497b) are provided with removable sections for convenience

FIG. 497a

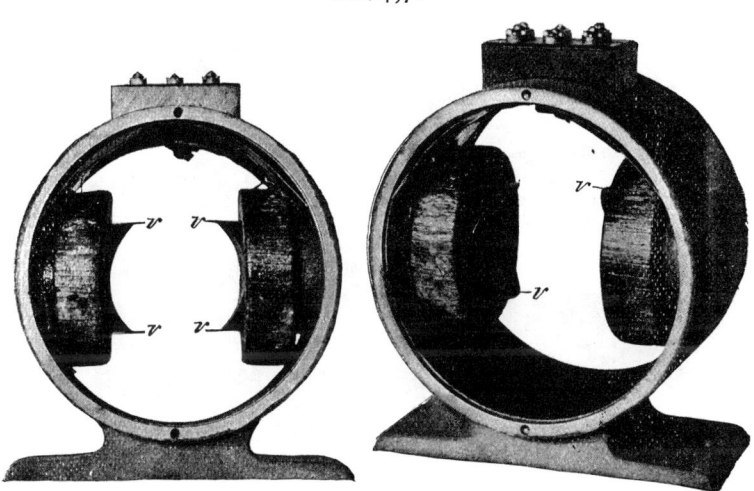

Field Magnet with Heads and Armature Removed.

in examining the interior of the field magnet. The commutator brushes consist of spring-pressed carbon rods inserted in the insulated brass sockets placed in horizontal holes bored in opposite sides of the head near one of the journals. The commutator revolves between and in contact with these carbon brushes. The electrical connections are made at the top of the casing, as will be presently described.

The steel armature shaft, which is supported by the self-oiling bearings in the heads, has a uniform diameter of ⅝ inch, and is 16 inches long. On this shaft is placed a cast iron sleeve $3\frac{1}{4}$ inches long and 1 inch in outside diameter, (Fig. 498), with a head $2\frac{1}{8}$ inches in diameter and $\frac{3}{16}$ inch thick formed on one end, and a wrought iron nut of the same size on the other end. The sleeve is secured by a key. On the sleeve between the head and nut are mounted the disks of which the armature core is formed. The end ones are $\frac{1}{16}$ inch thick; the intermediate ones, No. 25, all are of soft sheet steel. The disks are varnished with thin shellac and dried before they are placed on the sleeve. These disks each have 18 notches, each of which is $\frac{3}{8}$ inch wide

FIG. 497b

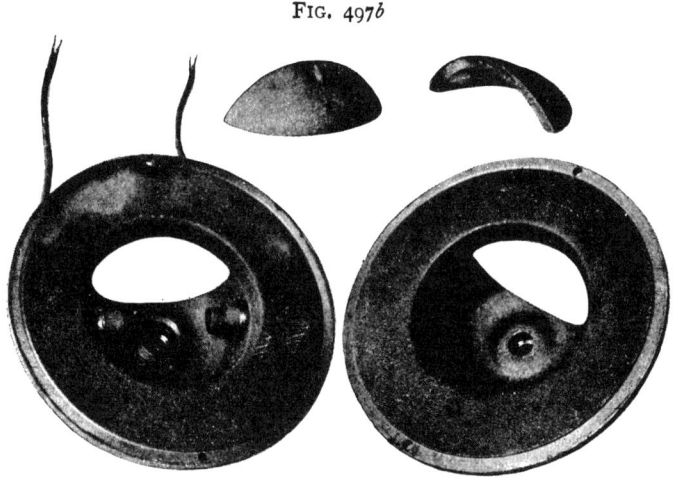

The Heads of the Field Magnet.

at the periphery, $\frac{1}{8}$ inch wide at the bottom and $\frac{3}{4}$ inch deep. These notches, when the disks are placed together, form grooves for receiving the armature winding. The grooves are lined with strips of leatherboard which completely cover the edges of the disks. A washer of vulcanized fiber $\frac{1}{32}$ inch thick is placed at each end, covering the end disks, and having notches of the same size as those in the steel disks and holes large enough to admit the nut and flange.

The nut and flange are each insulated by a washer of canvas provided with a number of radial slits to enable them to form down over the edges of the flange and nut. These canvas washers are coated with shellac varnish.

Tubes of vulcanized fiber are placed on each end of the armature shaft adjoining the cast iron sleeve, so that no electrical connection can be made by the winding with the shaft. Every portion of the steel of the armature is thus protected, so that it cannot come into electrical contact with the winding. All of the insulating material is held in place by thick shellac varnish, which is allowed to dry thoroughly before the winding is done.

The armature is wound with No. 22 (A. W. G.) single

FIG. 498.

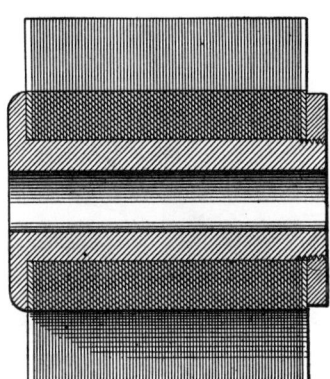

Armature Core, Half Size.

silk-covered copper wire. There are eighteen coils on the armature, with fifty-eight convolutions in each coil. The winding is done while the armature shaft is supported on lathe centers. To begin, 3 inches of wire are left projecting from the commutator end of the armature, as shown in Fig. 498a, and to avoid mistakes, the armature groove in which the beginning of the winding is made is marked 18-1 on opposite sides. Nine grooves are counted off and 8 is marked on one side of the ninth groove and 9 on the other side. The wire is then carried along in the groove 18-1 to the back end of the armature, thence over the end and past the shaft to the groove 8-9; along this groove to the commutator end of the armature, past the shaft to groove 18-1, along this groove to the rear end of the armature, and so on until grooves 18-1 and 8-9 contain fifty-eight turns; then carry the wire to groove 9-10, and back across the commutator end of the armature to groove 1-2, when a loop of about 4 inches long is formed outside of the groove, and the winding is continued in groove 9-10 and groove 1-2 as before, filling in fifty-eight turns, and so on until the grooves are half filled and the winding is one-half way around the armature; but

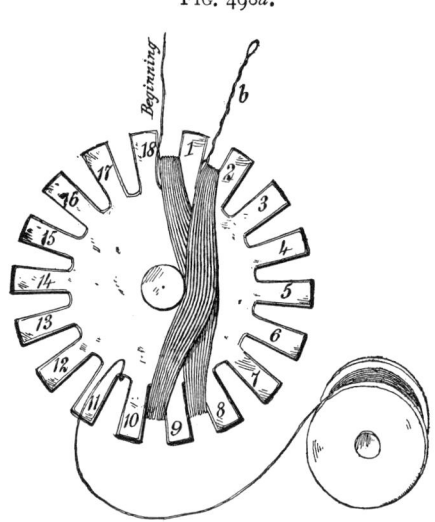

FIG. 498a.

Beginning of the Winding.

each time a new coil is started the previous one must be protected by small pieces of oiled silk applied to the ends of the armature where the wires cross. After the coils are one-half on, i. e., after the winding has been carried around to the point of starting, the looped and twisted ends designed for commutator connections are thoroughly protected with tape close up to the coils, and the winding is continued the same way as before, winding fifty-eight turns of wire on top of the wire first wound, until every groove is filled, and the loop is brought out of every space, when these loops are protected by tape, as already described, and the commutator is placed on the shaft. The doubled and twisted ends are cut off at the extreme end of the loop for attachment to the commutator bars.

The commutator (Fig. 498e) is a vital part of the motor, needing great care in its construction. It has a hub or sleeve, b, 1½ inches long, provided with an undercut collar, c, formed on its inner end, and is provided with a removable collar, d, having an undercut portion, c', corresponding

FIG. 498c.

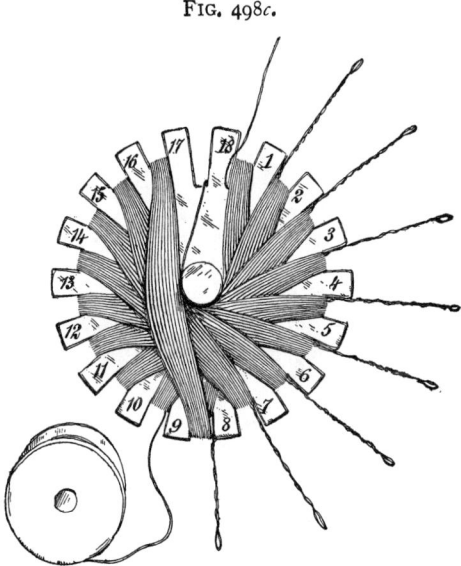

Winding the Last Coil of the First Series.

to the undercut flange, c, but oppositely arranged. Between these undercut portions are clamped the commutator bars, e. The latter are made of cast phosphor-bronze, when only a few are required. These are milled on the ends and inner side, the outside being turned off after the commutator is assembled.

Another way to make the commutator is to cast a cylinder of phosphor-bronze, having the cross section shown in Fig. 498e; afterward sawing the cylinder up into eighteen bars as shown. Each bar has a short, slotted arm extending in a radial direction for receiving a pair of terminals of the armature coil. The commutator sleeve or hub, *b*, is provided with a wrapping, *f*, and a washer, *g*, of mica for thoroughly insulating the bars from the sleeve, and with a mica disk, *h*, for insulating them from the collar, *d*. This collar is held in place by four screws passing through it into the sleeve, *b*. To economize space the inner end of the sleeve, *b*, is chambered, as shown. The commutator is prevented from turning on the shaft by a

Fig. 498*d*.

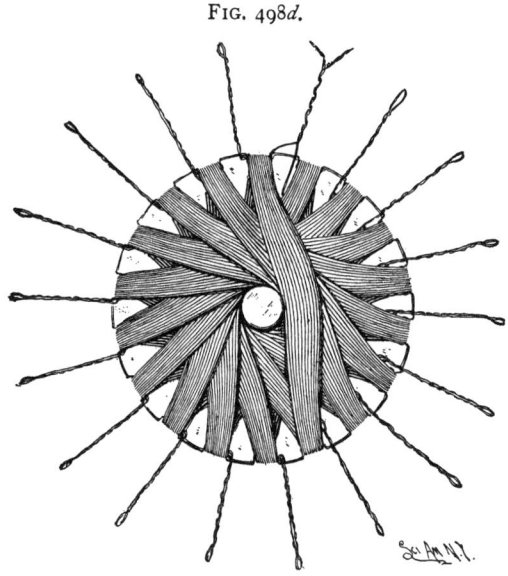

The Finished Winding.

lug pin, *i*, which is inserted in the shaft, and is received in a short slot, *j*, in the sleeve, *b*.

The commutator is $2\frac{7}{16}$ inches in diameter, the bars are 1 inch long, and $\frac{7}{16}$ wide on the outer face. They are separated evenly by strips of mica, which are turned off when the commutator is turned. The dimensions can be obtained from the engraving, which is one-half size.

The looped ends of the armature coils are all carried to one side to the fourth commutator bar, and cleaned off and inserted in the slots of the arms projecting from the com-

A QUARTER-HORSE POWER ELECTRIC MOTOR. 517

mutator bars, the wires being soldered in the slots. This arrangement of the wires admits of placing the carbon commutator brushes in a horizontal position, which is desirable.

The terminals of the armature winding are all thoroughly separated by pieces of oiled silk. They are wrapped with stout twine to resist centrifugal force and are varnished with shellac.

In each slot of the armature is placed a strip of leatherboard to cover the coil, and in peripheral grooves turned at three points in the length of the armature, are placed windings of No. 18 piano wire, which are carefully soldered

FIG. 498e.

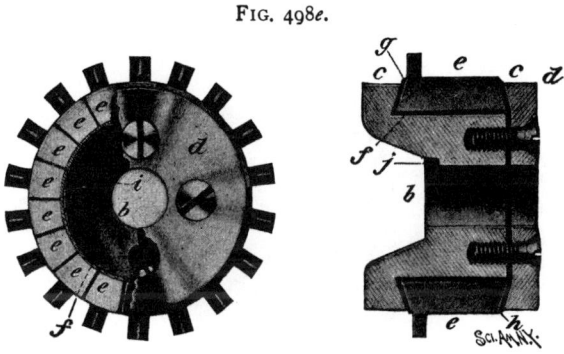

Sectional Views of the Commutator.

(Fig. 499). These windings prevent the bursting of the armature by centrifugal force. The winding and insulation of the armature is now provided with two or three coats of shellac varnish to exclude moisture and bind the wires and insulators.

The armature is now carefully balanced by adding solder to the peripheral bands, or by increasing the solder on the arms of the commutator bars. The armature is tested by rolling it on parallel bars arranged in a perfectly horizontal plane.

The field magnet, which has internal poles, v, is now bored to receive the armature and is faced off and bored to receive the heads. The latter are accurately fitted, so that each may be retained in place by two tap bolts as shown. Accuracy of position is secured by facing the inner side of the head and providing it with a shoulder which fits the finished edge of the field magnet. While the head is still in position in the lathe the support for the journal box is bored to insure proper alignment.

The journal boxes, aa^1 (Fig. 499a), each consist of a bronze sleeve bored internally to fit the shaft and turned off externally in two diameters, one to fit the hub of the magnet head and the other to make room for the oil chamber, k, which is further enlarged by the chambering out of the hub. The journal box is provided with an annular groove, m, to receive the oil from the end of the shaft, and a small duct, n, conveys the oil back to the reservoir. The journal outside

FIG. 499

The Finished Armature.

of the groove, m, is counter-bored to prevent it from coming into contact with the shaft.

A slot is cut across the upper portion of the journal box to loosely admit the ring, l, which rests upon the shaft and is revolved by frictional contact. The ring, l, carries up sufficient oil to keep the journal lubricated. The surplus oil falls from the inner end of the journal box directly into the oil reservoir. Collars are secured to the shaft at each end adjoining the journal boxes. The armature shaft is thus mounted so that the armature revolves within one-thirty-second inch of the polar extremities of the field magnet.

The commutator brushes, A (Fig. 499b), which form the electrical contacts with the commutators, each consist of a brass tube, q, having a thick head for receiving a binding

A QUARTER-HORSE POWER ELECTRIC MOTOR. 519

screw, r, for holding the lead wire, s. In the brass tube is placed a spiral spring, t, which presses the cylindrical carbon rod, u, forward into contact with the commutator cylinder. Each field magnet pole, v, is furnished with a coil, which is wound on a wooden form between two wooden collars on a strip of vulcanized fiber one-sixteenth inch thick wrapped around the form. After winding each coil is wound with adhesive tape and varnished with shellac. There is a total of $3\frac{3}{4}$ pounds of No. 28 single cotton-covered wire in the coils.

The wooden form for these coils should be a little larger than the magnet pole to insure the fitting of the magnet coils to the poles.

The coils are both wound in the same direction and connected in series, the outside ends being connected together,

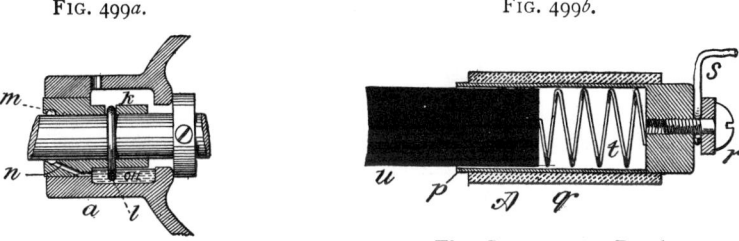

FIG. 499a.   FIG. 499b.

The Journal Box.   The Commutator Brush.

while the inner ends are connected one with the binding post, L, and the other with the binding post, F, both mounted on the block secured to the top of the field magnet. The binding post, L, is also connected with one of the commutator brushes, and the binding post, A, is connected with the commutator brush, but with no other part of the motor.

A rheostat or starting box is required to start the motor with safety. A diagram of the connections is given in Fig. 500, which is a plan view of the motor and a diagrammatic view of the rheostat. In this view the binding posts, F, A, are shown, the binding post, L, being connected with the line and one of the commutator brushes; the binding post, A, being connected with the other commutator brush and with the middle binding post of the rheostat. The binding post, F, of the rheostat is connected with the center binding post, F, of the motor, and the binding post, L, of the rheostat is connected with the line and also with the switch arm 1. The latter is capable of making a contact with either of the buttons 2, 3, 4, 5, 6, 7. The switch arm 1 has a spring which tends to throw it around into contact with the button 2, which has no electrical connection. The buttons 3 to 7 inclusive are connected with the resistance coils. When

FIG. 500

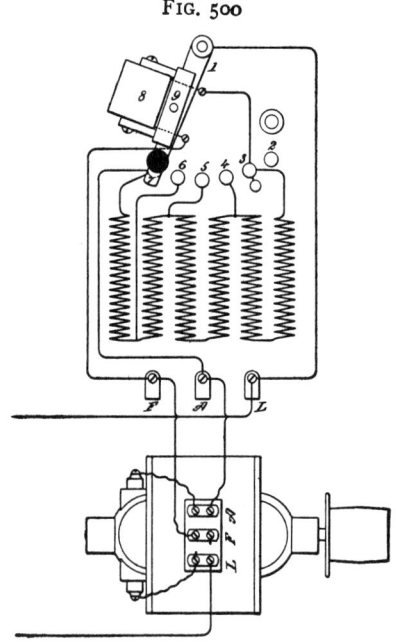

The Rheostat or Starting Box.

the arm 1 is brought into contact with the button 3, it throws all of the resistance into the armature circuit, and also throws the switch magnet, 8, into the field circuit, in which it remains with more or less resistance as long as the motor runs.

The switch arm cuts out the resistance gradually until the current is all on, when the armature, 9, carried by the arm, 1, is held by the magnet, 8. The parts remain in this position so long as the current passes. When the cir-

A QUARTER-HORSE POWER ELECTRIC MOTOR.

cuit is broken at any point, the magnet 8 releases the switch arm, and the spring causes it to fly back and break the circuit, thereby avoiding the danger of throwing a heavy current suddenly on the motor.

This motor will readily run on current taken from an ordinary lamp socket. It is so proportioned that the counter electro-motive force keeps the speed down to 1,600 revolutions per minute, and controls the current so as to render the motor uniform in its action.

By connecting the field magnet with the armature the machine may be run by power as a generator. When run at 1,850 revolutions per minute, it is able to supply three 16 candle power 110-volt lamps.

The following is a table of the dimensions of the motor:

Length of armature shaft	16	inches
Diameter " "	$5/8$	"
Diameter of the armature	$3\tfrac{3}{4}$	"
Diameter of the commutator	$2\tfrac{3}{8}$	"
Width of commutator	$1\tfrac{13}{16}$	"
Number of commutator bars, 18.		
Width of commutator bars	$\tfrac{5}{16}$	"
Length of bearing surface	$3/4$	"
Bore of the field magnet	$3\tfrac{3}{16}$	"
Outside diameter of field magnet drum	$8\tfrac{11}{16}$	"
Inside " " " "	$7\tfrac{7}{8}$	"
Width of " " " "	$5\tfrac{1}{4}$	"
Base	4 x $9\tfrac{1}{2}$	"
Height of drum above bottom of base	1	"
Diameter of inside of convex portion of head	4	"
Diameter of hole for receiving commutator brushes	1	"
Outside diameter of carbon brush holder	$1\tfrac{3}{16}$	"
Carbon rod	$\tfrac{11}{16}$	"
Length of carbon	$1\tfrac{5}{8}$	"

Wire on the field magnet, $3\tfrac{3}{4}$ pounds of single cotton-covered, No. 28.
Wire on armature, $2\tfrac{1}{4}$ pounds No. 22 single silk-covered.
Normal speed of motor, 1,600 revolutions per minute.

This little motor is a marvel of mechanical construction. It does not possess a single unnecessary part, and everything is plain, direct and simple.

TABLE OF TANGENTS.

Degrees.	Tangents.	Degrees.	Tangents.	Degrees.	Tangents.	Degrees.	Tangents.
1.	.0175	18.	.3249	35.	.7002	52.	1.279
1.5	.0262	18.5	.3346	35.5	.7133	52.5	1.303
2.	.0349	19.	.3443	36.	.7265	53.	1.327
2.5	.0437	19.5	.3541	36.5	.7400	53.5	1.351
3.	.0524	20.	.3640	37.	.7536	54.	1.376
3.5	.0612	20.5	.3739	37.5	.7673	54.5	1.401
4.	.0699	21.	.3839	38.	.7813	55.	1.428
4.5	.0787	21.5	.3939	38.5	.7954	55.5	1.455
5.	.0875	22.	.4040	39.	.8098	56.	1.482
5.5	.0963	22.5	.4142	39.5	.8243	56.5	1.510
6.	.1051	23.	.4245	40.	.8391	57.	1.539
6.5	.1139	23.5	.4348	40.5	.8541	57.5	1.569
7.	.1228	24.	.4452	41.	.8693	58.	1.600
7.5	.1317	24.5	.4557	41.5	.8847	58.5	1.631
8.	.1405	25.	.4663	42.	.9004	59.	1.664
8.5	.1495	25.5	.4770	42.5	.9163	59.5	1.697
9.	.1584	26.	.4877	43.	.9325	60.	1.732
9.5	.1673	26.5	.4986	43.5	.9490	60.5	1.767
10.	.1763	27.	.5095	44.	.9657	61.	1.804
10.5	.1853	27.5	.5206	44.5	.9827	61.5	1.841
11.	.1944	28.	.5317	45.	1.	62.	1.880
11.5	.2035	28.5	.5430	45.5	1.0176	62.5	1.921
12.	.2126	29.	.5543	46.	1.035	63.	1.962
12.5	.2217	29.5	.5658	46.5	1.053	63.5	2.005
13.	.2309	30.	.5774	47.	1.072	64.	2.050
13.5	.2401	30.5	.5890	47.5	1.091	64.5	2.096
14.	.2493	31.	.6009	48.	1.110	65.	2.144
14.5	.2586	31.5	.6128	48.5	1.130	65.5	2.194
15.	.2679	32.	.6249	49.	1.150	66.	2.246
15.5	.2773	32.5	.6371	49.5	1.170	66.5	2.299
16.	.2867	33.	.6494	50.	1.191	67.	2.355
16.5	.2962	33.5	.6619	50.5	1.213	67.5	2.414
17.	.3057	34.	.6745	51.	1.234	68.	2.475
17.5	.3153	34.5	.6873	51.5	1.257	68.5	2.538

TABLE OF TANGENTS—*Continued.*

Degrees.	Tangents.	Degrees.	Tangents.	Degrees.	Tangents.	Degrees.	Tangents.
69.	2.605	74.5	3.605	80.	5.671	85.5	12.706
69.5	2.674	75.	3.732	80.5	5.975	86.	14.300
70.	2.747	75.5	3.866	81.	6.313	86.5	16.349
70.5	2.823	76.	4.010	81.5	6.691	87.	19.081
71.	2.904	76.5	4.165	82.	7.115	87.5	22.903
71.5	2.988	77.	4.331	82.5	7.595	88.	28.636
72.	3.077	77.5	4.510	83.	8.144	88.5	38.188
72.5	3.171	78.	4.704	83.5	8.776	89.	57.290
73.	3.270	78.5	4.915	84.	9.514	89.5	114.588
73.5	3.375	79.	5.144	84.5	10.385	90.	
74.	3.487	79.5	5.395	85.	11.430		

INDEX.

Absorption, 6
Absorption of gases, 64
Absorption of light, 217
Achromatic objective, telescopic, 315
Acid, lithic, 273
Acid, stearic, 273
Acid, sulphuric, and water, 3
Actions, molecular, 56
Adhesion, 58
Adjustable sound reflector, 159
Adjustment, fine, 281
Aerial top, 109
Air compressed, 99
Air, effect of centrifugal force on, 12
Air, inertia of, 109
Air, pressure of, 89
Air pressure, weight lifted by, 89
Air pump, inexpensive, 91
Air pump receiver, 96
Air pump, treadle for, 95
Air, reduction of volume of, by pressure, 6
Air, resistance of, on falling bodies, 38
Air thermometer, 184
Air, weight of, 90
Air, withdrawing, from microscope slides, 99
Airy's spirals, 258
Alcohol, cotton in, 1
Alcohol and water, mixture of, 3
Alum cell, 189
Amalgamation of zincs, 402
Ammeter, 453
Ampere, 427
Analysis and synthesis of light, 213
Analysis of sounding flames, 148
Analyzer, 235
Anchor plants, 290
Ancient inventions, 106
Anhydrite, 260
Aniline and water, 4
Annealed glass, 240
Annunciator, 487
Ano-kato, 363
Aquatic plants, 287
Aragonite hemitrope, 258
Arborescent magnetic figures, 356
Armature, 518
Armature core, 513
Armature, drum, 475
Armature, Gramme, 474
Arrangement of battery cells, 429
Arrangement of polarizer and analyzer, 238
Art, decorative, suggestions in, 272
Ascensional power of heated air, 196
Aspirators, 101
Aspirators, blast produced by, 105
Aspirators, Chapman's, 106
Aspirators, receiver for, 105
Atmosphere, crushing force of, 89
Atmospheric pressure, 91
Atomizing petroleum burner, 101

Bage plate holder, 320
Balance, 54
Balance, readily made, 86
Balance, thermoscopic, 185
Ball bearings, 10
Ball experiment, 99
Balloon, dilatation of in a vacuum, 85
Balls, dropped and projected, 43
Bar, compound, 182
Barometer, 91
Barrett, Prof., 165
Barry, Philip, 167
Batteries, two-fluid, 409
Batteries, polarization of, 395

Battery and galvanometer, experiments with, 394
Battery, caustic potash, 408
Battery cells, arrangement of, 429
Battery, chromic acid, 412
Battery, Daniell, 409
Battery, dry, Dr. Gassner's, 407
Battery, Fuller, 413
Battery, gravity, 410
Battery, Grenet, 399
Battery, Grove, 411
Battery, large plunge, 401
Battery plates, roughening, 418
Battery, secondary, formation of, 420
Battery, simple plunge, 400
Battery, Smee's, 398
Battery, thermo-electric, 422
Beach pyro-potash developer, 321
Bee's wing, doubling hooks of, 291
Bell, Alexander Graham, 190
Bell in vacuo, 98
Bell, over-tones of, 144
Bell, polarized, 482
Bell telephone, principle of, 477
Bichromate of potash, 260
Bicycles, pedals and shafts of, 10
Bifilar suspension, 53
Bird, mechanical, Penaud's, 111
Black glass polarizer, 238
Black tones on photographs, 327
Blackened glass, polarization by reflection from, 240
Blast produced by aspirator, 105
Blood, circulation of, 195
Bodies, falling, 38
Bohnenberger machine, 21
Boiling water in vacuo, 98
Bologna flask, 56
Bomb, candle, 195
Boomerang, 113
Border, composite, 276
Border, dado and frieze, 274
Bouquet, phantom, 211
Box, musical, 119
Branch circuits, joint resistance of, 448.
Branchipus, 290.
Brazilian pebbles, 269.
Bridge key and connections, 447.
Bridge resistance box, 444.
Bridge, Wheatstone's, 443.
Bridges, railroad, vibration of, 126.
Bromine box, 342.
Brown tones on photographs, 327.
Buffing the plate, 339.
Bugle, 121.
Bunsen's filter pump, 102.
Burner, petroleum, atomizing, 101.
Bursting fly-wheels, 34.
Buzz, 117.

Cadmium, sulphate of, 272, 274
Calcite, 258
Calcium sulphide, 198.
Camera, adjustment of, 319.
Camera, photographic, 318.
Camera, pocket, 328.
Candle bomb, 195.
Capillarity, 60.
Capillary depression, 60.
Capillary elevation, 60.
Capitol at Washington, 159.
Carbonic acid, absorption of, by charcoal, 64, 66.
Carbonic acid, for experiment, a preparation of, 65.
Carbons, paraffine, 402.
Card experiment, 100.
Cartesian diver, 79.

INDEX. 525

Cast iron field magnet, 508.
Caustics, 208.
Caution about illumination, 283.
Cell, gutta-percha, 403.
Cell, wax, making, 300.
Centrifugal force, 11.
Centrifugal force, action of, on air, 12.
Centrifugal force, action of, on liquids, 12, 17.
Centrifugal force, top for showing the action of, 13.
Centrifugal railway, 11.
Centrifugal siren, 171.
Chapman's aspirator, 103.
Charcoal, absorption of carbonic acid by, 64
Chemical thermoscope, 198.
Chime, electrical, 384.
Chloride of silver cell, 405.
Choral top, 12.
Chromic acid battery, 412.
Chromotrope, 217.
Chrysoberyl, 261.
Chrysolite, 260.
Church windows, rattling of, 125.
Ciliated organisms, microscopic examination of, by intermittent light, 292.
Circuit of simple electric motor, 505.
Circuits, branch, 448.
Circular polarization, 253.
Circulation of blood in fish's tail, 295.
Circulation of blood in frog's foot, 295.
Clamond's thermo-electric battery, 422.
Clappers, 116.
Cleaning glass for polariscope, 262.
Clock, one hundred year, 54.
Clock, Wheatstone's polar, 269.
Clocks, application of pendulum to, 49.
Coating the plate, 340.
Cohesion, 56.
Cohesion, demonstration of, 56.
Cohesion, force of, on liquids, 56.
Collimation of telescopes, 313.
Color, 214.
Communicating vessels, equilibrium in, 74.
Commutator, 517
Commutator brush, 519
Compact telescope, 316.
Comparison of sound and light waves, 200.
Composite border, 276
Composition of vibrations, 136.
Compound bar, 183.
Compound microscope, 281.
Compounding rectangular vibrations, 140.
Compressed air, 99, 100.
Compressibility, 6.
Compressing pump, 96.
Compression, 4.
Compression, heat due to, 194.
Compressor, 292.
Concave cylindrical mirror, 208.
Concave reflectors, 192.
Concave spherical mirror, 211.
Concentration of sound, 158, 162.
Condenser, electrical, 381.
Condenser, sub-stage, 285
Conduction of heat, 193.
Conduction of sound, 125.
Conductivity of metals, 194.
Cone, mica, 248.
Conical pendulum. 46.
Conjugate foci. 204.
Connections of plating dynamo, 496.
Consequent pole, 353.
Contact prints, 325.
Converging rays, convex lens, 206.
Converter, 358.

Convex cylinder mirror, 208.
Copying, photographic, 319.
Cotton in alcohol, 1.
Coulomb, lines of torsion, 53.
Course of light through Iceland spar, 234
Course of light through a prism, 203.
Cricket, 116.
Crooke, W. A., 189.
Cross, Maltese, 249.
Crushing force of the atmosphere, 89.
Cryophorus, Wollaston's, 188.
Crystallization, examples of, 274.
Crystals, 211.
Crystals, leaves, stalks and flowers, 277.
Crystals, panel, with ornaments of, 274.
Crystals, polariscope for, 255.
Crystals, wide-angled, 255.
Current generator, simple, 468.
Curves, magnetic, formation of, 354.
Curves traced by vibrating rods, 304.
Cutting prints, 325.
Cycloid, 40.
Cycloidal curve, 52.
Cycloid, method of describing, 42.
Cyclops, 290.
Cylindrical mirror, 134.

Daguerre, 337.
Daguerreotype fixing, 344.
Daguerreotype gallery, 342.
Daguerreotype gilding or toning, 344.
Daguerreotype plate, coating, 340.
Daguerreotype plate, developing, 344.
Daguerreotype plate, scouring, 337.
Daguerreotype, 337.
Daguerre's discovery, 344.
Daniell battery, 409.
Daphnia, 291.
Dark room, 341.
Davy, 518.
Dead-beat galvanometer, 434.
Decorative art, suggestions in, 272.
Depolarization of electrode by mechanical agitation, 416.
Deprez-D'Arsonval galvanometer, 434.
Designs on wire cloth, 62.
Destruction of life by removal of air, 98.
Detail, to bring out in a photographic negative, 323.
Determining speed by resonance, 169.
Developer, Beach, pyro-potash, 321.
Developer, hydrochinon, 323.
Development of plates, 321
Diamagnetism, 464.
Diaphragm cell, 179.
Diaphragm iris, 285.
Diaphragm microscope, 281.
Diaphragms,' vibration of, 642.
Diatoms, markings on, 230.
Diffusion of gases, 66
Diffusion of gases, law of, 71.
Diffusion of gases, simple way of showing, 66, 67.
Dilatation of a balloon in a vacuum, 85
Dimensions of motor, 521
Discharge, electric, over finely divided metal, 375.
Discharger, jointed, 38.
Disk, with silvered beads, 175.
Disruptive discharge, 384.
Distribution of electricity on the plates of the Wimshurst machine, 374
Diver, Cartesian, 79.
Diverging rays, concave lens, 206.
Diversion of electric discharge by moisture, 377.
Divisibility, 4.
Divisibility, extreme, 4.

Double mouthpiece, 180.
Double polarization by single plate, 245.
Double refraction, 233.
Doubler, Norremberg, 245, 253
Doubling hooks of bee's wing, 290
Draper, 338
Driving gear, friction, for a gyroscope, 20.
Dropped and projected balls, 43, 44.
Drops, Prince Rupert's, 57.
Drum armature, 475.
Dry objects, quick method of mounting, 297.
Dutch tears, 57.
Dynamic electricity, 392.
Dynamo, electro-plating, 495.
Dynamo, hand power, 487.
Dynamo, principles of, 465.
Dynamo, Westinghouse, 335.

Earth, magnetism by induction from, 347.
Earth's rotation shown by pendulum, 47.
East River bridge, 126.
Eaton, Prof. A. K., 317.
Edison listening to the first phonogram from England, 151.
Edison's new phonograph, 151.
Effect of armature on permanent magnet, 350.
Effects of magnetic induction, 351.
Egg and brine experiment, 81.
Elasticity, 6.
Elasticity of gases, 7.
Electric chime, 384.
Electric condenser, 380.
Electric discharge through vacuum tube, 380.
Electric discharge, various phases of, 374.
Electric fly, 385.
Electric machine, Wimshurst, 370.
Electric machine, Winter's, 369.
Electric motor, 390, 470.
Electric motor, fifty-cent, 451.
Electric pendulum, 360.
Electrical gyroscope, 24, 27.
Electrical gyroscope for showing the rotation of the earth, 28, 29.
Electrical measurements, 442.
Electrical perfection of glass, 384.
Electrical units, 425.
Electricity, frictional, 359.
Electricity, masked, 362.
Electricity, vitreous and resinous, 361
Electrified threads, 364.
Electro-magnet, 457.
Electro-magnet, experiments with, 464.
Electro-magnet, inexpensive, 461.
Electrodes, mechanical depolarization of, 414.
Electro-plating dynamo, 495.
Electrophorus, 368.
Electroscope, 361.
Electroscope, experiments with, 362
Elliptical polarization, 253.
Endosmometer, 71.
Endosmose, pressure by, 70.
Escapement, 50.
Equatorially mounted electrical indicator, 33.
Equilibrium in communicating vessels, 74.
Euler, 200.
Exhausting Geissler tube, 104.
Exosmose, 67
Exosmose, partial vacuum by, 70.
Expansion, 4, 181.
Expansion of gases, 85.
Experiment with scientific top, 153.

Extension, 1.
Extraordinary ray, 233.
Eye-pieces for telescope, 312.

Falling bodies, 38.
Falling bodies, law of, 39.
Fifty-cent electric motor, 471.
Field magnet of motor, 511
Films, mica, 253.
Films, thin, 255.
Filter pump, Bunsen's, 102.
Filtration, 6.
Fixing bath for photos, 377
Fixing daguerreotype, 344.
Flageolet, 123.
Flame, manometric, 135.
Flame, speaking, 132.
Flame, vibrating, 134.
Flames, musical, 145.
Flames, sounding, 145.
Flames, sounding, analysis of, 145.
Flames, vibrating, 130.
Flask, Bologna, 57.
Flexure, elasticity of, 7.
Floating magnet, 357.
Flowers, sensitive, 165.
Fly, electric, 385.
Fly-fly, 111.
Flying pendulum, 55.
Fly-wheels, 89.
Fly-wheel, bursting of, 34.
Fly-wheels, flexible, 35.
Focus, principle of concave lens, 205.
Focusing camera, 319.
Fogging, 319.
Foraminifera, 290.
Force, 8.
Force, lines of, 354.
Force of cohesion in liquids, 56.
Force of steam, 195.
Formation of secondary battery, 420.
Forms of lenses, 204.
Foucault's experiments, 46.
Fountain, Hero's rotary, 18.
Franklin, 184
Franklin's plate, 381.
Freezing by rapid evaporation, 188.
Fresnel, 200.
Friction, 9.
Friction, heat due to, 194.
Friction, rolling, 10.
Friction, sliding, 10.
Frictional driving gear for gyroscopes, 20.
Frictional electricity, 359.
Frog plate, 296.
Fuller cell, 413.
Fusion, 181.

Galileo's discovery, 46
Galleries, whispering, 159.
Gallery, daguerreotype, 343.
Galvanometer and battery, experiments with, 394.
Galvanometer, Deprez-D'Arsonval, 334.
Galvanometer, tangent, 440.
Galvanometers, 433.
Gas pressure, apparatus for producing, 167.
Gas wheel, 87.
Gases, 6, 85.
Gases, absorption of, 64.
Gases, diffusion of, 66.
Gases, diffusion of, simple way of showing, 67.
Gases, expansion of, 85.
Gases, weighing of, 86.
Gassiot's cascades, 390.
Gassner's dry battery, 407.
Gathering microscope objects, 287.

INDEX.

Gathering microscopic objects, implement for, 287.
Geissler tube, exhausting, 104.
Geissler tube, self-exciting, 365.
Generator of currents, 392.
Gilding and toning, 345.
Girder, breaking of by pith balls, 127.
Glass, hand, 88.
Glass, perfectly elastic, 7.
Glass, pulse, 184.
Glass, strained, 241.
Glass, strained by heat, 243.
Glass, strained by pressure, 243.
Glass top, 12.
Gloucester cathedral, 159.
Glow of the negative collectors, 378.
Glow of the positive collectors, 378.
Gold crystals, 291.
Goldfish, circulation of blood in tail of, 297.
Gramme armature, 474.
Gramme machine for illustration, 472.
Gravity battery, 410.
Green, John, 317.
Grenet battery, bottle form, 399.
Grenet battery with air tubes, 415.
Grindstones, bursting of, 37.
Grove battery, 411.
Gun, Quaker, 43.
Gutta-percha-lined cells, 403.
Gyroscope, 19.
Gyroscope, electrical, 24, 27, 31.
Gyroscope, friction driving gear for, 20.
Gyroscope, pneumatic, 21.
Gyroscope, steam, 21.

Hammer, water, 39.
Hand glass, 88.
Hand power dynamo, 48.
Harmonica, 120.
Harmonica vibrations, 126.
Harp, Marloye's, 122.
Heads of field magnet, 512.
Heat, 181.
Heat, conduction of, 193.
Heat due to friction, 194.
Heat due to pressure, 194.
Heat, latent, 185.
Heat, a mode of motion, 181.
Heat, radiant, Tyndall's experiment on, 189.
Heat, reflection of, 193.
Heat, reflection and concentration, 192.
Heated air, ascensional power of, 196.
Heating tool, 299.
Hemitrope, 258.
Hero's fountain, rotary, 18.
Herschel, 212.
Hoffman, 223.
Holder for soap film, 203.
Hooke, 200.
Hooke's invention, 54.
Hot air motor, 196.
Hundred year clock, 54.
Huyghens, 200.
Huyghens' inventions, 49.
Hydraulic ram, 80.
Hydrochinon developer, 223.
Hydrostatic press, hypothetical, 76.
Hydrostatic press, principle of, 75.
Hydrostatic pressure, 72.
Hygrometry, 196.
Hygroscope, 197.
Hygroscopic roses, 198.
Hygroscopic and luminous roses, 198.
Hypothetical lens, 203.

Iceland spar, 233.
Illumination, caution about, 283.
Illusion, disk, 175.

Illusion, optical, Rapieff's, 230
Illusions, 228.
Illusions, optical, 223.
Impenetrability, 1, 2.
Implements for gathering microscopic objects, 287.
Incandescent lamp, 221.
Inclined plane, 40.
Indicator, electric, 33.
Induced current from induced magnetism, 467.
Induction machine, Wimshurst, 370.
Induction, magnetic, 473.
Inertia, 8.
Inertia of air, 109.
Inertia locomotive, 9.
Inexpensive air pump, 91.
Infusoria, 290.
Insects, 278
Instantaneous photography, 319.
Instruments, stringed, 124.
Insulating stool, 387.
Intensification of photographic negatives, 234.
Intensity of light, 222.
Intermittent light, examination of ciliated objects by, 292
Interrupter, light, for microscope, 293.
Inventions, ancient, 106.
Iodine box, 342.
Iodine cell, 189.
Iris diaphragm, 285.
Irradiation, 221.
Isochronism, 51.

Jew's harp, 143.
Joint resistance of branch circuits, 448.
Jointed discharger, 386.
Journal box, 519
Jupiter, 314.

Kaleidoscope, 213.
Kater's reversible pendulum, 48.
Kent's trough, 295.
Key, bridge, 447.
Kits, photographic, 319.
Koenig's manometric flames, 178.

Lamp, incandescent, 221.
Lantern slides, 325.
Latent heat, 185.
Lateral pressures, 79.
Latour, Cagniard, 172.
Law controlling gyroscopic movement, 32
Law of diffusion of gases, 71.
Law of falling bodies, 39.
Law, Pascal's, 73.
Law, Pascal's, demonstration of, 72.
Leaf, sensitive, 107.
Leclanche battery, 406.
Le Conte, Dr., 165.
Length of resonant tube, 145.
Lengthening the spark, 378.
Lens, hypothetical, 204.
Lens, sound, 165.
Lenses, 204.
Lenses, forms of, 204.
Leyden jar, 382.
Leyden jar, attachment to Wimshurst machine, 373
Light, 200.
Light, analysis and synthesis of, 213
Life, destruction of by removal of air, 98.
Light, intensity of, 222.
Light interrupter for microscope, 293.
Light modifier, 285.
Light, polarized, 233.
Light, polarized, experiments in, 239.

INDEX.

Light and sound, reflection of, 159.
Lines of force, 354.
Liquids, 72.
Liquids, compressibility of, 72.
Liquids, pressure exerted by, 72.
Liquids, top for showing centrifugal action on, 13.
Lissajous' experiments, 136.
Lithic acid, 273.
Lockyer, J. Norman, 214.
Locomotive, inertia, 9.
Lubricant, friction lessened by, 9.
Luminous paint, 198.
Luminous roses, 198.

Magnet, electric, 457.
Magnet. electric, for experimentation, 458.
Magnet, field, cast iron, 508.
Magnet forming the field, 503.
Magnet and rolling armature, 358.
Magnets, floating, 357.
Magnetic curves, arborescent, 356.
Magnetic curves, formation of, 354.
Magnetic curves in relief, 355.
Magnetic induction, 466, 473.
Magnetic induction, effects of, 351.
Magnetic key, 478.
Magnetic pole, neutralizing effects of an opposing, 352.
Magnetic top, 358.
Magnetism, 347.
Magnetism by induction from the earth, 347.
Magnetism by torsion, 348.
Magnetization of bars, 348.
Magneto-electric machine, 476, 479, 481
Magnifier, water bulb, 208
Magnus' experiment, 33
Maltese cross, 249
Manometric flames, 135
Manometric flames, Koenig's, 178
Markings on diatoms, 230
Marloye's harp, 122
Mars, 314
Masked electricity, 362
Mayer, Prof. A. M., 357
Mayer's floating needles, 357
Measurement of time by pendulum, 49
Measurements, electric, 442
Measuring jar, 383
Mechanical bird, Penaud's, 111
Mechanical depolarization of electrodes, 414.
Mercurial column supported by atmosphere, 91
Mercurial shower, 5
Mercury bath, 344
Metals, conductivity of, 194.
Metallic thermometer, 182
Metallophone, 119
Mica cone, 249
Mica films, 253
Mica objects for polariscope, 247
Mica plates, 203
Mica semi-cylinders, 248
Mica semi-cylinders, crossed, 248
Mica star, fan and crossed bars, 251
Mica wheel, 250
Microscope, modern, 284
Microscope, simple polariscope for, 306
Microscope slides, withdrawing air from, 99
Microscope, water lens, 279
Microscopic examination of ciliated organisms, 292
Microscopic exhibition of vibrating rods, 304
Microscopic objects, gathering, 287

Microscopic objects, simple polariscope for, 262
Microscopic objects, various, 287
Microscopy, 278
Mill, Barker's, 79
Mirror, convex spherical, 211
Mirror, cylindrical, 134
Mirror, rotating, 132
Mirror, spherical, 210
Mirrors, 20
Modifier, light, 284
Molecular actions, 56
Molecular forces, 4
Molecules, adhesion and cohesion of, 56
Moon, 315
Mortar, 390
Motion, 8
Motion produced by permanent magnet, 349
Motion, vortex, 114
Motor, electric, 470
Motor, electric, simple, 497
Motor, hot air, 196
Mouth organ, 120
Mouth used as a resonator, 143
Mouth vacuum apparatus, 106
Mouthpiece, double, 180
Movement, gyroscopic, law of, 32
Multiple reflection, 212
Music box, 119
Musical flames, 145
Musical instruments, stringed, 7
Musical top, 116

Nebulæ, 315
Neutralizing effect of opposing poles, 352
Newton, Sir Isaac, 200
Newton's rings, 302
Nicol prism, 269
Niepce, 337
Niter, 258
Noise, 116
Norremberg doubler, simple, 245, 251, 253

Objects, gathering microscopic, 287
Objects, mica, for polariscope, 247
Objects, microscopic, various, 287
Objects for polariscope, 244
Objects for simple polariscope, 307
Objects, telescopic, 314
Oblate spheroid, 17
Oblique lines, apparent deviation by, 225
Ocarina, 123
Ohm's law, 427
Optical illusions, 223
Optical illusions, curious, 226
Optical illusions, Thompson, 228
Ordinary ray, 233
Organ, mouth, 120
Oscillating and conical pendulums, 45
Over-development of photographic plate, 323
Over-exposure of photographic plate, 322
Over-tones of a bell, 144

Paint, luminous, 199
Pandean pipes, 122
Panel with ornaments of crystals, 274
Parabolic reflections, 159
Paraffining carbons, 402
Pascal's experiment, 73, 91
Pascal's law, 73
Pascal's law, demonstration of, 72
Pebbles, Brazilian, 269
Penaud's mechanical bird, 111
Pendulum with audible beats, 47
Pendulum, application of to clocks, 50

INDEX.   529

Pendulum, calculating of length of, 46
Pendulum, flying, 55
Pendulum, Kater's reversible, 48
Pendulum, length of at Hammerfest, 45
Pendulum, length of at St. Thomas, 45
Pendulum, measuring time by, 49
Pendulum, oscillating and conical, 45
Pendulum, seconds, 45
Pendulum, simple, 45
Pendulum, torsion, 52
Perforation of glass, 384.
Permanent magnet, effect of, on armature, 350
Permanent magnet, motion produced by, 347
Persistence of vision, 220
Persistent rotation, 9
Phantom bouquet, 211
Phonogram, Edison listening to, 151
Phonograph, Edison's, 151
Phonograph, first audience of, 151
Phonograph, perfected, 150
Phonograph simple, 150
Phonograph, test of, 156
Phonographic records, 156
Phosphorescence, 199
Photographic camera, 318
Photographic clearing, solution, 323
Photographic copying, 13, 19
Photographic negative, to bring out detail in, 323
Photographic negatives, washing, 325
Photographic plate, over-exposure of, 321
Photographic print, toning solution for, 327
Photographic shutters, 321
Photographs, black tones on, 327
Photographs, brown tones on, 327
Photographs, fixing bath for, 327
Photographs, instantaneous, 319
Photography, 318
Photography, best season for, 319
Photography, development of, 321
Photometer, 222
Photo-micrographic apparatus, 332
Photo-prints, coating, 325
Pipes, 122
Pipes, closed, 123
Pipes, open, 123
Pipes, reed, 121
Pith ball, electrical attraction and repulsion of, 359
Pith ball electroscope, 391
Pith balls, dancing, 390
Plane of rotation, change in, 14
Plant hairs, 290
Plante's secondary battery, 418
Plants, aquatic, 287
Plate holders, 319
Plate holders, bag, 330
Plating dynamo, connections of, 496
Plates, tourmaline, 236
Pleurosigma angulatum, 232
Plunge battery, 401
Pneumatic gyroscope, 24, 31
Pneumatic syringe, 6
Polar clock, 269
Polariscope for determining temperatures, 269
Polariscope for large objects, 262
Polariscope, mica objects for, 247
Polariscope for microscopic objects, 262
Polariscope, objects for, 244
Polariscope, simple, for microscope, 306
Polariscope for wide-angled crystals, 255
Polariscopes, 251
Polarization and analyzation by a bundle of plates, 244
Polarization of batteries, 395

Polarization, circular, 259
Polarization, elliptical, 253
Polarization by reflection, 237
Polarization by refraction, 237
Polarization by single plate, 246
Polarized bell, 485
Polarized light, 233
Polarized light, experiments in, 239
Polarizer, black glass, 238
Pollen of marshmallow, 290
Polyprism, 203
Pond life, 278
Popgun used as a pneumatic syringe, 6
Pores, physical, 4
Pores, physical and sensible, 4
Pores, sensible, 4
Porosity, 4
Power, storage of, 9
Practical application of the polariscope, 268
Press, hydraulic, simple, 77
Press, hydrostatic, principle of, 75
Pressure of air, 89
Pressure of endosmose, 67
Pressure, heat due to, 194
Pressure, hydrostatic, 73
Pressure, lateral, 79
Pressure for sensitive flames, 166
Prince Rupert's drops, 57
Principal focus of a convex lens, 204
Principle of tangent galvanometer, 439
Printing photographs, 325
Prints, contact, 325
Prism, course of light through a, 203
Prism, Nicol, 234
Prism, rocking, 214
Prisms, 202
Proboscis of blow-fly, 290
Properties of bodies, 1
Pulse glass, 184
Pump, filter, Bunsen's, 102
Pump, Wirtz's, 109

Quaker gun, 43
Quarter H.P. electric motor, 510.
Quartz, 259
Quartz polarized circular, 259
Quick method of mounting dry objects, 297

Radial disks, 176
Radiant heat, Tyndall's experiment in, 180.
Radiometer, 189
Railroad bridges, vibration of, 126
Railway, centrifugal, 11
Railway, spiral, 11
Ram; hydraulic, 80
Rapieff's optical illusions, 230
Ray, extraordinary, 233
Ray, ordinary, 233
Reaction, 80
Reactionary apparatus, 79
Real image, 206
Receiver, air pump, 96
Receiver for aspirator, 105
Record, phonographic, 156
Recording voltmeter, 454
Rectangular vibrations, compounding of, 140
Rectilinear motion, conversion of into rotary, 9
Reduction of volume of alcohol and water mixture, 3
Reeds, 121
Reflecting telescope, 212
Reflection and concentration of heat, 192
Reflection and concentration of sound, 158
Reflection of light and sound, 159

INDEX.

Reflection, multiple, 212
Reflectors, parabolic, 159
Reflection, polarization by, 237
Reflectors, concave, 192
Refraction, 202
Refraction, double, 233
Refraction, polarization by, 237
Refraction of sound, 164
Re-enforcement of sound, 141
Resistance box, bridge, 444
Resonance, determining speed by, 169
Resonant, mouth used as, 143
Resonant tubes, 145
Resonant tubes, length of, 146
Resonant vessel, selected power of, 142
Rest, 8
Rest, relative, 8
Reversible pendulum, Kater's, 47
Revolving tables, substitute for, 282
Rocker, Trevelyan, 163
Rocking car, 164
Rocking prism, simple, 214
Rods, vibrating, 118
Rods, vibrating, microscopic, exhibition of, 304
Rolling armature, 358
Rolling friction, 10
Roses, hygroscopic and luminous, 198
Rotating mirror, 132
Rotation of the earth, gyroscope for showing, 33
Rotation, persistent, 9
Rotifer, 290
Rotifer, exhibited by intermittent light, 295
Roughening battery plates, 419
Ruby light, 319

Sails, as concentrators and reflectors of sound, 159
Salicine crystals, 272
San Salvador, 159
Santonine, 273
Saturn, 314
Savart's wheel, 174
Scientific top, 14, 177
Scouring the plate, 337
Secondary battery, Planté, 418
Seconds pendulum, 45
Seeds, 290
Selected power of resonant vessels, 141
Selenite, 253
Self-exciting Geissler tube, 365
Self-luminous buoy, 366
Sensible pores, 4
Sensitive flames, 165
Sensitive flames, burner for, 166
Sensitive flames with gas at ordinary pressure, 168
Sensitive flames, producing gas pressure for, 167
Sensitive leaf, 197
Shower, mercurial, 5
Shutter, photograph, simple, 331
Shutters, photographic, 321
Silver chloride cell, 405
Silver crystals, 291
Simple air pump, 91
Simple electric motor, **497**
Simple pendulum, 45
Simple phonograph, 149
Single fluid battery, 398
Single refracting bodies, 233
Siren, 174
Siren, centrifugal, 171
Siren for measuring velocities, 170
Slides, lantern, 325
Sliding friction, 10
Smee's battery, 398
Smoke wreaths, 113

Soap film, 203
Sound, 116
Sound, concentration of, 162
Sound, conduction of, 125
Sound lens, 160
Sound and light waves compared, 200
Sound receiver, simple, 128
Sound, reflection and concentration of, 158
Sound reflector, adjustable, 159
Sound, re-enforcement of, 141
Sounding flames, 146
Sounds by changes of temperature, 190
Spar, Iceland, 233
Speaking name, 132
Spectrum, 215
Spectrum, apparatus, 217
Spectrum, simple method of producing, 215
Spectroscope, 215
Sphere, insulated, 387
Spherical mirror, 210
Spheroid, oblate, 17
Spicules, 290
Spinning device, frictional, 14
Spiral railway, 11
Spirals, Airy's, 258
Sponges, spicules of, 290
Springs, 7
Stand, telescope, 315
Stars, single and double, **315**
Starting box, 520
Steam engine, fifty-cent, 195
Steam, force of, 195
Steam gyroscope, 23, 25
Stearic acid, 277
Stentor, 290
Stevens, Prof W. Le Conte, 212
Stewart, Balfour, 237
Stile's wax cell, 300
Storage of power, 9
Strained glass, 241
String telephone, 125
Stringed instruments, 7, 124
St. Paul's, 159
Sub-stage condenser, 285
Sugar, solution of in water, 2
Suggestions in decorative art, 272
Sulphate of cadmium, 272
Sulphate of nickel, 260
Sulphuric acid and water, mixture of, 3
Sun, 315
Surface tension, 59
Swiftest descent apparatus, 39, 40
Switch connections, tangent galvanometer, 441
Synapia inherens, 290

Table of tangents, 522
Tangent galvanometer, 440
Tangent galvanometer, principle of, 439
Tangent galvanometer, switch connections, 441
Telephone, Bell, principle of, 477
Telephone, string, 124
Telescope, achromatic objective for, 313
Telescope, collimation of, 313
Telescope, eye-pieces for, 311
Telescope, reflecting, 212
Telescope stand, 313
Temperatures, polariscope for determining, 269
Tension, 7
Telescope, inexpensive, 309
Terrestrial eye-piece for telescope, 311
Testing simple air pump, 91
Thermo-electric battery, 423
Thermo-electric current, 422
Thermo-electric series, 422
Thermometer, air, 184

INDEX. 531

Thermoscope, chemical, 198
Thermoscopic balance, 185
Thermostat, metallic, 182
Thermostat, simple, 183
Thin films, 255
Thompson, optical illusion, 228
Thompson, Prof. Silvanus P., 228
Threads, electrified, 364
Toepler's experiment, 149
Tolles, R. B., 316
Tongs, tourmaline, 255, 260
Toning solutions, 327
Top, aerial, 109
Top, chameleon, 177
Top, choral, 12
Top, experiments with, 16
Top, glass, 13
Top, magnetic, 358
Top, a scientific, 14, 178
Top for showing centrifugal action, 13
Top, spinning device for, 14
Topaz, 260
Torricelli's experiment, 91
Torsion, 7
Tourmaline crystals, action of, 234
Tourmaline tongs, 255
Toys, musical, 116
Trajectory, 43
Transferring objects to slide, 291
Treadle for air pump, 95
Trevelyan rocker, 163
Tricycles, 10
Tube with interrupted conductor, 308
Tuning forks and resonant tubes, 145
Two-fluid batteries, 409
Tyndall, John, 181
Tyndall's experiment with radiant heat, 189
Tyrol, 124

Unannealed glass, 241
Units, electrical, 425
Universal discharger, 386
Universal microscope, 284

Vacuum apparatus, mouth, 106
Vacuum, dilatation of a balloon in, 85
Vacuum produced by exosmose, 70
Vaporization, 181
Velocities, siren for showing, 170
Verre-trempe, 244
Vibrating flames, 130
Vibrating flame apparatus, 134
Vibrating rods, 118

Vibrating rods, microscopic exhibition of, 304
Vibrations, composition of, 136
Vibrations, compounding rectangular, 140
Vibration, longitudinal, of rods, 121
Vibrations, harmonic, 126
Vial for four liquids, 81
Virtual image, 207
Vision, persistence of, 220
Vitreous and resinous electricity, 361
Vocal sounds, re-enforcement of, 141
Voltaic arc, 518
Voltmeter, expansion, 449
Voltmeter, recording, 454
Volume, reduction of by mixture, 3
Vortex motion, 113
Vorticella, 290

Washing the negative, 325
Watch balance, 54
Water, boiling in vacuo, 98
Water bulb magnifier, 208
Water colored by aniline, 4
Water hammer, 39
Water lens microscope, 279
Wax cell, Dr. Stiles', 300
Wax patterns, 700
Webster, G. Watmough, 228
Weighing gases, 86
Weight of air, 88
Weight lifted by air pressure, 90
Wheatstone's bridge, 443
Wheatstone's polar clock, 269
Wheel, gas, 86
Wheel, mica, 250
Whispering galleries, 159
Winding, 514
Wide-angled crystals, 255
Wide angled crystals, polariscope for, 255
Wimshurst induction machine, a modified, 370
Wire cloth, designs on, 63
Wirtz's pump, 109
Wollaston's cryophorus, 188
Wreaths of smoke, 133

Young, 200

Zoetrope, 220
Zylophone, 118

The Scientific American Cyclopedia of Receipts

Notes and Queries

15,000 Receipts

734 PAGES

Price, $5.00

Mailed to Any Part of the World

Leather Bindings as follows:

Sheep, $6.00
Half Morocco, $6.50

MUNN & CO.
PUBLISHERS

REVISED EDITION

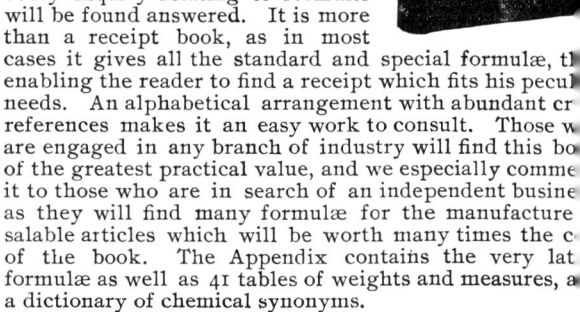

THE SCIENTIFIC AMERICAN CYCLOPEDIA OF RECEIPTS, NOTES AND QUERIES, first published in the autumn of 1891. It was well received by the press, came quickly into the favor of purchasers and has had an unprecedented sale. It has been used by chemists technologists, and those unfamiliar with the arts, with equal success, and has demonstrated that it is a book which is useful in the laboratory, factory or home. It consists of a careful compilation of the most useful receipts, and information germane to the scope of the book, which has appeared in the SCIENTIFIC AMERICAN for more than half a century. The Publishers now take pleasure in offering the Twenty-fifth Revised Edition, which has been brought up to the latest requirements by the insertion of 900 new formulæ, making it the latest and most complete volume on the subject of receipts ever presented. Over 15,000 selected formulæ are here collected, nearly every branch of the useful arts being represented. Many of the principal substances and raw materials used in the arts are described, and almost every inquiry relating to formulæ will be found answered. It is more than a receipt book, as in most cases it gives all the standard and special formulæ, thus enabling the reader to find a receipt which fits his peculiar needs. An alphabetical arrangement with abundant cross references makes it an easy work to consult. Those who are engaged in any branch of industry will find this book of the greatest practical value, and we especially commend it to those who are in search of an independent business as they will find many formulæ for the manufacture of salable articles which will be worth many times the cost of the book. The Appendix contains the very latest formulæ as well as 41 tables of weights and measures, and a dictionary of chemical synonyms.

SEND FOR FULL TABLE OF CONTENTS.

SCIENTIFIC AMERICAN OFFICE **361 Broadway, New York**

SCIENTIFIC AMERICAN REFERENCE BOOK

12mo; 516 pages; illustrated; 6 colored plates.   Price $1.50, postpaid

¶ The result of the queries of three generations of readers and correspondents is crystallized in this book, which has been in course of preparation for months. It is indispensable to every family and business man. It deals with matters of interest to everybody. The book contains 50,000 facts, and is much more complete and more exhaustive than anything of the kind which has ever been attempted.

The "Scientific American Reference Book" has been compiled after gauging the known wants of thousands. It has been revised by eminent statisticians. Information has been drawn from over one ton of Government reports alone. It is a book for everyday reference—more useful than an encyclopedia, because you will find what you want in an instant in a more condensed form. The chapter relating to patents, trademarks and copyrights is a thorough one and aims to give inventors proper legal aid. The chapter on manufactures deals with most interesting figures, admirably presented for reference. The chapter dealing with Mechanical Movements contains nearly three hundred illustrations, and they are more reliable than those published in any other book—they are operative. Weights and measures occupy a considerable section of the book, and are indispensable for purposes of reference. Sixty years of experience alone have made it possible for the publishers of the Scientific American to present to the purchasers of this book a remarkable aggregation of information. The very wide range of topics covered in the "Scientific American Reference Book" may be inferred by examining the table of contents on back. The first edition of this work is 10,000 copies. Remit $1.50, and the book will be promptly mailed. Send to-day.

REDUCED FACSIMILE PAGE 118.

MUNN & CO., Publishers
Scientific American Office
361 Broadway, New York City

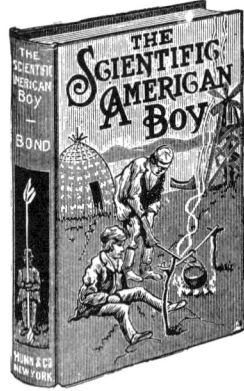

Just Published

The Scientific American Boy

By
A. RUSSELL BOND

12mo.   320 Pages.   340 Illustrations.
Price $2.00 Postpaid

THIS is a story of outdoor boy life, suggesting a large number of diversions which, aside from affording entertainment, will stimulate in boys the creative spirit. In each instance complete practical instructions are given for building the various articles.

¶ The needs of the boy camper are supplied by the directions for making tramping outfits, sleeping bags and tents; also such other shelters as tree houses, straw huts, log cabins and caves.

¶ The winter diversions include instructions for making six kinds of skate sails and eight kinds of snow shoes and skis, besides ice boats, scooters, sledges, toboggans and a peculiar Swedish contrivance called a "rennwolf."

¶ Among the more instructive subjects covered are surveying, wigwagging, heliographing and bridge building, in which six different kinds of bridges, including a simple cantilever bridge, are described.

¶ In addition to these, the book contains a large number of miscellaneous devices, such as scows, canoes, land yachts, windmills, water wheels and the like. A complete table of contents sent on request

MUNN & COMPANY
361 Broadway   PUBLISHERS OF "SCIENTIFIC AMERICAN"   New York City

HOME MECHANICS
FOR
AMATEURS

By GEORGE M. HOPKINS

Author of "Experimental Science"

2mo, 370 Pages, 320 Illustrations.   Price, $1.50, Postpaid

¶ The book deals with wood-working, household ornaments, metal-working, lathe work, metal spinning, silver working; making model engines, boilers and water motors; making telescopes, microscopes and meteorological instruments, electrical chimes, cabinets, bells, night lights, dynamos and motors, electric light, and an electrical furnace. It is a thoroughly practical book by the most noted amateur experimenter in America.

¶ Every reader of "Experimental Science" should possess a copy of this most helpful book. It appeals to the boy as well as the more mature amateur. Holidays and evenings can be profitably occupied by making useful articles for the home or in building small engines or motors or scientific instruments.

MUNN & COMPANY
Publishers of "Scientific American"
61 BROADWAY   -   -   -   -   NEW YORK

The Most Popular Scientific Paper in the World
Established 1845.   Weekly, $3.00 a Year; $1.50 Six Months.

THIS unrivaled periodical is now in its *sixtieth year*, and, owing to its ever-increasing popularity, it enjoys the largest circulation ever attained by any scientific publication. Every number contains sixteen large pages, beautifully printed, handsomely illustrated; it presents in popular style a descriptive record of the most novel, interesting and important developments in Science, Arts and Manufactures. It shows the progress of the World in respect to New Discoveries and Improvements, embracing Machinery, Mechanical Works, Engineering in all its branches, Chemistry, Metallurgy, Electricity, Light, Heat, Architecture, Domestic Economy, Agriculture, Natural History, etc. It abounds in fresh and interesting subjects for discussion, thought or study. To the inventor it is invaluable, as every number contains a complete list of all patents and trademarks issued weekly from the Patent Office. It promotes Industry, Progress, Thrift and Intelligence in every community where it circulates.

The SCIENTIFIC AMERICAN should have a place in every Dwelling, Shop, Office, School, or Library. Workmen, Foremen, Engineers, Superintendents, Directors, Presidents, Officials, Merchants, Farmers, Teachers, Lawyers, Physicians, Clergymen—People in every walk and profession in life will derive satisfaction and benefit from a regular reading of the SCIENTIFIC AMERICAN.

If you want to know more about the paper send for *"Fifteen Reasons Why You Should Subscribe to the Scientific American,"* and for *"Five Reasons Why Inventors Should Subscribe to the Scientific American."* Fifty-two numbers make 832 large pages, equal to 3,328 ordinary magazine pages, and 1,000 illustrations are published each year. Can you and your friends afford to be without this up-to-date periodical which is read by every class and profession? Remit $3.00 by postal order or check for a year's subscription, or $1.50 for six months.

MUNN & CO., Publishers, 361 Broadway, New York City

Established 1876.

THIS journal is a separate publication from the SCIENTIFIC AMERICAN, and is designed to extend and amplify the work carried on by the parent paper. In size and general make-up it is uniform therewith, covering sixteen pages of closely printed matter, handsomely illustrated. It has no advertising pages, and the entire space is given up to the scientific, mechanical and engineering news of the day. It differs from the SCIENTIFIC AMERICAN in that it contains many articles that are too long to be published in the older journal, or of a more technical nature.

The price of the SUPPLEMENT is $5.00 a year, but where subscribers take both the SCIENTIFIC AMERICAN and the SCIENTIFIC AMERICAN SUPPLEMENT a special combined rate of $7.00 for both is made if the papers are mailed to one address. Remit by postal order or check. All copies of the SUPPLEMENT since January 1, 1876, are in print and can be supplied at the uniform price of 10 cents each, thus enabling readers to obtain access to a most valuable source of information on almost every subject at the most moderate price. A large *Supplement Catalogue* giving a list of nearly 15,000 valuable papers will be mailed free to any one. Address

MUNN & CO., Publishers, 361 Broadway, New York City